Vitamins in Human Biology and Medicine

Editor

Michael H. Briggs, D.Sc., Ph.D.

Professor of Human Biology
Deakin University
Geelong, Victoria
Australia

CRC Press
Taylor & Francis Group
Boca Raton London New York

CRC Press is an imprint of the
Taylor & Francis Group, an **informa** business

First published 1981 by CRC Press
Taylor & Francis Group
6000 Broken Sound Parkway NW, Suite 300
Boca Raton, FL 33487-2742

Reissued 2018 by CRC Press

A Library of Congress record exists under LC control number: 80019887

ISBN 13: 978-1-138-55053-7 (hbk)
ISBN 13: 978-1-315-12137-6 (ebk)

Visit the Taylor & Francis Web site at http://www.taylorandfrancis.com and the CRC Press Web site at http://www.crcpress.com

PREFACE

The early decades of the twentieth century saw an amazing series of discoveries. Diseases which were widespread in many countries, and affecting millions of people, both adults and children, were suddenly revealed as caused by dietary deficiencies of biochemical substances that the human body either could not synthesize at all, or could synthesize in only limited, usually insufficient, amounts. Over the space of only a few years, conditions such as scurvy, rickets, pellegra, pernicious anaemia, together with a less well-defined group of skin and eye conditions, became responsive to successful and rational treatment by vitamin supplements.

It is no wonder that the very word 'vitamin' was carried into popular parlance as the epitome of a wonder drug, and still, to this day, occupies a position in the public mind that has resisted competitions from such rivals as 'antibiotic' and 'tranquillizer'. There is no doubt that, in the early years of the present century, vitamins did, indeed, perform medical miracles. It is entirely understandable that, to the general man-in-the-street, they should continue to do so.

But the days when vitamins produced dramatic cures are over in developed countries, for modern food supplies adequate vitamins to the majority of people, although this could be improved by education aimed at reducing vitamin losses during food storage and preparation. Vitamin supplemention should be considered only when there are definite clinical signs of deficiency, or in special circumstances where a deficiency state may arise unless a supplement is used. These special cases include the chronic use of certain drugs, malabsorption and other gastrointestinal disorders, the exclusive use of a vegan diet, and pregnancy.

It is the purpose of the present book to try to offer a modern perspective on the importance of vitamins in human biology and medicine.

There have been recent revivals of attempts to use vitamins as therapeutic agents by administering doses many times the recommended allowance. This use of 'megavitamin therapy' is reviewed by Dr. Reynold Spector, with particular emphasis on vitamin entry to the central nervous system. Of related interest is the chapter by Dr. John Blass discussing central nervous system manifestations of thiamin deficiency, and also the contribution by Ms. Fiona Cumming and colleagues reviewing adverse effects and clinical toxicity of vitamin supplementation.

Changes in plasma triglycerides and cholesterol concentrations related to vitamin C are described by Dr. Emil Ginter, while Dr. Mark Wahlqvist reviews effects of nicotinic acid, and related compounds, on plasma cholesterol. Dermatological studies of vitamin A acid (retinoic acid) are presented in detail by Dr. William Cunliffe.

Developments and improvements in laboratory tests of vitamin status are reviewed by Dr. Mahtab Bamji, while Ms. Sue Tonkin describes the many studies which demonstrate significant changes in many of these tests in women using oral contraceptives. There is also a presentation by Ms. Delia Flint and Professor Derek Prinsley on changes in vitamin status in elderly persons.

It is hoped that the present volume will have something of relevance for all interested in human nutrition and foods. It is, however, intended primarily for postgraduate reading, especially for those conducting research in the wide, but fascinating field of the vitamins.

THE EDITOR

Michael H. Briggs is Professor of Human Biology at Deakin University, Geelong, Victoria, Australia. Currently he is also Dean of Sciences.

Born in Manchester, England, in 1935, Dr. Briggs was educated at the University of Liverpool, then undertook postgraduate studies in biochemistry and nutrition at Cornell University and the University of New Zealand. He is a Fellow of the Royal Society of Chemistry, a Fellow of the Royal Institute of Biology, and a Member of the Royal College of Pathologists. Dr. Briggs is also a member of a wide range of national and international scientific and medical societies. Since 1970 he has been a Consultant to the World Health Organization, and in this capacity has visited 15 different countries.

Immediately prior to his present position, Dr. Briggs was Director of Biochemistry at the Alfred Hospital, Melbourne, a major teaching hospital of Monash University. In 1971 and 1972, Dr. Briggs was at the University of Zambia as Head of the Department of Biochemistry, where a Food and Agriculture Organization nutrition survey team formed part of his unit.

Dr. Briggs also spent seven years in the pharmaceutical industry working as a research and development director on therapeutic substances and nutrient additives for humans and farm animals.

Among his more than 200 publications are contributions to textbooks, monographs, and encyclopedias, as well as articles in scientific and medical periodicals on biochemistry, pathology, toxicology, pharmacology, enzymology, and reproductive biology. Dr. Briggs was editor of seven volumes of *Advances in Steroid Biochemistry and Pharmacology* and is co-editor of the new series *Progress in Hormone Pharmacology and Biochemistry*. Among other recent volumes he has written or edited are: *Oral Contraceptives* (4 volumes, 1977-1980, Eden Press), *Biochemical Contraception* (Academic Press 1975), *Chemistry and Metabolism of Drugs and Toxins* (Heinemann Medical, 1974), *Pharmacological Models in Contraceptive Development* (W.H.O., 1974), *Implications of Steroid Hormones in Cancer* (Heinemann Medical, 1971), *Steroid Biochemistry and Pharmacology* (Academic Press 1970), *Advances in the Study of the Prostate* (Heinemann Medical, 1970), and *Urea as a Protein Suppliment* (Pergamon Press, 1968).

Dr. Briggs is married to a medical practitioner (Dr. Maxine Briggs) who has been his research collaborator for many years and is co-author of many of their publications.

Michael H. Briggs

CONTRIBUTORS

Mahtab S. Bamji, Ph.D.
Deputy Director
National Institute of Nutrition
Hyderabad, India

John P. Blass, M.D., Ph.D.
Professor of Neurology and Medicine
Cornell University Medical College
Burke Rehabilitation Center
White Plains, New York

Maxine Briggs, M.B., Ch.B., D.P.H., D.O.H., F.R.A.C.M.A.
Director of Rehabilitation Services
Geelong, Victoria
Australia

Michael H. Briggs, D.Sc., Ph.D., F.R.S.C., F.I.Biol., M.R.C. Path.
Professor of Human Biology
Deakin University
Geelong, Victoria
Australia

Fiona Cumming, B.Sc., Cert. Diet., R.D.
Deakin University
Geelong, Victoria
Australia

W. J. Cunliffe, B.Sc., M.D., F.R.C.P.
Consultant Dermatologist
General Infirmary
Senior Lecturer in Dermatology
University of Leeds
Yorkshire, England

Delia M. Flint, M.Sc., Cert. Diet., R.D.
Lecturer, Human Nutrition
Deakin University
Geelong, Victoria
Australia

Emil Ginter, Ph.D.
Senior Scientist
Institute of Nutrition
Bratislava, Czechoslovakia

Derek M. Prinsley, M.D., F.R.C.P., F.R.A.C.P., F.R.S.H.
Professor of Gerontology and Geriatric Medicine
University of Melbourne
Parkville, Victoria
Australia

Reynold Spector, M.D.
Chief, Division of Clinical Pharmacology
Professor of Medicine and Pharmacology
University of Iowa
Iowa City, Iowa

Suzanne Y. Tonkin, B.Sc., M.Sc.
Tutor in Biochemistry
Deakin University
Geelong, Victoria
Australia

Mark L. Wahlqvist, B.Med. Sc., F.R.A.C.P.
Professor of Human Nutrition
Deakin University, Geelong
Consultant Physician
Prince Henry's Hospital, Melbourne
Victoria, Australia

TABLE OF CONTENTS

Chapter 1
Laboratory Tests for the Assessment of Vitamin Nutritional Status 1
Mahtab S. Bamji

Chapter 2
Oral Contraceptives and Vitamin Status 29
Suzanne Y. Tonkin

Chapter 3
Vitamin Status of the Elderly 65
Delia M. Flint and Derek M. Prinsley

Chapter 4
Effects on Plasma Cholesterol of Nicotinic Acid and its Analogues 81
Mark L. Wahlqvist

Chapter 5
Plasma Lipids and Vitamin C 95
Emil Ginter

Chapter 6
Thiamin and the Wernicke-Korsakoff Syndrome 107
John P. Blass

Chapter 7
Megavitamin Therapy and the Central Nervous System 137
Reynold Spector

Chapter 8
Retinoic Acid in Dermatology 157
W. J. Cunliffe

Chapter 9
Clinical Toxicology of Vitamins... 187
Michael H. Briggs, Maxine Briggs, and Fiona Cumming

Index 245

Chapter 1

LABORATORY TESTS FOR THE ASSESSMENT OF VITAMIN NUTRITIONAL STATUS

Mahtab S. Bamji

TABLE OF CONTENTS

I. Introduction ... 2
- A. Are Laboratory Tests Essential? ... 2
- B. What is an Ideal Biochemical Test? ... 3
- C. Laboratory Methods ... 3
  1. Urinary Excretion of Vitamins ... 3
  2. Blood Levels of Vitamins ... 4
  3. Functional Tests ... 4
  4. Standardization of Methods, Quality Control, and Guidelines for Interpretation of the Methods ... 5
  5. Factors which Lead to Nutritional Deficiencies ... 5

II. Vitamin A ... 6

III. Thiamin ... 8
- A. Urinary Thiamin ... 8
- B. Blood Thiamin ... 9
- C. Functional Tests ... 9
  1. Glucose Load Test ... 9
  2. Erythrocyte Transketolase Test (ETK) ... 10

IV. Riboflavin ... 11
- A. Urinary Riboflavin ... 11
- B. Red Blood Cell Riboflavin ... 11
- C. Erythrocyte Glutathione Reductase Test ... 12

V. Pyridoxine ... 13
- A. Urinary Pyridoxine ... 13
- B. Blood Levels of Pyridoxine ... 14
- C. Functional Tests ... 14
  1. Tryptophan Load Test ... 14
  2. Enzymic Tests — Blood Transaminase Activities ... 15
  3. Methionine Load Test ... 16

VI. Niacin ... 16

VII. Folic Acid (Folate) ... 16
- A. Serum and Red Blood Cell Folate ... 17
- B. Functional Tests ... 18
  1. Histidine Load Test — Urinary Excretion of Formiminoglutamate and Urocanate ... 18
  2. Deoxyuridine Suppression Test ... 18

VIII. Vitamin B_{12} 19
A. Serum Vitamin B_{12} 19
B. Valine Load Test — Methylmalonic Acid (MMA) Excretion 20

IX. Pantothenic Acid 20

X. Ascorbic Acid (Vitamin C) 20
A. Serum and Leukocyte Ascorbic Acid Levels 20
B. Functional Tests 21

XI. Vitamin D 21

XII. Vitamin E 22

XIII. Vitamin K 22

XIV. Summary and Conclusions 23

Acknowledgments 23

References 24

I. INTRODUCTION

A. Are Laboratory Tests Essential?

Assessment of nutritional status can be done in a number of ways such as (a) physical examination of the subject for the presence or absence of signs and symptoms of vitamin deficiency, (b) inquiry into dietary history, and (c) laboratory tests on easily available body fluids such as blood and urine. Each of these methods has its own advantages and disadvantages.

In the development of any disease, biochemical changes precede clinical changes and hence biochemical tests help to identify the disease at the subclinical stage. They also help to confirm the clinical diagnosis, since clinical signs are often not specific.[1] For instance, lesions of the mouth, such as angular stomatitis and glossitis, can be due to deficiency of riboflavin, pyridoxine, niacin, and iron or combinations of these. They may also arise from nonnutritional causes like chewing of betal nut preparations containing an irritant like lime. Peripheral neuropathy can be due to thiamin, pyridoxine, or pantothenic acid deficiency or some other unidentified causes, and folicular hyperkeratosis has been attributed to deficiency of vitamin A,[1] B-complex, and essential fatty acids,[2,3]

This is not altogether surprising, since the biochemical machinery of the body is very complex and the deficiency of more than one nutrient may lead to the molecular event(s) which may ultimately be the cause of the morphological lesion(s). In this connection it should be mentioned that though more than six decades have passed since the discovery of vitamins, and a great deal is known about the biochemical and clinical lesions of specific deficiencies, for not a single nutritional deficiency disease is the exact molecular basis understood. Nor are the very early symptoms of deficiency recognized.

Diet surveys are useful, particularly at the community level, but often it is difficult to obtain reliable data and even when available, they provide at best a rough estimate of the amount consumed, but not the amount absorbed or utilized. Hence the need for good biochemical tests is obvious.

B. What is an Ideal Biochemical Test?

An ideal biochemical test is one which can be carried out on easily available body fluids such as blood and urine. It should be specific, simple, inexpensive, and should reveal tissue depletion at an early stage, and not merely reflect recent intake of the nutrient. With so many criteria to satisfy, it is not surprising that we do not have an ideal biochemical test for any of the vitamins. The choice of the method depends on the situation in which it is applied and often more than one test may have to be used. In community surveys, one needs a method which is simple to perform, inexpensive, and not too sophisticated in terms of equipment and skill, even if it means sacrificing to some extent the other criteria like specificity. For the diagnosis of an individual patient, specificity is more important. With increasing expertise becoming available, greater numbers of health surveys are employing even the more sophisticated methods. In some instances samples are sent from the field to the laboratories in the same country or even other countries. As a result of these developments, it is possible that in a few years from now we would have either gained greater confidence in the methods in use or become more disillusioned with their usefulness.

C. Laboratory Methods

Basically there are two types of biochemical tests — those based on the measurement of either the vitamin or its metabolite in blood or urine and those based on the measurement of one or more of their functions.

1. Urinary Excretion of Vitamins

Urinary excretion of a vitamin or its metabolites is measured either in a random sample of urine, in the first voided morning sample, or in a 24-hr specimen. Collection of a 24-hr specimen though desirable is difficult. A random sample, though most convenient to collect, is undependable due to large fluctuations that can occur between samples even when the values are related to creatinine. Creatinine excretion shows considerable variation between days and between timed specimens within a day.[4] Early morning sample has the advantage of being standardized at least for fluid intake and physical activity. Its usefulness has been demonstrated for some constituents such as sodium, potassium, and calcium.[5] Whether it is so for vitamins needs to be ascertained.

One of the major criticisms against the use of urinary excretion as a biochemical index is that it reflects the recent intake of the vitamin more than the state of tissue saturation. To overcome this objection, load retention tests have been described in which urinary excretion of the vitamin is measured in a timed sample of the urine collected after an oral or a parenteral load of the vitamin. Deficient subjects would be expected to retain more of the administered load and thus excrete less as compared to normal subjects. In recent years, load tests are rarely employed and little if any attempt has therefore been made to compare the results of vitamin load return tests with the more sophisticated functional tests.

Under certain circumstances such as infections, the use of antibiotics, and in conditions which lead to negative nitrogen balance or failure of utilization, the urinary excretion of vitamins is known to increase. Such individuals may actually be at risk of developing hypovitaminosis, but the urinary excretion data may be misleading since they suggest a state of adequate vitamin status.

2. Blood Levels of Vitamins

Whole blood, plasma, erythrocytes, and leukocytes have all been used for nutritional evaluation. Blood levels, particularly plasma levels, may also reflect recent dietary intake (even if care is taken to obtain specimens under basal conditions), rather than the true state of tissue concentration. Blood cells are regarded by some workers as convenient biopsy material, but whether vitamin levels in erythrocytes or leukocytes reflect the state of the other tissues even in uncomplicated deficiency states is a moot point. Under some circumstances the distribution of vitamins between tissues is known to alter. One such example which has been studied recently is oral contraceptive use.

The choice of the blood component is guided by the concentration of the vitamin in it and its sensitivity to deficiency states. The change in vitamin concentration should not be so rapid as to reflect only the recent dietary intake, nor so slow as to appear just before the appearance of clinical manifestations. Thus, vitamin B_{12} measurements are generally made in the serum or plasma, whereas for most other water-soluble vitamins, plasma as well as blood cells are used. For the fat-soluble vitamins, only plasma is used because the concentrations of these vitamins or their metabolites in blood cells are very low.

3. Functional Tests

It has been possible to develop functional tests, in the case of the B-complex vitamins, since most of their biochemical functions are well established. These tests are of two types: enzymatic tests and metabolic tests. In the enzymatic tests the activity of an enzyme which requires the vitamin as a coenzyme is measured with and without saturating amounts of the coenzyme added in vitro. The in vitro stimulation of the enzyme by the coenzyme is taken as a direct measure of unsaturation of the enzyme protein with respect to the coenzyme and hence a measure of vitamin deficiency. In vitro stimulation index can be expressed as activation coefficient (AC) or percent stimulation. The relationship between the two is as follows:

$$AC = \frac{\text{Enzyme activity with added coenzyme}}{\text{Enzyme activity without added coenzyme}}$$

$$\text{Percent stimulation} = AC \times 100 - 100$$

When enzymatic tests were first described, it was assumed that apoenzyme levels were not affected by vitamin deficiencies and it was therefore believed that the coenzyme stimulation index would be a better parameter for assessing vitamin nutrition status than enzyme activity measurements. However, there are now some indications that apoenzyme levels are also altered in deficiency states. It is therefore advisable to take into account the basal enzyme activity as well as its activation with coenzyme in the interpretation of the results.

At present, all the factors which alter apoenzyme levels are not recognized. They may vary from one enzyme to another. Vitamin deficiency or excess, administration of hormones, drugs, and disease states may all increase or decrease apoenzyme levels and make the interpretation of enzymatic tests difficult.

Examples of enzymatic tests are the transketolase test for thiamin, glutathione reductase test for riboflavin and alanine aminotransferase, and aspartate aminotransferase tests for pyridoxine. These will be discussed at greater length in later sections.

In the metabolic tests, rise in the concentration of a metabolite in blood or urine as a result of a vitamin deficiency-mediated enzymatic lesion is measured, preferably after administering a load of an appropriate precursor. Examples of such tests are the glucose load test for thiamin (blood levels of pyruvate and lactate), tryptophan load test for pyridoxine (urinary levels of xanthurenic acid and other metabolites), histidine load test for folic acid (formiminoglutamic acid excretion in urine), and valine load test for vitamin B_{12} (methylmalonic acid excretion in urine).

4. Standardization of Methods, Quality Control, and Guidelines for Interpretation of the Values

An important problem in clinical chemistry is standardization of methods and suitable quality control. While standard quality control sera are available for most of the routine assays (including serum enzyme assays) done in clinical chemistry laboratories, suitable quality control materials are not available for vitamin assays.

Vitamins being fairly sensitive compounds, their stability in quality control materials poses a major problem. Preparation of a reference material for erythrocyte enzymes which would show consistent coenzyme stimulation is very difficult. Quality control material for serum vitamin A and C is available from the Center for Disease Control, Atlanta, Georgia. An erythrocyte preparation for erythrocyte enzymes is being developed by the Wolfson Research Laboratory in Birmingham, U.K. International agencies like the World Health Organization and the International Union of Nutritional Sciences — Committee on Vitamins have recognized this as being a priority research area. Internal quality control using pooled erythrocytes (frozen at −20°C in small aliquots) is employed by some laboratories.

Though desirable, it would be impractical to expect all laboratories to use the same procedures. For instance, kinetic assays for the measurement of erythrocyte enzymes can be used only by laboratories which possess sophisticated recording spectrophotometers with temperature control. Other laboratories would have to use colorimetric two-point assays. However, some degree of standardization based on optimization of the methods is necessary between laboratories that use similar methods.

In the absence of standardized methods and quality control, it would be difficult to develop suitable interpretive guidelines which can be used with confidence by all laboratories. Even with standardized methods, some differences between populations can be expected and hence it is desirable that each investigator tries to establish the normal values for the population that is under study. This exercise is essential at least until such time that methods are better standardized and it is possible to find out if differences between populations with regard to normal values exist.

A practical question that frequently comes up relates to the basis for fixing guidelines. Is good health synonymous with maximum tissue saturation or can a person with suboptimal tissue saturation still be healthy? This is an important question not only from the scientific point of view but also from the practical point of view for those dealing with populations with limited resources. The need for imaginative and well-planned studies to answer it is obvious. Clinical indices now available represent the extreme end of the spectrum and hence are not sensitive enough. Between the appearance of biochemical changes and overt morphological changes, there must be a spectrum of subtle clinical and physiological manifestations which perhaps indicate varying grades of inadequate health. Until some insight into these is gained, it may be better to err on the safe side and develop guidelines aimed at achieving tissue saturation, a difficult enough job. Sauberlich et al.[5] have described tentative guidelines for the interpretation of laboratory tests currently available for assessing vitamin nutrition status from the information available in the literature. These will be referred to throughout this chapter.

5. Factors Which Lead to Nutritional Deficiencies

Vitamin deficiency in man can arise because of the following reasons: (1) inadequate intake due to poverty, ignorance, lack of incentive, anorexia, or dental problems; (2) poor digestion and absorption due to diarrhea and use of certain drugs; and (3) increased requirement associated with certain physiological states like rapid growth, pregnancy (fetal requirement and hormone effects), and lactation.

Certain drugs and hormone therapies can raise the requirement, either by interfering with the metabolism of the vitamins (antivitamins)[6] or by raising the levels of carrier proteins, tissue binders, and apoenzymes, which bind to the vitamins. The latter can lead to a redistribution of vitamins between tissues and also between enzymes within a tissue leading to pockets of deficiency and excess. A notable example of this kind of effect is the action of oral contraceptives.[7] Women taking oral contraceptives have raised levels of retinol and retinol-binding protein in serum.[8] This is believed to be due to mobilization of the vitamin from the liver into circulation. In animals, oral contraceptives produce a fall in liver vitamin A.[7] Contrarily, administration of oral contraceptives to women or female rats leads to a fall in blood levels of riboflavin[9,10] and folate,[11,12] but in the rat an increase in liver levels of these vitamins is observed[10,12] due to a rise in some apoenzymes or other vitamin-binding proteins.

Drugs can selectively inhibit or stimulate enzymes. For instance, pregnant women and women taking oral contraceptives show marked increase in urinary excretion of tryptophan metabolites because estrogens induce tryptophan oxygenase but inhibit kynurinase of the tryptophan-niacin pathway.[13] Estrogens also raise the levels of certain transaminases.[9,13]

In situations such as these, evaluation of nutritional status using either the functional tests or blood and urine levels becomes extremely difficult. How does one interpret a situation where tissue levels rise and serum levels fall? Should attempts be made to supplement high doses of vitamins to restore serum levels? High doses of vitamins can also create imbalance by raising the levels of certain apoenzymes.

Vitamin-responsive inborn errors of metabolism have been described.[14] These manifest as functional defects despite adequate tissue levels of the vitamin.

II. VITAMIN A

Vitamin A deficiency is a global nutritional problem. A considerable degree of blindness in developing countries is due to vitamin A deficiency. The Ten-State Nutrition Survey[15] in the U.S. shows that over 40% Spanish Americans, 20% blacks, and 10% whites have low plasma vitamin A levels.

Since the functional role of vitamin A except in the rhodopsin cycle is not understood, serum vitamin A measurement continues to be the method of choice for assessing vitamin A nutrition status, even though it is not the ideal method. Vitamin A is not excreted in human urine. Several metabolites of vitamin A have been recognized in the urine but they do not seem to reflect vitamin A status.[16] Measurements of dark adaptation to identify night blindness are too complex to be of value in routine assessment.

Many investigators advocate the measurement of liver vitamin A since 90% of the body's vitamin A reserves are in the liver. However, several investigators have shown that liver reserves do not correlate well with serum vitamin A levels. In a recent study on autopsy specimens from accident victims in urban Thailand, Suthutvoravoot and Olson[17] observed that liver reserves of vitamin A rise with age. Women tend to have higher serum vitamin A but lower liver levels compared to men.This may be a hormonal effect.

Extrapolating from animal studies and available human data on the rate of depletion of liver and body reserves of vitamin A, these workers feel that liver reserves of 20 μg/g wet liver in preschool children and 10 μg/g wet liver in adults should afford protection for a period of 100 days. In fact, Olson[18] has shown that this approach can be adopted for assessing the vitamin A status of a community at large.

Measurements of liver reserves of vitamin A are routinely beset with practical problems. Recently a new approach to determine liver vitamin A in vivo without having to do a liver biopsy has been described by Rietz et al.[16] The method is based on the principle that when tritium- or deuterium-labeled vitamin A is administered, it equilibrates with the total body vitamin A. Measurement of specific activity of tritium: or deuterium:hydrogen ratio in a freshly taken sample of blood would reflect body reserves. Since most of the body's vitamin A is in the liver, it would reflect liver reserve. This method has been applied to pigs and sheep but its usefulness in humans has not yet been ascertained.

A dose response relationship between dietary or administered vitamin A and serum vitamin A has been observed in controlled studies on human volunteers[19] as well as field trials. Subjects with clinical evidence of vitamin A deficiency invariably have serum values below 10 μg/100 mℓ, but it is not uncommon to find individuals with clinical signs of vitamin A deficiency having serum values greater than 20 μg and individuals without clinical evidence having values less than 10 μg.

Serum vitamin A values less than 10 μg/100 mℓ are regarded as deficient (high risk), 10 to 19 μg/100 mℓ low (medium risk), and greater than or equal to 20 μg/100 mℓ acceptable (low risk) (Table 1). Pregnant women tend to have higher values for serum vitamin A.

Hypovitaminosis A is regarded to be a public health problem when serum vitamin A levels are less than 20 μg/100 mℓ in 15% or more of the population and less than 10 μg/mℓ in 5% or more of the population. Serum β-carotene measurements are of limited value for nutritional evaluation, but in populations which receive most of their vitamin A through vegetable sources, serum β-carotene concentration reflects the intake of foods rich in vitamin A precursors.

Several diseases and drugs affect serum vitamin A levels.[5] Chronic infections, liver diseases, sprue, cystic fibrosis, and protein malnutrition tend to lower serum vitamin A levels. In cystic fibrosis and protein-calorie malnutrition, the transport of retinol from the liver is affected and hence serum levels may be low despite adequate liver reserves.[5,20] As mentioned earlier, oral contraceptive use is associated with raised levels of serum vitamin A and retinol-binding protein (RBP) presumably due to increased mobilization of the vitamin from the liver.[7,8]

Since serum vitamin A, particularly retinol-bound to the transport protein, RBP determines the availability of the vitamin to the tissues, it would appear that regardless of what the liver reserves are, serum levels should determine the state of health.

Several methods have been described for the measurement of serum vitamin A in lipid extracts of serum. These include (a) spectrophotometric method,[21,22] (b) colorimetric procedures based on measurement of the blue color formed by complexing vitamin A with oxidizing agents like antimony trichloride,[23] trifluoroacetic acid,[24] and trichloroacetic acid,[25] and (c) fluorometric methods based on direct measurement of fluorescence in serum extracts[22,26] or measurement after removal of other fluorescing impurities by column chromatography.[27] In a recent study, Sivakumar[22] compared a direct microfluorometric procedure with spectrophotometric as well as trifluoroacetic acid-colorimetric procedures. The three methods compared well except with sera obtained from pregnant women, in which the fluorometric procedure gave values which were several-fold higher. This precludes the use of direct fluorometric procedure for the measurement of serum vitamin A in pregnancy.

Table 1
TENTATIVE GUIDELINES FOR THE INTERPRETATION OF BLOOD VITAMIN LEVELS AND ENZYME TESTS[a]

	Deficient (high risk)	Low (medium risk)	Acceptable (low risk)
Serum vitamin A (μg/100 mℓ)	<10	10—19	≥20
Erythrocyte transketolase — AC	>1.25	1.15—1.25	<1.15
Erythrocyte riboflavin (μg/100 mℓ)	<10	10—14.9	≥15
Erythrocyte glutathione reductase — AC	>1.4	1.2—1.4	<1.2
Erythrocyte aspartate aminotransferase — AC	≥1.5-2.0[a]		<1.5—2.0
Erythrocyte alanine aminotransferase — AC	≥1.25		<1.25
Serum folate (ng/mℓ)	<3.0	3.0—5.9	≥6.0
Erythrocyte folate (ng/mℓ)	<140	140—159	≥160
Serum vitamin B_{12} (βg/mℓ)	<150	150—200	≥200
Serum ascorbic acid (mg/100mℓ)	<0.2	0.2—0.29	≥0.3
Leucocyte ascorbic acid (mg/100 mℓ)	0—7	8—15	15

[a] Adapted from Sauberlich, Dowdy, and Skala,[5] and Salkeld, Knörr, Kärner, and Thurnham.[77,78]

III. THIAMIN

The incidence of beriberi has decreased remarkably in recent years, but the disease is still seen in the populations of Southeast Asian countries where foods rich in thiaminase (raw fish) or antithiamin substances (fermented tea leaves) are consumed. Milder manifestations of thiamin deficiency are common among pregnant women and the elderly. Wernicke's encephalopathy among alcoholics is believed to be due to thiamin deficiency.

Thiamin nutritional status can be evaluated by measuring urine or blood levels of the vitamin or by functional tests.

A. Urinary Thiamin

There appears to be a correlation between dietary intake of thiamin and urinary levels of the vitamin.[5,28,29] This is so not only in controlled studies among human volunteers wherein 24-hr specimens are collected, but also in nutrition surveys where random urine samples are collected and thiamin excretion related to creatinine excretion. The latter, however, is useful only in community surveys involving large population groups, but is totally unsatisfactory for assessment of individual cases for reasons discussed in Section IIIC1. Load return tests using 1 mg thiamin given parenterally or 5 mg given orally have been described.[5]

Adults receiving adequate dietary thiamin excrete more than 100 μg in 24 hr.[5,29] According to another guideline, urinary excretion greater than or equal to 66 μg/g creatinine indicates low risk[5] (Table 2). Children excrete more thiamin per gram creatinine than adults. It is, however, desirable that each investigator establish the range of normal values for the population he studies. For instance, limited data on Indian subjects[28] suggest that for a given dietary intake, urinary excretion of thiamin in Indians tends to be higher than that reported by Sauberlich et al.[5] based on ICNND data.

Table 2
TENTATIVE GUIDELINES[5] FOR ACCEPTABLE URINARY EXCRETION OF VITAMINS OR METABOLITES IN ADULTS

	Per day	Per gram creatinine
Thiamin (μg)	≥100	≥66
Riboflavin (μg)	≥120	≥80
Pyridoxine (μg)	—	≥20
4-Pyridoxic acid (mg)	≥0.8	—
Xanthurenic acid (mg)	<25	—
2-Pyridone/N′ methyl nicotinamide	1.0—4.0	—
FIGLU mg/8 hr	<30	—
Methylmalonic acid (mg)	<24	

thiamin requirement is linked with calorie intake, urinary excretion of thiamin can be expected to be influenced by calorie intake. A rapid linear increase in urinary thiamin occurs at intakes exceeding 0.5 to 0.6 mg/day or 0.3 to 0.4 mg/1000 calories.[5,28] At lower intakes, the rate of excretion increases very slowly, most of the excretion being in the form of metabolites.

Urinary thiamin can be estimated by the thiochrome method after one step purification on an ion exchange resin like Decalso.[30] A direct procedure in which thiamin is measured before and after destruction of fluorescence due to thiamin by the addition of benzylsulfonyl chloride has been described.[31]

B. Blood Thiamin

The concentration of thiamin in blood is very low and difficult to measure. Opinion with regard to the usefulness of blood thiamin for the purpose of nutritional evaluation is divided. According to Burch et al.[32] blood thiamin levels do not change even when dietary intake of the vitamin is very low. Baker and Frank,[33] on the other hand, claim that blood thiamin concentration is the best index of thiamin nutrition status. They have suggested the use of a unicellular organism, *Ochromonas danica,* for the measurement of blood thiamin levels, and have reported a range of 25 to 75 ng/mℓ for whole blood and 15 to 42 ng/mℓ for serum. The reproducibility of this method in the hands of other investigators is yet to be established.

Recently, Warnock et al.[34] have developed a method for the specific measurement of thiamin pyrophosphate (TPP) in blood using yeast apo pyruvate decarboxylase. TPP concentration in normal human erythrocytes was found to range from 50 to 150 μg/mℓ of packed cells. Based on experiments in rats these authors claim that erythrocyte TPP concentration is a more sensitive index of thiamin status than is erythrocyte transketolase activity (to be discussed in Section IIIC2).

C. Functional Tests

In mammalian tissues, thiamin is required as a cofactor for oxidative decarboxylation of alpha-keto acids and for transketolase activity. Both these functions are depressed in thiamin deficiency.

1. Glucose Load Test

Horwitt et al.[35] have described a glucose load test in which pyruvate and lactate levels in blood are measured 1 hr after administration of glucose (1.8 g/kg body weight) followed by a standard exercise. Though the test gives useful information, it is difficult to carry out and hence has found limited application.

2. Erythrocyte Transketolase Test (ETK)

Transketolase is a TPP-dependent enzyme. It catalyzes the following two reactions of the HMP-shunt pathway.

1) Xylulose-5-P + Ribose-5-P $\rightleftharpoons$ Sedoheptulose-7-P + glyceraldehyde-3-P

2) Xylulose-5-P + Erythrose-4-P $\rightleftharpoons$ Fructose-6P + glyceraldehyde-3-P

In thiamin deficiency ETK activity falls, but its in vitro stimulation with TPP, referred to as TPP effect, increases.[36] TPP effect less than 15% (AC,1.15) is regarded as being normal. Values between 15 and 25% (AC,1.15 − 1.25) are indicative of marginal deficiency and values greater than 25%, (AC, 1.25) severe deficiency (Table 1).

The method has been found to work well in controlled human studies[5,28] where thiamin deficiency is induced experimentally. Experiments in rats as well as humans show that prolonged deficiency of thiamin lowers the apo-ETK[28,36] and, hence, in such situations the TPP effect would fail to assess the true magnitude of thiamin deficiency. Failure of in vitro added TPP to bind with apo-ETK could also lead to low TPP effect, despite low enzyme activity. It is therefore advisable to be guided not just by the TPP effect, but also by enzyme activity, and suspect thiamin deficiency when the enzyme activity is low. The fallacies in relying completely on the TPP effect have been discussed by Warnock.[37]

Studies in rats show that deficiencies of other nutrients such as protein, riboflavin, pyridoxine, folic acid, and vitamin C do not affect the measurement of TPP effect.[36,38] According to a recent study in rats, undernutrition lowers the enzyme activity but not the TPP effect, and hence those authors regard TPP effect as a more reliable index of thiamin nutrition status than ETK activity.[39]

Low values of ETK activity have been observed in patients suffering from liver diseases,[40,41] uremic neuropathy,[42] cancer,[43] early onset diabetes,[44] and disorders of the gastrointestinal tract.[45] In cancer patients there is a concomitant rise in TPP effect, despite dietary adequacy of thiamin, suggesting an impaired conversion of thiamin to TPP. On the other hand, diabetic patients show a fall in unstimulated as well as stimulated ETK activities with no change in TPP effect, indicating an apoenzyme deficiency. ETK activity tends to be raised in pernicious anemia[44-46] but not in other anemias.[44,46] Very often patients with cirrhosis, despite low enzyme activity, do not show a high TPP effect.[40] However, recently it was reported that administration of high doses of thiamin (200 mg/day) to patients with chronic liver disease brought the TPP effect to normal and stimulated the synthesis of the enzyme.[41]

Transketolase activity can be measured by colorimetric procedures involving measurement of ribose disappearance or the formation of sedoheptulose or hexose. Ribose-5-phosphate is used as the substrate. Xylulose-5-phosphate is formed rapidly from ribose-5-phosphate by the action of pentose phosphate epimerase and pentose phosphate isomerase which are abundantly present in the erythrocytes. A refinement of the colorimetric procedure which distinguishes transketolase activity from overall ribose-5-phosphate utilization by measuring ribose disappearance and sedoheptulose formation in the same color reaction has been described.[37]

Recently a kinetic procedure based on the measurement of NADH oxidation was described.[47] The coefficient of variation of this method appears to be lower than that of the colorimetric procedures.

Occasionally, addition of TPP has been found to inhibit rather than stimulate the transketolase activity. The reason for this is not clear. High concentration of TPP has been reported to inhibit ETK activity.[48]

The enzyme activity has been expressed on the basis of mℓ RBC, g hemoglobin, or number of red blood cells by different workers.

Though the erythrocyte transketolase test has proved to be quite useful in nutrition surveys as well as individual assessment, correlations between dietary intake of thiamin, presence or absence of clinical signs of deficiency, and biochemical deficiency are not always seen.

Since erythrocytes are nonnucleated and have a long half-life, it has been suggested that leukocyte transketolase may be a better measure of thiamin nutrition status than erythrocyte enzyme.[49] The usefulness of leukocyte transketolase for assessing thiamin nutrition status has been verified in animals, but the technical difficulties involved in obtaining sufficient leukocytes in man may offset the advantages.

In rats leukocyte pyruvate decarboxylase has been found to be less sensitive than ETK to thiamin deficiency.[50]

IV. RIBOFLAVIN

Riboflavin deficiency is widely prevalent in developing countries, particularly among pregnant women, nursing women, and children in whom the incidence of clinical signs may be as high as 20 to 25% and biochemical deficiency over 70%. Biochemical deficiency of riboflavin is seen among pregnant women and children even in developed countries. The typical clinical symptoms are the triad or orolingual lesions — angular stomatitis, glossitis, and cheilosis.

Riboflavin nutrition status has been assessed by measuring urinary excretion of the vitamin in fasting or random or 24-hr specimens, load return tests, measurement of erythrocyte riboflavin concentration, and the erythrocyte glutathione reductase test.

A. Urinary Riboflavin

Riboflavin is not metabolized extensively by the body and hence metabolites (except those formed by bacteria) are not detected in the urine. Correlations between urinary riboflavin expressed per gram creatinine and dietary intake have been observed in controlled experiments on human volunteers as well as in population surveys,[5,29,51,52] but the limitations discussed earlier for urinary thiamin apply to riboflavin as well. The method is useful only for population surveys and not for individual assessment. Load return tests may prove to be more useful for this purpose. On adequate dietary intake of riboflavin, urinary excretion is more than 120 μg/day. According to another guideline it is greater than or equal to 80 μg/g creatinine (Table 2). Children excrete more riboflavin per gram creatinine than adults. Each investigator should attempt to obtain his own guidelines.

Conditions causing negative nitrogen balance and ingestion of antibiotics increase urinary riboflavin due to tissue depletion rather than improvement in nutrition status.[5,53] Use of oral contraceptives lowers urinary excretion of riboflavin in women[54] and rats.[10] This appears to be due to an increased retention of the vitamin by tissues such as liver.[10]

Measurement of urinary riboflavin can be done by fluorometric as well as microbiological procedures.

B. Red Blood Cell Riboflavin

Use of erythrocyte riboflavin for nutrition evaluation was suggested by Bessey et al.,[55] but this is not a very sensitive index since the changes observed are of a small magnitude. Riboflavin in erythrocytes, plasma, or whole blood has been measured by fluorometric and microbiological assays.[5,33,55,56] A good correlation between fluorometric and microbiological assays using *Lactobacillus casei* as the test organism has been observed.[56] The ICNND guidelines[5] for interpreting erythrocyte riboflavin concentration per 100 dℓ cells are: less than 10 μg, deficient; 10 to 14.9 μg, low; and 15 μg and above, normal (Table 1). However, in a recent study, these criteria did not seem to be valid for rural schoolboys in India, where even the clinically and enzymically deficient children had erythrocyte riboflavin levels higher than 15 μg/100 dℓ cells[57] with no clear-cut difference between those with clinical lesions and those without lesions. In that study all the boys were found to suffer from biochemical riboflavin deficiency as judged by the erythrocyte glutathione reductase test (Section IV).

In rats, pyridoxine deficiency has been shown to elevate erythrocyte riboflavin concentration,[58] but in the Indian study cited no correlation seemed apparent between erythrocyte riboflavin and pyridoxine nutritional status. Children with erythrocyte riboflavin concentration less than 22 μg/100 mℓ were invariably found to be deficient by the enzymatic test, but high levels of riboflavin did not always mean sufficiency of riboflavin.

C. Erythrocyte Glutathione Reductase Test

Most flavin enzymes are mitochondrial and are therefore absent in erythrocytes. However, erythrocytes do contain the flavin enzyme — erythrocyte glutathione reductase (EGR) — which catalyzes the reduction of oxidized glutathione (GSSG) by the following reaction:

$$\text{GSSG} + \text{NADPH} + \text{H}^+ \rightarrow \text{NADP}^+ + 2\,\text{GSH}$$

Experimental studies in animals and man have shown that in riboflavin deficiency, EGR activity falls and that its in vitro stimulation by FAD is raised.[5,59] The latter is either expressed as activation coefficient (AC) (activity with added FAD/activity without added FAD) or as FAD effect (AC × 100 − 100).[5,59,60]

In a controlled human study it was observed that the EGR activity showed a linear increase with increasing intake of dietary riboflavin, but began to plateau at intakes close to 0.5 mg/1000 calories.[60] At this level of riboflavin intake, EGR-AC was less than 1.25 indicating that the cutoff level for EGR-AC to differentiate riboflavin inadequacy and sufficiency should be around 1.25.[60] Other studies also suggest that when riboflavin intake is adequate, AC is less than 1.25 (FAD effect <25%).[5,59] Tentative guidelines for interpreting EGR-AC are as follows[5]: <1.20, acceptable; 1.20 to 1.4, low; and >1.4, deficient (Table 1).

Studies in rats suggest that deficiency of other vitamins such as thiamin, folic acid, and vitamin C do not affect the EGR-test.[39,58] Undernutrition[39] and pyridoxine deficiency[58] produced a fall in EGR activity but no change in AC, suggesting a decrease in apo-EGR.

An inverse correlation between urinary excretion of riboflavin on the one hand and EGR-AC on the other has been reported in population surveys, but the relationship is weak and does not appear to be a simple one. According to Brubacher,[51] while thiamin deficiency per se tends to decrease the level of apo-ETK, a similar decrease in apo-EGR is not seen in riboflavin deficiency, thus increasing the confidence in EGR-AC test as an index of riboflavin nutrition status. However, recently we have observed that prolonged supplementation with riboflavin does raise apo-EGR levels as indicated by an increase in total enzyme activity.[57]

EGR test does not correlate very well with erythrocyte riboflavin concentration[57,60] even though most of the riboflavin in blood is in the form of FAD, and EGR is the major flavoprotein in erythrocytes.

In the author's experience, subjects with clinical evidence of riboflavin deficiency generally show EGR-AC greater than 1.25, but the reverse is not true. Several subjects with severe riboflavin deficiency, as judged by the EGR test, do not show clinical evidence of riboflavin deficiency. Thus, the test has proved to be of great value in identifying subclinical riboflavin deficiency.

Increased EGR activity with a high degree of saturation with FAD has been reported in the red cells of patients with severe uremia, cirrhosis of the liver, and glucose-6-phosphate dehydrogenase deficiency.[61] Higher activity has also been reported in iron deficiency anemia.[62]

EGR activity can be assayed either by measuring rate of NADPH oxidation (kinetic method) or by colorimetric estimation of the reduced glutathione formed. The former method requires a high-performance spectrophotometer with temperature control to permit measurements of optical density change in dense solutions, since high concentration of NADPH is used for the assay. Hemoglobin also contributes to absorption at 340 nm. The kinetic procedure has been optimized.[63] Using the optimized method we have observed good correlations between the colorimetric method[60] and the kinetic method.[63] High concentrations of FAD and low ratio of GSSG to FAD have been found to inhibit EGR activity, giving rise to AC ratios less than 1.0.[64]

Glatzle has demonstrated that whole blood glutathione reductase activation with FAD (BGR-AC) can also be used for assessment of riboflavin nutrition status.[65] The method can be applied to capillary blood specimens. An automated Auto Analyzer method involving measurement of reduced glutathione formed with 5,5′-dithiobis-2-nitrobenzoic acid (DTNS) has been described recently for BGR assay.[66] Using this procedure, BGR-AC was found to be between 1.00 and 1.35 in 617 normal children. BGR-AC tends to be lower than corresponding EGR-AC,[67] suggesting that the method might be less sensitive, though more convenient.

At present, EGR-AC test is regarded as the best laboratory test for assessing riboflavin nutrition status.

V. PYRIDOXINE

Vitamin B_6 occurs in several biologically active forms such as pyridoxol, pyridoxal, pyridoxamine, and phosphorylated derivatives of each of these. The major biologically inactive metabolite of vitamin B_6 in urine is 4-pyridoxic acid. The presence of 4-pyridoxic acid phosphate has also been reported in urine.

Pyridoxal phosphate (PLP) catalyzes a variety of biochemical reactions. While most of them are related to protein metabolism, pyridoxine is also required for the metabolism of carbohydrates and fats. Biochemical assessment of pyridoxine status is rather difficult due to high sensitivity of PLP-enzymes to hormones and drugs. This leads to alterations in the distribution of the vitamin between tissues and between its numerous biochemical functions within a tissue. Symptoms of pyridoxine deficiency in man are not fully established. Experimentally induced deficiency leads to oral lesions and peripheral neuritis similar to those seen in other B-complex vitamin deficiencies. Vitamin B_6 deficiency in infancy can lead to central nervous system disorders such as hyperactivity, behavioral changes, convulsive seizures, and possible mental retardation. Anemia and dermatitis may also occur. PLP synthesis in mammalian tissues is dependent on a flavin enzyme and there is a close interrelationship between these two vitamins.[68]

A. Urinary Pyridoxine

A correlation between pyridoxine intake and urinary excretion per gram creatinine has been reported in controlled studies on human volunteers. Urinary excretion of less than 20 μg/g creatinine is indicative of marginal or inadequate dietary intake of the vitamin[5] (Table 2). The limitations regarding the interpretation of urinary excretion data discussed earlier for other vitamins apply to pyridoxine as well. Pyridoxine requirement is generally related to dietary protein, but dietary protein does not affect urinary vitamin B_6 significantly. Like thiamin and riboflavin, children tend to excrete more pyridoxine per gram creatinine than do adults.

The most convenient method for measuring urinary pyridoxine is the microbiological assay using yeast, *Saccharomyces carlsbergensis,* also known as *S. uvarum* (ATCC No. 9080). A protozoological assay using *Tetrahymena pyriformis* has also been described.[69]

Between 20 to 50% of dietary pyridoxine is excreted as 4-pyridoxic acid. A number of other unidentified metabolites are also excreted. Excretion of pyridoxic acid in urine correlates with ingested pyridoxine. Its measurement, however, is difficult and the method is seldom used. On normal diet, 4-pyridoxic acid excretion is more than 0.8 mg/day (Table 2).

B. Blood Levels of Pyridoxine

Unlike the other water-soluble vitamins whose concentrations in red blood cells are higher than in plasma, the concentration of pyridoxine is higher in plasma than in red cells. In one study involving large number of subjects, vitamin B_6 levels for whole blood, red cells, and plasma of normal subjects were reported to be 37 ± 6 (mean ± SD), 20 ± 3, and 59 ± 13 ng/mℓ, respectively, by the protozoological assay.[69] Lower levels are, however, observed with the yeast assay.

Blood pyridoxine levels show some correlation with dietary vitamin, but due to wide range of reported values, guidelines for detecting deficiency have not been set up. Measurement of serum pyridoxal phosphate may prove to be of greater value. The most reliable and sensitive procedures for measuring serum pyridoxal phosphate are methods based on apoenzyme-coenzyme coupling, using apotyrosine decarboxylase[70] or apotryptophanase[71] and labeled tyrosine or tryptophan, respectively. These methods can be employed only by laboratories that have sensitive radioactive counters. In a recent study on pregnant women, Lumeng et al.[72] found that measurements of plasma pyridoxal phosphate are more reliable for assessing vitamin B_6 status than are enzymic tests involving stimulation of erythrocyte transaminases with PLP. The same group of workers has observed a good correlation between plasma PLP and tissue PLP levels in rats.[73]

C. Functional Tests

1. Tryptophan Load Test

The enzyme kynurinase of the tryptophan-niacin pathway requires pyridoxal phosphate as a cofactor. Its activity is affected by pyridoxine deficiency and hence urinary excretion of xanthurenic acid after a tryptophan load is higher in pyridoxine deficient subjects even though the formation of xanthurenic acid from 3-OH kynurenine also involves PLP-dependent transamination (Figure 1). Excretions of other metabolites such as kynurenic acid and 3-hydroxykynurenine are also increased.

In the tryptophan load test, 2 or 5 g of tryptophan is given orally and urine collections made for a 6- to 8-hr period or preferably 24-hr period after the load. According to Luhby et al.[74] normal subjects excrete less than 32 μmol of xanthurenic acid in 8-hr collection after 2 g tryptophan load. This value is lower than the guidelines given by Sauberlich et al.[5] (less than 25 mg/day acceptable) (Table 2) assuming that 63% of xanthurenic acid excreted in 24 hr is recovered in 8-hr collection, as shown by Luhby et al.[74]

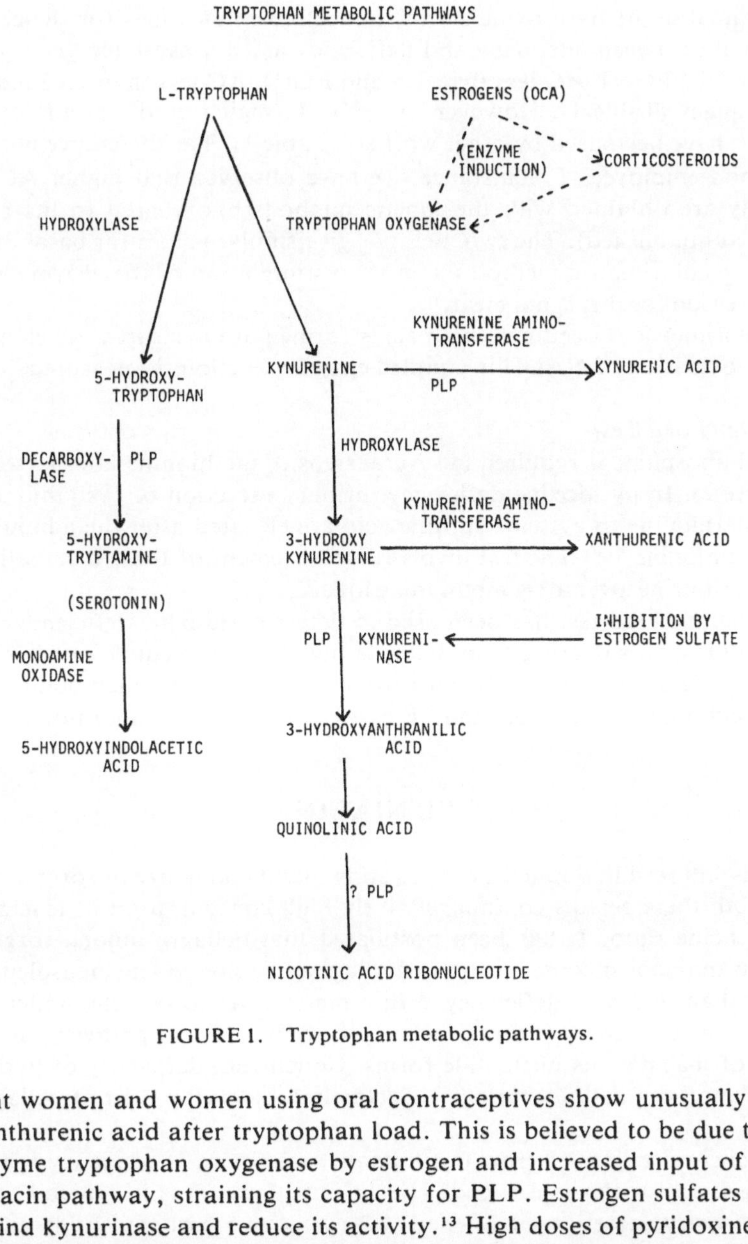

FIGURE 1. Tryptophan metabolic pathways.

Pregnant women and women using oral contraceptives show unusually high excretion of xanthurenic acid after tryptophan load. This is believed to be due to induction of the enzyme tryptophan oxygenase by estrogen and increased input of tryptophan into the niacin pathway, straining its capacity for PLP. Estrogen sulfates are also believed to bind kynurinase and reduce its activity.[13] High doses of pyridoxine are needed to correct this abnormality in pregnant women and oral contraceptive users. Mentally retarded children[75] and cancer patients[76] show increase in urinary excretion of xanthurenic acid and other metabolites of tryptophan after tryptophan load.

2. *Enzymatic Tests — Blood Transaminase Activities*

Activities of erythrocyte glutamate oxaloacetate transaminase (EGOT) as well as glutamate pyruvate transaminase (EGPT), also known as aspartate aminotransferase (EASPAT) and alanine aminotransferase (EALAT), respectively, are regarded to be sensitive indices of pyridoxine nutrition status. In pyridoxine deficiency, the activities of these enzymes fall, but in vitro stimulation of the enzymes with PLP (activation coefficients) shows a rise.

Enzymatic tests based on activation of EGOT and EGPT with PLP have been widely used for evaluation of pyridoxine status, but reliable guidelines for the cutoff points to differentiate between adequacy and deficiency are not available yet. According to Sauberlich et al.[5] EGOT-AC less than 1.5 and EGPT-AC less than 1.25 indicates vitamin B_6 adequacy (Table 1). However, for EGOT, higher cutoff points ranging from 1.7[77] to 2.0[78] have been used by other workers (Table 1). The difference may be related to the method employed. For instance, we have observed that higher AC values for EGOT assay are obtained with the kinetic methods[63] compared to the colorimetric procedure[79] (unpublished). The two methods give similar values for basal EGOT activity, but in the colorimetric method the in vitro stimulation of the enzyme with PLP is lower. The reason for this is not clear.

In the colorimetric procedures, keto acids formed are measured, whereas in kinetic procedures, oxidation of NADH in coupled enzyme reactions is measured.

3. Methionine Load Test

Pyridoxal phosphate is required in several steps of methionine homocysteine metabolism (Figure 2). In pyridoxine deficiency, urinary excretion of cystathionine and the ratio of cystathionine to cysteine sulfinic acid are elevated after the administration of a load of methionine.[80,81] The test involves measurement of these metabolites in a 24-hr collection of urine after a 3-g methionine load.

The methionine load test has been used to detect pyridoxine deficiency in pregnant women,[82] adult men suffering from neuropathy,[83] and patients treated with isonicotinic acid hydrazide.[84] Since analysis of urine for methionine metabolites is done on an amino acid analyzer, the test can be carried out by only laboratories having this facility.

VI. NIACIN

Pellagra is still seen in populations subsisting mainly on maize or sorghum (*Sorghum vulgare*). Both these cereals contain relatively high concentration of leucine and high leucine:isoleucine ratio. It has been postulated that pellagra among sorghum eaters belonging to the poor income groups in India may be due to leucine:isoleucine imbalance rather than to niacin deficiency.[85] In a manner yet to be fully understood, high leucine in jowar diet appears to affect the tryptophan-niacin pathway, as well as the conversion of niacin to its nucleotide forms. Concurrent deficiency of pyridoxine appears to be another possible etiological factor in the development of pellagra in Indians.[86]

Since niacin is also derived from tryptophan, assessment of adequacy of this vitamin based upon diet histories is difficult. Niacin nutrition status is generally assessed by measuring urinary excretion of N′-methyl-2-pyridone-5-carboxylamide and N′-methylnicotinamide. The ratio 2-pyridone/N′-methylnicotinamide is regarded to be the best index of niacin nutrition status.[5,87] It is independent of age and creatinine excretion.[5] Tentatively, a ratio of less than 1.0 is regarded as indicating niacin deficiency.[5] In normal subjects its value ranges between 1.0 to 4.0[5] (Table 2).

Measurement of niacin or its nucleotide in blood is not very reliable for the purpose of evaluating niacin status. The erythrocytes of pellagrins have been reported to have higher amounts of nicotinamide mononucleotide, but lower levels of NAD and NADP than the erythrocytes of normal subjects.[88]

VII. FOLIC ACID (FOLATE)

The earliest hematological manifestation of folic acid deficiency (eventually leading to megaloblastic anemia) is hypersegmentation of the nuclei of the neutrophilic poly-

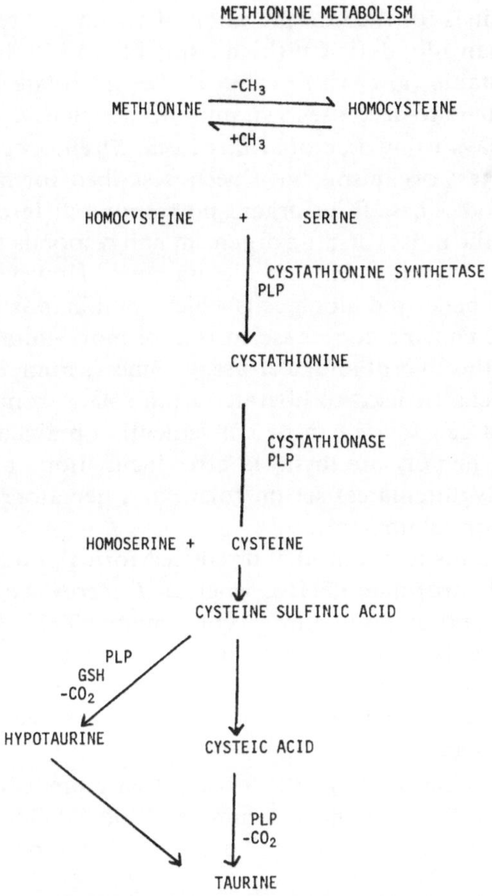

FIGURE 2. Methionine metabolism.

morphonuclear leukocytes (PMN). Megaloblastic changes also occur in epithelial tissues from a variety of organs such as cervical epithelium.

Folic acid deficiency is frequently seen in pregnant women and children in the developing countries, though the most severe clinical manifestations may be masked by concomitant presence of iron deficiency in them. In developed countries, 2.5 to 5% of pregnant women may suffer from megaloblastic anemia.[5] The segmentation of the neutrophilic PMN nuclei tends to be reduced in pregnancy and hence this hematological parameter fails to show a good correlation with biochemical indices of folate deficiency in pregnant women.

Folic acid nutrition status is generally assessed by measuring serum folate, red blood cell folate, or urinary excretion of formiminoglutamic acid (FIGLU) after a histidine load. Urinary excretion of folate has not been of much value in nutritional evaluation.

A. Serum and Red Blood Cell Folate

Experimental folate deficiency in man was found to reduce serum folate after 22 days on the deficient diet, whereas red cell folate showed a fall only after 123 days.[89] While serum folate tends to reflect recent dietary intake of the vitamin, red cell folate can be expected to reflect folate status at the time of formation of the red blood cells. Nevertheless, good correlations have been observed between serum and red cell folate, serum and liver folate, and red cell and leukocyte folate, indicating that serum as well as red cell folate values reflect tissue folate status.

The tentative guidelines for the interpretation of serum and red cell folate are serum folate (ng/mℓ), less than 140, deficient (high risk); 140 to 159, low (medium risk); and 160 and above, acceptable (low risk) (Table 1). Serum folate levels tend to increase but erythrocyte levels tend to decrease in vitamin B_{12} deficiency.

Microbiological assays using *Lactobacillus casei, Streptococcus faecalis,* and *Pediococcus cerevisiae* as test organisms have been described for measuring the different forms of folate in blood. These three organisms respond differently to different forms of folate.[90] *L. casei* is the most versatile organism and responds to all forms of oxidized and reduced folate monoglutamates and diglutamates. However, it does not respond to polyglutamates and hence red blood cells which contain mostly polyglutamates have to be treated with the enzyme conjugase to release monoglutamates from polyglutamates, before doing the microbiological assay. Since serum is a rich source of the enzyme conjugase, it can be used to liberate serum folate from polyglutamates in red cells. Estimates of red cell folate can be conveniently obtained from values for total folate in whole blood hemolysate (assayed after incubation at 37°C to liberate monoglutamates from polyglutamates) serum folate and hematocrit. Serum folate is entirely in the form of monoglutamates.

S. faecalis also responds to oxidized and reduced forms of mono- and diglutamates, except 5-methyl tetrahydrofolate (THF), whereas *L. cerevisae* responds only to fully reduced folate mono- and diglutamates except 5-methyl THF. For nutritional evaluation, *L. casei* is the organism of choice since it measures total folate activity. This assay, though very sensitive, requires meticulous care to prevent high blanks and spuriously high or low values. Presence of antibiotics and sulfa drugs in blood may prevent the growth of the organism.

Recently isotopic protein-binding assays using folate binders in milk have been described for measuring folate content of serum and blood.[91] Good correlations between isotopic assays and microbiological assays have been observed.

B. Functional Tests

1. Histidine Load Test — Urinary Excretion of Formiminoglutamate and Urocanate

Formiminoglutamate (FIGLU) and urocanate are products of histidine metabolism in mammals (L-histidine→urocanic acid→imidazolonepropionic acid→FIGLU). In folic acid deficiency urinary excretion of FIGLU and urocanate is markedly elevated, particularly after an oral load of 2 to 15 g of histidine. Generally only FIGLU excretion is measured in 5-, 8-, or 24-hr collections of urine after the histidine load, since measurement of urocanate does not provide any additional information. Normal adults excrete less than 30 mg FIGLU in 8 hr after the load[5,90] (Table 2). Folate-deficient subjects excrete five to ten times this level.[5] FIGLU excretion is raised even in vitamin B_{12} deficiency and hence the test is not specific for folate deficiency. Urinary FIGLU has been found to show better correlation with red cell folate than with serum folate values.

2. Deoxyuridine Suppression Test

Objective methods for defining early megaloblastic change are not available. PMN segmentation is modified by pregnancy and infections and hence this method is not very reliable for detecting folic acid deficiency. Recently a deoxyuridine suppression test was described.[92,93] In normal bone marrow, ^{3}H-thymidine incorporation is substantially depressed by prior incubation of the bone marrow with deoxyuridine — a precursor of thymidine. This suppression has been found to be markedly reduced in bone marrows of patients suffering from deficiencies of folic acid and vitamin B_{12}. In the former, the defect can be corrected by the in vitro addition of pteroyl glutamic acid. Similar correction by the in vitro addition of vitamin B_{12} is not as marked in the latter. This test can be used to differentiate between folic acid and vitamin B_{12} deficiencies. However, it is difficult to carry out. Bone marrow biopsy is an invasive procedure and can be applied only in special cases admitted to the hospital.

VIII. VITAMIN B_{12}

Dietary deficiency of vitamin B_{12} is not very common but it may be seen in vegetarians who do not consume animal products due to economic or other reasons, e.g., Vagans. Dietary deficiency of vitamin B_{12} has also been reported among the elderly.[94]

Milder forms of vitamin B_{12} deficiency may manifest as anorexia, mild neuropathies, and soreness of tongue. More severe deficiency can lead to megaloblastic anemia, which is indistinguishable from that due to folic acid deficiency, and severe neurological lesions.

Pernicious anemia is due to lack of intrinsic factor in gastric juice. This condition can be identified by carrying out radioactive vitamin B_{12} absorption test (Schilling test or its modifications). In this test, a tracer dose (0.5 to 2.0 μg) of radiolabeled vitamin B_{12} is administered orally and a flushing dose (1 mg) of unlabeled vitamin B_{12} is administered intramuscularly. Urinary excretion of vitamin B_{12} is measured in the following 24 hr. A normal person excretes 10 to 40% of the administered radioactivity in 24 hr, whereas a patient of pernicious anemia excretes less than 7% of the administered dose. In severe undernutrition, vitamin B_{12} absorption may be affected.[111]

Vitamin B_{12} nutrition status is generally assessed by measurement of serum vitamin B_{12} concentration or by measuring methylmalonic acid excretion in urine after a load of valine.

A. Serum Vitamin B_{12}

According to the guidelines provided by the World Health Organization,[95] serum vitamin B_{12} values less than 150 pg/mℓ indicate deficient — high-risk status; 150 to 200 pg/mℓ, low — medium-risk status; and values above 200 pg/mℓ, acceptable — low-risk status (Table 1). Other workers have used slightly different guidelines — less than 100 pg deficient, 100 to 149 or 159 pg low, and greater than 150 or 160 pg adequate.[5]

Serum vitamin B_{12} levels can be estimated by microbiological assays. The organisms most commonly used are *Lactobacillus leichmanni* and *Euglena gracilis*. The *L. leichmanni* technique requires initial extraction of protein and it has been suggested that this may be a source of inaccuracy. Numerous isotope dilution assays which use vitamin B_{12} binders from serum have also been described. While good correlations between microbiological and isotope dilution assays have been observed, the latter tend to give slightly higher values. The isotopic methods are simple, reliable, and not affected by the presence of antibiotics in the serum, but they require isotopes, expensive counting equipment, and larger quantities of serum.

Low values of serum vitamin B_{12} in the absence of folate or iron deficiency suggest reduction of liver levels of vitamin B_{12}.[96] In the presence of folate or iron deficiency, fall in serum vitamin B_{12} is associated with reduction in tissue vitamin B_{12} levels but not of the same magnitude. According to a recent report, the fall in serum vitamin B_{12} in pregnancy may not have a functional significance.[97] Low levels of serum vitamin B_{12} are observed in certain iatrogenic conditions. Serum vitamin B_{12} levels may be normal despite tissue vitamin deficiency, if pernicious anemia is associated with chronic myeloid leukemia, but this is a rare circumstance.[96] Liver diseases tend to increase serum vitamin B_{12} levels.

B. Valine Load Test — Methylmalonic Acid (MMA) Excretion

Vitamin B_{12} is required for the isomerization of methylmalonyl coenzyme A to succinyl coenzyme A. The urinary excretion of MMA by normal subjects is less than 12 mg in 24 hr (Table 2). In pernicious anemia it may exceed 500 mg in 24 hr. Since valine is a precursor for MMA, an oral load of 5 to 10 g valine is given to raise the MMA excretion. Urinary MMA is not affected by folate deficiency and hence the test can be used to differentiate between megaloblastic anemia of folate and vitamin B_{12} deficiency. MMA excretion may be elevated in vitamin E deficiency (Section XII).

IX. PANTOTHENIC ACID

Pantothenic acid is ubiquitously distributed in natural foods and therefore its deficiency is not regarded as a public health problem. A syndrome called "burning feet syndrome" or "nutritional melalgia" has been found to respond to treatment with pantothenic acid.[98]

Assessment of pantothenic acid nutrition status on the basis of dietary intake is difficult because estimates of its daily requirement for man are not available. However, urinary excretion in 24-hr specimens and blood levels of the vitamin have been found to reflect the dietary intake of pantothenic acid. Most of the pantothenic acid in red blood cells is in the form of coenzyme A, but it is rapidly converted to free vitamin if blood is stored. Serum contains mostly free pantothenic acid. For nutritional evaluation measurement of total blood pantothenate is recommended.[5] Whole blood pantothenic acid value less than 100 μg/100 mℓ and urinary excretion of less than 1.0 mg/day are regarded to be indicative of pantothenic acid deficiency.

Urinary pantothenic acid excretion has been found to be low in acute alcoholism but it is raised in chronic alcoholism. The former may be due to reduced food intake, whereas the latter may indicate impaired utilization.[99] Indian women of low income groups who are known to suffer from deficiency of several nutrients were recently reported to have normal values for pantothenic acid in urine, blood, and milk, suggesting that pantothenic acid deficiency may not be a widespread problem even in malnourished communities.[100]

Functional tests based on acetylation of sulfanilamide by red cells[101] and urinary excretion of acetylated p-aminobenzoic acid (PABA) after a load of PABA[102] have been described, but their usefulness is not established.

X. ASCORBIC ACID (VITAMIN C)

The clinical manifestations of scurvy are well recognized, but the biochemical role of vitamin C is yet to be elucidated. Ascorbic acid appears to be required as a reductant in some hydroxylation reactions such as hydroxylation of p-hydroxy phenylpyruvic acid (a metabolite of tyrosine) and hydroxylation of lysine and proline residues in the protocollagen molecule. However, according to a recent report, other reductants can partially replace ascorbic acid in the prolyl hydroxylase reaction.[103]

Assessment of ascorbic acid nutrition status is generally done by measuring leukoycte and serum levels of the vitamin. Urinary excretion and red blood cell values are not regarded as being very specific and useful indices.

A. Serum and Leukocyte Ascorbic Acid Levels

Serum levels of ascorbic acid show a linear increase with dietary intake of the vitamin up to the serum concentration of 1.4 mg/100 mℓ. Beyond this threshold for serum, urinary excretion shows rapid increase. A temporary increase beyond this level may however occur after the ingestion of high doses of ascorbic acid. The guidelines provided by ICNND for the interpretation of serum ascorbate levels have been revised by Sauberlich et al.[5] The revised guidelines are as follows: serum levels (mg/100 mℓ) less than 0.2, deficient — high risk; 0.2 to 0.29, low — medium risk; and levels exceeding 0.3, acceptable — low risk (Table 1). Children having low values for vitamin C in the serum often suffer from iron deficiency.[104]

Leukocyte ascorbic acid levels are more representative of tissue stores of the vitamin than are serum levels. Measurement of leukocyte ascorbic acid is, however, more difficult and requires large volumes of blood. Nutrition surveys often depend on serum or plasma levels of the vitamin. The guidelines for the interpretation of leukocyte ascorbic acid (mg/100 mℓ) are: values between 0 to 7, deficient — high risk; 8 to 15, medium risk; values greater than 15, acceptable — low risk (Table 1).

Ascorbic acid levels in tissues can be estimated by colorimetric procedures employing 2.6-dichlorophenol (DCP) or dinitrophenylhydrazine reagents. Ascorbate sulfate tends to interfere with the latter if the reaction is carried out at temperatures exceeding 37°C.

B. Functional Tests

Vitamin C deficiency results in increased urinary excretion of some metabolites of tyrosine. However, this abnormality is not specific for vitamin C deficiency. Tests based on disappearance of intradermally injected solution of DCP or lingually applied solution of DCP have been described but have not proved to be of much value.[105]

XI. VITAMIN D

Rickets can arise out of deficient intake of vitamin D, inadequate exposure to sunlight, metabolic defects leading to inadequate synthesis of the biological active metabolite 1,25-dihydroxy D_2 or D_3, and abnormality of renal tubular function leading to renal hypophosphatemia despite normal vitamin D metabolism.[106] Nutritional vitamin D deficiency rickets is uncommon in countries where food supply is fortified with irradiated ergosterol, but it is seen in developing countries despite abundant sunshine. Rickets has become a clinical problem among the Asian immigrants in the U.K., presumably due to inadequate exposure to sunlight and dietary deficiency of the vitamin. High phytate content of the diet, though once suggested, is an unlikely etiological factor.

Increase in serum alkaline phosphatase activity and decrease in inorganic phosphorus levels have long been used as biochemical indicators of rickets. While this is generally true in undernourished children despite radiological evidence of rickets, alkaline phosphatase activity is normal or only marginally raised. A variety of diseases alter alkaline phosphatase activity. Changes in serum calcium and phosphorus are also nonspecific.

25-(OH) D_3 is the major metabolite of vitamin D in circulation. However, serum levels of 25-(OH) D_2 may also rise following therapy with vitamin D_2. Protein-binding assays using binding proteins from serum or kidney have been developed for the assay of these metabolites. Small amounts of 1,25-$(OH)_2$ D_2 and D_3 are present in the blood. A radioreceptor assay using cytosol receptors from chick intestine has been described for estimating 1,25$(OH)_2$ D_2 and D_3. The sensitivity of this assay is 1.4 times higher for 1,25-$(OH)_2$ D_3 than for 1,25-$(OH)_2$ D_2. Hausler and colleagues have applied the chick receptor assay for the measurement of 1,25-$(OH)_2$ as well as 25-(OH) derivatives of vitamins D_2 and D_3 after chromatographic separation of the metabolites.[107] The normal plasma level of 25-OH D and 1,25-$(OH)_2$ D measured in Tucson adults were 25 to 40 ng/mℓ and 2.1 to 4.5 ng/100 mℓ, respectively.[107] More than 90% of both were present as vitamin D_3 metabolites. Hypervitaminosis D produced a marked increase in 25-OH D, particularly the D_2 derivative.

Measurement of these metabolites in blood promises to be of value in the identification of vitamin D deficiency and disorders related to vitamin D metabolism.

XII. VITAMIN E

Dietary deficiency of vitamin E is seldom seen in man. Low serum levels have been observed in conditions of malabsorption, e.g., cystic fibrosis, in premature infants due to low reserves and transient malabsorption, in some children suffering from protein calorie malnutrition, and in patients with a rare genetic disorder, d-β lipoproteinemia.

Though tocopherol-deficient red cells show enhanced in vitro susceptibility to peroxides and shortened survival time, hemolytic anemia is not a manifestation of vitamin E deficiency. Reports claiming treatment of anemia in premature infants and children with PCM are controversial. Unlike animals, in humans, muscle performance is not known to be affected by vitamin E deficiency, though ceroid accumulation does occur in a number of organs. The subject of vitamin E nutrition in man was recently reviewed.[108]

Vitamin E nutrition status has been assessed by either measuring serum vitamin E levels or by measuring the in vitro hemolysis of red cells by hydrogen peroxide or dialuric acid. The correlation between the two parameters is poor. The latter has been criticized as being nonspecific. Serum tocopherol levels show direct correlation with total lipids, cholesterol, and β-lipoproteins, and hence it has been suggested that plasma tocopherol concentrations should be related to total lipid content of plasma. According to Horrowit and colleagues (quoted by Beiri and Farrell),[108] 0.8 mg of tocopherols per gram of total lipids is indicative of adequate nutritional status in infants. However, more work relating these two components is necessary to obtain reliable interpretive guidelines.

Serum or plasma tocopherols can be measured by colorimetric procedures based on the Emmerie Engel reaction or by gas liquid chromatography.[5]

Premature infants suffering from vitamin E deficiency have been reported to excrete large amounts of methylmalonic acid — an abnormality similar to that seen in vitamin activity of vitamin B_{12} requiring enzyme methylmalonyl coenzyme A mutase, presumably due to impaired conversion of cyanocobalamine into its coenzyme adenosyl cobalamine.[109]

XIII. VITAMIN K

Dietary deficiency of vitamin K is very rare in adults, but vitamin K-responsive blood clotting disorders can arise in malabsorption syndromes, in ulcerative colitis, and in liver diseases. Use of antibiotics which destroy the intestinal flora and antivitamin K compounds such as coumarin and indanedione can also lead to deficiency symptoms.

Significant advances in our understanding of the molecular mode of action of vitamin K in the synthesis of blood clotting factors have been made in recent years.[110] The vitamin seems to be required for γ-carboxylation of certain glutamic acid residues in the proteins whose synthesis depends on vitamin K. Ca^{2+}-dependent binding of prothrombin to phospholipid surface seems to require the γ-carboxyglutamic acid residues in prothrombin.

Biochemical tests for the evaluation of vitamin K status are not available and hence laboratory tests for detecting vitamin K deficiency depend on the measurement of blood clotting time and prothrombin time. If the clotting disorder is due to vitamin K deficiency, it responds rapidly to treatment with the vitamin.

XIV. SUMMARY AND CONCLUSIONS

The goal of finding ideal biochemical tests which would satisfy all the criteria described in Section I is elusive and perhaps not attainable in the near future. Despite limitations, biochemical tests provide useful information regarding tissue stores of vitamins, and future research should be aimed at identifying the constraints underlying each of these tests and use this knowledge for the interpretation of data.

Correlation between biochemical and clinical indices is not aways seen in cross-sectional studies, mainly because during development of deficiency, biochemical changes precede clinical changes, and during recovery, biochemical lesions get corrected before morphology reverses to normal state. Thus in a population survey, one may find an individual who has biochemical but not clinical deficiency and vice versa. Different biochemical indices also change at different rates. There is a considerable amount of individual variation with regard to susceptibility to a given state of tissue depletion. Besides, clinical manifestations usually have a complex etiology.

In the case of B-complex vitamins, the conventional tests based on the measurement of tissue levels of the vitamins have been replaced by functional tests. Interpretation of enzymatic tests is easy in situations where enzyme activity and its in vitro stimulation by coenzyme are inversely related as in primary vitamin deficiency. However, there are conditions including chronic vitamin deficiency where this relationship does not hold good due to changes in the levels of apoenzymes. At present it is not clear if good health should be equated with enzyme activity or percent saturation of the enzyme with respect to its coenzyme. Can increased sequestration of coenzyme by one of the enzymes create deficiency at other functional points? Increased requirement in the tissues due to increase in specific enzymes or binding proteins may raise the tissue concentration of the vitamin but reduce serum levels. Is this to be regarded as deficient state? If so, serum levels of vitamins or coenzymes and load return tests may be more representative of body vitamin economy than some of the other tests.

Some workers have attempted to derive guidelines for the interpretation of laboratory tests. To find out if these are applicable to all populations, methodology has to be standardized and monitored by the use of appropriate quality control measures. Even under optimal conditions, it is uncertain if it will be possible to obtain precise estimates of cutoff points to differentiate between adequacy and deficiency, and ascertain the magnitude of biochemical insult an organism can suffer before succumbing to disease because this threshold is likely to have considerable degree of individual variation and capacity for adaptation. However, future research should be aimed at understanding the functional implications of marginal malnutrition as revealed through biochemical tests.

ACKNOWLEDGMENTS

The author is very grateful to Dr. S. G. Srikantia, Director, National Institute of Nutrition for a critical scrutiny of the contents of this chapter, and to Dr. Faizy Ahmed for the assistance given in the preparation of the manuscript.

REFERENCES

1. **Jelliffe, D. B.,** *The Assessment of Nutritional Status of the Community,* World Health Organization, Geneva, 1966, chap. 2.
2. **Bhat, K. S. and Belavady, B.,** Biochemical studies in phrynoderma (follicular hyperkeratosis). III. Thiamine, riboflavin and nicotinic acid nutritional status of children suffering from phrynoderma, *Indian J. Med. Res.,* 58, 753, 1970.
3. **Bhat, K. S. and Belavady, B.,** Biochemical studies in phrynoderma (follicular hyperkeratosis). II. Polyunsaturated fatty acid levels in plasma and erythrocytes of patients suffering from phrynoderma, *Am. J. Clin. Nutr.,* 20, 386, 1967.
4. **Mohan Ram, M. and Reddy, V.,** Variability in urinary creatinine, *Lancet,* 2, 674, 1970.
5. **Sauberlich, H. E., Dowdy, R. P., and Skala, J. H.,** Laboratory tests for the assessment of nutritional status, *CRC Crit. Rev. Clin. Lab. Sci.,* 4, 215, 1973.
6. **Roe, D. A.,** *Drug Induced Nutritional Deficiencies,* Avi Publishing, Westport, Conn., 1966, chap. 1.
7. **Bamji, M. S.,** Implications of oral contraceptive use on vitamin nutritional status, *Indian J. Med. Res.,* Suppl. 68, 80, 1978.
8. **Briggs, M. H.,** Vitamin A and teratogenic risks of oral contraceptives, *Br. Med. J.,* 3, 170, 1974.
9. **Ahmed, F., Bamji, M. S., and Iyengar, L.,** Effect of oral contraceptive agents on vitamin nutrition status, *Am. J. Clin. Nutr.,* 28, 606, 1975.
10. **Ahmed, F. and Bamji, M. S.,** Biochemical basis for the "riboflavin defect" associated with the use of oral contraceptives, *Contraception,* 14, 297, 1976.
11. **Pietarinen, G. L., Leichter, L., and Pratt, R. F.,** Dietary folate intake and concentration of folate in serum and erythrocytes in women using oral contraceptives, *Am. J. Clin. Nutr.,* 30, 375, 1977.
12. **Lakshmaiah, N. and Bamji, M. S.,** Effect of oral contraceptives on folate economy — study in female rats, *Hormones and Metabolic Research,* in press.
13. Oral contraceptive agents and vitamins, *Nutr. Rev.,* 30, 229, 1972.
14. **Scriver, C. R.,** Vitamin responsive inborn errors of metabolism, *Metabolism,* 22, 1319, 1973.
15. Ten-State Nutrition Survey Reports, I-V, Center for Disease Control, Atlanta, 1972.
16. **Rietz, P., Wiss, O., and Weber, F.,** Metabolism of vitamin A and the determination of vitamin A status, *Vitam. Horm.,* 32, 237, 1974.
17. **Suthutvoravoot, S. and Olson, J. A.,** Plasma and liver concentrations of vitamin A in a normal population of urban Thai, *Am. J. Clin. Nutr.,* 27, 883, 1974.
18. **Olson, J. A.,** A steady-state approach to vitamin A nutriture, in *11th Int. Congr. Nutrition,* Rio de Jenairo, Aug. 27 to Sept. 1, 1978, Abstr. No. 189.
19. **Hodges, R. E. and Kolder, H.,** *Summary of Proceedings of Workshop on Biochemical and Clinical Criteria for Determining Human Vitamin A Nutriture,* Food and Nutrition Board, National Academy of Sciences, National Research Council, Washington, D. C., 1971, 10.
20. **Patwardhan, V. N.,** Hypovitaminosis A and epidemiology of xerophthalmia, *Am. J. Clin. Nutr.,* 22, 1106, 1969.
21. **Bessy, O. A., Lowry, O. H., Brock, M. J., and Lopez, J. A.,** The determination of vitamin A and carotene in small quantities of blood and serum, *J. Biol. Chem.,* 166, 177, 1946.
22. **Sivakumar, B.,** Evaluation of a microfluorimetric method for serum vitamin A, *Clin. Chim. Acta,* 79, 189, 1977.
23. **Moore, T.,** *Vitamin A* Elsevier, Amsterdam, 1957, 586.
24. **Roels, O. A. and Trout, M.,** in *Standard Methods of Clinical Chemistry,* Vol. 7, Cooper, G. R. and King, J. S., Jr., Eds., Academic Press, New York 1972, 215.
25. **Bayfield, R. F.,** Colorimetric determination of vitamin A with trichloroacetic acid, *Anal. Biochem.,* 39, 282, 1971.
26. **Drujan, B. D., Castillon, R., and Guerreo, E.,** Application of fluorometry in the determination of vitamin A, *Anal. Biochem.,* 23, 44, 1968.
27. **Garry, P. J., Pollack, J. D., and Owen, G. M.,** Plasma vitamin A assay by fluorometry and use of silicic acid column technique, *Clin. Chem.,* 16, 776, 1970.
28. **Bamji, M. S.,** Transketolase activity and urinary excretion of thiamin in the assessment of thiamin nutrition status of Indians, *Am. J. Clin. Nutr.,* 23, 52, 1970.
29. **Pearson, W. N.,** Blood and urinary vitamin levels as potential indices of body stores, *Am. J. Clin. Nutr.,* 20, 514, 1967.
30. Interdepartmental Committee on Nutrition for National Defense, *Manual for Nutrition Survey,* 2nd ed., U.S. Government Printing Office, Washington, D.C., 1963.
31. **Leveille, G. A.,** Modified thiochrome procedure for the determination of urinary thiamin, *Am. J. Clin. Nutr.,* 25, 273, 1972.

32. **Burch, H. B., Bessey, O. A., Love, R. H., and Lowry, O. H.**, The determination of thiamine and thiamine pyrophosphate in small quantities of blood and blood cells, *J. Biol. Chem.*, 198, 477, 1952.
33. **Baker, H. and Frank, O.**, *Clinical Vitaminology*, Interscience, New York, 1968, chap. 2.
34. **Warnock, L. G., Prudhomme, C. R., and Wagner, C.**, The determination of thiamin pyrophosphate in blood and other tissues and its correlation, with erythrocyte transketolase, *J. Nutr.*, 108, 421, 1978.
35. **Horwitt, M. K. and Kreisler, O.**, The determination of early thiamin-deficient states by estimation of blood lactic and pyruvic acids after glucose administration and exercise, *J. Nutr.*, 37, 411, 1949.
36. **Brin, M.**, Erythrocyte transketolase in early thiamine deficiency, *Ann. N. Y. Acad. Sci.*, 98, 528, 1962.
37. **Warnock, L. G.**, A new approach to erythrocyte transketolase measurement, *J. Nutr.*, 100, 1057, 1978.
38. **Bamji, M. S.**, Changes in hepatic and erythrocyte transketolase activity and thiamin concentration of liver in experimental deficiency of water soluble vitamins, *Int. J. Vitam. Nutr. Res.*, 42, 184, 1972.
39. **Vo-khactu, K. P., Sims, R. L., Clayburgh, R. H., and Sanstead, H. H.**, Effect of simultaneous thiamin and riboflavin deficiencies on the determination of transketolase and glutathione reductase, *J. Lab. Clin. Med.*, 87, 741, 1976.
40. **Fennely, J., Frank, O., Baker, H., and Leevy, C. M.**, Red blood cell transketolase activity in malnourished alcoholics with cirrhosis, *Am. J. Clin. Nutr.*, 20, 946, 1967.
41. **Rossouw, J. E., Labadarios, D., Krasner, N., Davis, M., and Williams, R.**, Red blood cell transketolase activity and the effect of thiamin supplementation in patients with chronic liver disease, *Scand J. Gastroenterol.*, 13, 133, 1978.
42. **Warnock, L. G., Cullum, U. X., Stouder, D. A., and Stone, W. J.**,Erythrocyte transketolase activity in patients with neuropathy, *Biochem. Med.*, 10, 351, 1974.
43. **Basu, T. K. and Kickerson, J. W.**, The thiamin status of early cancer patients with particular reference to those with breast and bronchial carcinomas, *Oncology*, 33, 250, 1976.
44. **Kjosen, B. and Sein, S. H.**, The transketolase assay of thiamin in some diseases, *Am. J. Clin. Nutr.*, 30, 1591, 1977.
45. **Markkanen, T. and Kallimaki, J. L.**, Transketolase activity of blood cells in various clinical conditions, *Am. J. Med. Sci.*, 252, 564, 1966.
46. **Wells, D. G., Baylis, E. M., Holoway, L., and Marks, V.**, Erythrocyte transketolase activity in megaloblastic anemia, *Lancet*, 2, 543, 1968.
47. **Smeets, E. H. J., Muller, H., and De Weal, J.**, A NADH dependent transketolase assay in erythrocyte hemolysates, *Clin. Chim. Acta*, 33, 379, 1971.
48. **Gilroy, J., Meyer, J. S., Bauer, R. B., Vulpe, N., and Greenwood, D.**, Clinical, biochemical and neurophysiological studies of chronic interstitial hypertrophic polyneuropathy, *Am. J. Med.*, 40, 368, 1966.
49. **Chang, M. C., Koch, M., and Shank, R. E.**, Leukocyte transketolase activity as an indicator of thiamin nutriture in rats, *J. Nutr.*, 106, 1978, 1976.
50. **Hathcock, J. N.**, Thiamin deficiency effects on rat leukocyte pyruvate decarboxylation rates, *Am. J. Clin. Nutr.*, 31, 250, 1978.
51. **Brubacher, G.**, Biochemical studies for assessment of vitamin status, *Bibl. Nutr. Dieta*, 20, 31, 1974.
52. **Kaufmann, N. A. and Gugenheim, K.**, The validity of biochemical assessment of thiamin, riboflavin and folacin nutriture, *Int. J. Vitam. Nutr. Res.*, 47, 40, 1977.
53. **Doiphode, N. G. and Bamji, M. S.**, Effect of aureomycin on the nutritional status of the B-vitamins in humans, *Int. J. Vitam. Res.*, 40, 58, 1970.
54. **Briggs, M. H. and Briggs, M.**, Oral contraceptives and vitamin nutrition, *Lancet*, 1, 1234, 1974.
55. **Bessey, O. A., Horwitt, M. K., and Love, R. H.**, Dietary deprivation of riboflavin and blood riboflavin levels in man, *J. Nutr.*, 58, 367, 1956.
56. **Bamji, M. S., Sharda, D., and Nadamuni Naidu, A.**, A comparison of fluorometric and microbiological assays for estimating riboflavin content of blood and liver, *Int. J. Vitam. Nutr. Res.*, 43, 351, 1973.
57. **Bamji, M. S., Rameshwar Sarma, K. V., and Radhaiah, G.**, Relationship between biochemical and clinical indices of B-vitamin deficiency. A study in rural school boys, *Br. J. Nutr.*, in press.
58. **Sharda, D. and Bamji, M. S.**, Erythrocyte glutathione reductase activity and riboflavin concentration in experimental deficiency of some water soluble vitamins, *Int. J. Vitam. Nutr. Res.*, 42, 43, 1972.
59. Erythrocyte glutathione reductase — a measure of riboflavin nutrition status, *Nutr. Rev.*, 30, 162, 1972.
60. **Bamji, M. S.**, Glutathione reductase activity in red blood cells and riboflavin nutritional status in humans, *Clin. Chim. Acta*, 26, 263, 1969.
61. **Yawata, Y. and Tanaka, K. R.**, Effect of metabolic stress on activation of glutathione reductase by FAD in human red blood cells, *Experientia*, 27, 1214, 1971.

62. **Ramachandaran, M. and Iyer, G. V. N.,** Erythrocyte glutathione reductase in iron deficiency anemia, *Clin. Chim. Acta,* 52, 225, 1974.
63. **Bayoumi, R. A. and Rosalki, S. B.,** Evaluation of methods of coenzyme activation of erythrocyte enzymes for detection of deficiency of vitamins B_1, B_2, B_6, *Clin. Chem.,* 22, 327, 1976.
64. **Hornbeck, C. L. and Bradley, D. W.,** Concentrations of FAD and glutathione as they affect values of erythrocyte glutathione reductase, *Clin. Chem.,* 20, 512, 1974.
65. **Glatzle, D., Vuilleumier, J. P., Weber, F., and Decker, K.,** Glutathione reductase test with whole blood, a convenient procedure for the riboflavin status in humans, *Experientia,* 30, 665, 1974.
66. **Garry, P. J. and Owen, G. M.,** An automated flavin adenine dinucleotide-dependent glutathione reductase assay for assessing riboflavin nutriture, *Am. J. Clin. Nutr.,* 29, 663, 1976.
67. **Boni, H., Wassmer, A., Brubacher, G., and Ritzel, G.,** Assessment of the vitamin B_2 status by means of the glutathione reductase test, *Nutr. Metab.,* 21 (Suppl. 1), 20, 1977.
68. The interrelationship between riboflavin and pyridoxine, *Nutr. Rev.,* 35, 237, 1977.
69. **Baker, H., Frank, O., Ning, M., Gellene, R. A., Hunter, S. H., and Leevy, C. M.,** A protozoological method for detecting clinical vitamin B_6 deficiency, *Am. J. Clin. Nutr.,* 18, 123, 1966.
70. **Sundaresan, P. R. and Coursin, D. B.,** Microassay of pyridoxal phosphate using L-tyrosine-1-^{14}C and tyrosine apodecarboxylase, in *Methods in Enzymology,* Vol. 18 (Part A), McCormick, D. M. and Wright, L. D., Eds., Academic Press, New York, 1970, 509.
71. **Okuda, K., Fujii, S., and Wada, M.,** Microassay of pyridoxal phosphate using tryptophan-^{14}C with tryptophanase, in *Methods in Enzymology,* Vol. 18 (Part A), McCormick, D. M. and Wright, L. D., Eds., Academic Press, New York, 1970, 505.
72. **Lumeng, L., Cleary, R. E., Wagner, R., Yu, P. L., and Li, T. K.,** Adequacy of vitamin B_6 supplementation during pregnancy. A prospective study, *Am. J. Clin. Nutr.,* 29, 1376, 1976.
73. **Kyan, M. P., Lumeng, L., and Li, T. K.,** Validation of plasma PLP measurements as an indicator of vitamin B_6 nutritional status, *Fed. Proc. Fed. Am. Soc. Exp. Biol.,* 35, 2508, 1976.
74. **Luhby, A. L., Brin, M., Gorden, M., Davis, P., Murphy, M., and Spiegel, H.,** Vitamin B_6 metabolism in users of oral contraceptive agents. I. Abnormal urinary xanthurenic acid excretion and its correction by pyridoxine, *Am. J. Clin. Nutr.,* 24, 684, 1971.
75. **Sabater, J. and Ricos, C.,** Abnormalities of tryptophan metabolism (kynurenine pathway) found in a group of 830 mentally retarded children, *Clin. Chim. Acta,* 56, 175, 1974.
76. **Rose, D. P. and Randall, Z. C.,** Influence of loading dose on the demonstration of abnormal tryptophan metabolism by cancer patients, *Clin. Chim. Acta,* 47, 45, 1973.
77. **Salkeld, R. M., Knorr, K., and Korner, W. F.,** The effect of oral contraceptives on vitamin B_6 status, *Clin. Chim. Acta,* 49, 195, 1973.
78. **Thurnham, D. I.,** The influence of breakfast habits on vitamin status in the elderly, *Proc. Nutr. Soc.,* 36, 97A (Abstr.), 1977.
79. **Caband, P., Leeper, R., and Wroblewski, F.,** Colorimetric measurement of serum glutamic oxaloacetic transaminase, *Am. J. Clin. Pathol.,* 26, 1101, 1956.
80. **Park, Y. K. and Linkswiler, H.,** Effect of vitamin B_6 depletion in adult man on the excretion of cystathionine and other methionine metabolites, *J. Nutr.,* 100, 110, 1970.
81. **Linkswiler, H.,** Tryptophan and methionine metabolism of adult females as affected by vitamin B_6 deficiency, *J. Nutr.,* 104, 1348, 1974.
82. **Krishnaswamy, K.,** Methionine load test in pyridoxine deficiency, *Int. J. Vitam. Nutr. Res.,* 42, 468, 1972.
83. **Krishnaswamy, K.,** Methionine metabolism in pyridoxine deficiency, *Nutr. Metab.,* 17, 55, 1974.
84. **Krishnaswamy, K.,** Isonicotinic acid hydrazide and pyridoxine deficiency, *Int. J. Vitam. Nutr. Res.,* 44, 457, 1974.
85. **Gopalan, C. and Jaya Rao, K.,** Pellagra and amino acid imbalance, *Vitam. Horm.,* 33, 505, 1975.
86. **Gopalan, C. and Krishnaswamy, K.,** Effect of excess leucine on tryptophan-niacin pathway and pyridoxine, *Nutr. Rev.,* 34, 318, 1976.
87. **de Lange, D. J. and Joubert, C. P.,** Assessment of nicotinic acid status of population groups, *Am. J. Clin. Nutr.,* 15, 169, 1964.
88. **Srikantia, S. G., Narsinga Rao, B. S., Raghuramula, N., and Gopalan, C.,** Pattern of nicotinamide nucleotides in the erythrocytes of pellagrins, *Am. J. Clin. Nutr.,* 21, 1306, 1968.
89. **Herbert, V.,** Biochemical and haematologic lesions in folic acid deficiency, *Am. J. Clin. Nutr.,* 20, 562, 1967.
90. **Blakley, R. L.,** The biochemistry of folic acid and related pteridines, in *North Holland Research Monographs, Frontiers of Biology,* Nenberger, A. and Tatum, E. L., Eds., North-Holland, Amsterdam, 1969, chap. 2.
91. **Rudzki, Z., Nazaruk, M., and Kimber, R. J.,** The clinical value of the radioassay of serum folate, *J. Lab. Clin. Med.,* 87, 859, 1976.
92. **Wickramsinghe, S. N. and Saunders, J. G.,** Letter: Deoxyuridine syppression test, *Br. Med. J.,* 2, 87, 1975.

93. **Zittoun, J.**, Effect of folate and cobalamin compounds on the deoxyuridine suppression test in vitamin B_{12} and folate deficiency, *Blood*, 51, 119, 1978.
94. **Elsborg, L., Lung, V., and Bastrup-Madsen, P.**, Serum vitamin B_{12} levels in the aged; *Acta Med. Scand.*, 200, 309, 1976.
95. Nutritional Anaemias, Report of a WHO Scientific Group, Tech. Rep. Ser. No. 405, World Health Organization, Geneva, 1968.
96. **Mollin, D. L., Anderson, B. B., and Burman, J. F.**, The serum vitamin B_{12} level: its assay and significance, *Clin. Haematol*, 5, 521, 1976.
97. **Kalamegham, R. and Krishnaswamy, K.**, Functional significance of low serum vitamin B_{12} levels in pregnancy, *Int. J. Vitam. Nutr. Res.*, 47, 52, 1977.
98. **Gopalan, C.**, The "burning feet" syndrome, *Indian Med. Gaz.*, 81, 22, 1946.
99. **Tao, H. G.**, Measurements of urinary pantothenic acid excretions of alcoholic patients, *J. Nutr. Sci. Vitaminol. (Tokyo)*, 22, 333, 1976.
100. **Srinivasan, V. and Belavady, B.**, Nutritional status of pantothenic acid in Indian pregnant and nursing women, *Int. J. Vitam. Nutr. Res.*, 46, 433, 1976.
101. **Ellestad, J. J., Nelson, R. A., Adson, M. A., and Palmer, W. M.**, Pantothenic acid and coenzyme A activity in blood and colinic mucosa from patients with chronic ulcerative colitis, *Fed. Proc. Fed. Am. Soc. Exp. Biol.*, 29 (Abstr.), 820, 1970.
102. **Sarma, P. S., Menon, P. S., and Venkatachalam, P. S.**, Acetylation in the laboratory diagnosis of "burning feet syndrome" (pantothenic acid deficiency), *Curr. Sci.*, 18, 367, 1949.
103. **Nolan, J. C., Cardinale, G. J., and Udenfriend, S.**, The formation of hydroxyproline in collagen by cells grown in the absence of serum, *Biochim. Biophys. Acta*, 543, 116, 1978.
104. **Dawson, K. P. and Dawney, J.**, The iron status of children with low leucocyte ascorbic acid levels, *N. Z. Med. J.*, 86, 479, 1977.
105. **Maclennan, W. J. and Hamilton, J.**, Quick assessment of vitamin C status, *Lancet*, 1, 585, 1976.
106. **Harrison, H. E. and Harrison, H. C.**, Rickets then and now, *J. Pediatr.*, 87, 1144, 1975.
107. **Hughes, M. R., Baylink, D. J., Jones, P. G., and Haussler, M. R.**, Radiological receptor assay for 25-hydroxy vitamin D2/D3 and alpha, 25-dihydroxy vitamin D2/D3, *J. Clin. Invest.*, 58, 61, 1976.
108. **Bieri, J. G. and Farrell, P. M.**, Vitamin E, *Vitam. Horm.*, 34, 31, 1976.
109. **Pappu, A. S., Fatterpaker, P., and Sreenivasan, A.**, Possible interrelationship between vitamins E and B_{12} in the disturbance in methyl malonate metabolism in vitamin E deficiency, *Biochem. J.*, 172, 115, 1978.
110. **Olson, R. G. and Suttie, J. W.**, Vitamin K and γ-carboxyglutamate biosynthesis, *Vitam. Horm.*, 35, 59, 1977.
111. **Krishnaswamy, K.**, unpublished results.

Chapter 2

VITAMINS AND ORAL CONTRACEPTIVES

Suzanne Y. Tonkin

TABLE OF CONTENTS

I. Introduction 30

II. Vitamin A 32

III. Vitamin B_1 33

IV. Vitamin B_2 35

V. Vitamin B_6 37

VI. Vitamin B_{12} 43

VII. Folic Acid 44

VIII. Vitamin C 47

IX. Vitamin D 49

X. Vitamin E 50

XI. Vitamin K 51

XII. Conclusion 52

References 53

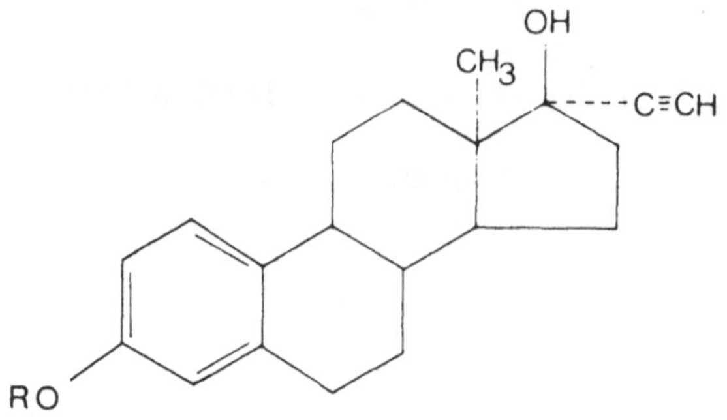

FIGURE 1. Estrogen structures: ethynyloestradiol (R = H); mestranol (R = CH_3); quinestrol (R = cyclopentyl).

I. INTRODUCTION

The realization that interest in nutritional status, in response to steroid contraception, has been largely concentrated in this decade makes it perhaps unexpected that such investigations have been extensive. Several reviews of the literature in the context of the metabolic consequences of contraceptives included nutritional implications,[1-5] but the subject of contraceptive steroids and vitamin status has also been documented previously on several occasions.[6-14] It is becoming increasingly apparent that the status of many vitamins is altered by steroid hormonal agents capable of regulating human fertility.

Of the oral contraceptives, the most widely used type contains a combination of an estrogen and a progestogen at fixed dose levels. Sequential and biphasic formulations are also available, wherein the dose of steroid varies according to the stage of the menstrual cycle. Contraception can also be maintained by a progestogen, received either orally or parenterally, the latter affording protection for an extended period of time. A large number of steroid compounds are used in conception control, yet only two types of estrogen have been employed commercially, e.g., ethinyl, estradiol, and mestranol (Figure 1). In contrast, the range of progestogens used is extensive (Figure 2), and thus, a large number of combinations exists, currently marketed at dose levels ranging from 30 to 100 μg estrogen and 0.15 to 2.5 mg progestogen. The trend in recent years has been to reduce the amount of steroid, particularly the estrogen component since this has been implicated in many of the complications described. In most of the studies in which the effects of steroid hormones on vitamin status was sought, reliance was placed on the commonly available products of the types described for the purpose of contraception, but in some studies steroids were administered to males and to post-menopausal females. By making comparisons, an impression has been gained as to the relative merits and disadvantages of each component, independently and in combination. Furthermore, the optimum dose for complete protection and minimal metabolic disturbance can be assessed. Although experience with the long-term depot progestogens is substantial, heavy reliance cannot be placed on some of the earlier results pertaining to metabolic effects, since the pharmacokinetics of compounds in this delivery system were not fully understood.[15]

An evaluation of the effects of contraceptive steroids relies on reference to a control group. Even if the latter are women who are carefully matched in many respects, the reason they are not using an oral contraceptive may be linked to the incidence of certain conditions. If the control subjects are using another type of contraceptive, such as

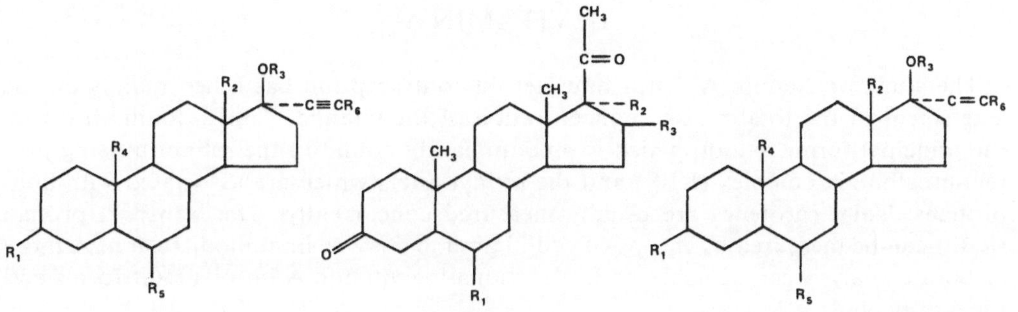

A B C

Compound	Formula	R_1	R_2	R_3	R_4	R_5	Δ
norethynodrel	A	O	Me	H	H	H	5 (10)
norethisterone	A	O	Me	H	H	H	4
norethisterone acetate	A	O	Me	Ac	H	H	4
ethynodiol diacetate	A	O.Ac	Me	Ac	H	H	4
lynestrenol	A	H_2	Me	H	H	H	4
norgestrienone	A	O	Me	H	H	H	4,9 (10), 11
norgestrel	A	O	Et	H	H	H	4
quingestanol acetate	A	O.CP	Me	Ac	H	H	3,5
dimethisterone	A	O	Me	Me	Me	α-Me	4
norgesterone	C	O	Me	H	H	H	5 (10)
chlormadinone acetate	B	Cl	O.Ac	H_2	–	–	4,6
medroxyprogesterone acetate	B	α-Me	O.Ac	H_2	–	–	4
megestrol acetate	B	Me	O.Ac	H_2	–	–	4,6
superlutin	B	H	O.Ac	CH_2	–	–	4,6

* This compound is a 17-vinyl, rather than a 17-ethynyl, derivative.
Abbreviations: Me = CH_3; Et = C_2H_5; CP = cyclopentyl; Ac = CH_3CO.

FIGURE 2. Progestogen structures.

an intrauterine device, this doubt is partially overcome, but the incidence of iron deficiency anemia may be higher. To use the woman as her own control eliminates many of these problems, but may introduce other variables such as those related to the passage of time, for example, the types of food available. Consequently, dietary vitamin sources fluctuate according to seasonal variation.

Specific laboratory procedures for the assessment of vitamin status have been described previously[16] and form the subject of another chapter.[17] The study of contraceptive steroids and nutritional status has relied heavily on chemical methods. Trends are becoming clear for a number of vitamins, although the biological significance of many of these changes has not been elucidated. The interpretation of laboratory tests for the assessment of nutritional status requires skill, particularly as the changes during steroid contraception are often more subtle than those encountered in the classical nutritional deficiences. In the event that a marginal or deficient state is demonstrated, the clinical implications are often elusive. Allied with these problems is the comparison of results which have been based on different laboratory techniques. Although suitable quality control material is lacking for some of the vitamins, standardization of methods for the assessment of nutritional status, currently being supervised by the World Health Organization, would appear to merit wider participation.

II. VITAMIN A

The study of vitamin A status and steroid contraception has relied mainly on the assessment of the total plasma concentration of the vitamin. This accounts for both the principal form, retinol, which is predominantly bound to the retinol-binding protein-prealbumin complex (RBP) and the retinyl esters which are associated with lipoproteins. Total carotenes are usually measured concurrently. The transport protein (RBP) can be measured by means of radioimmunoassay or immunodiffusion. A more elaborate analysis can be achieved by fractionating vitamin A into its ester forms and the free alcohol.

Females who receive combined oral contraceptive medication incur a change in vitamin A status. In 1971, Gal et al.[18] reported that the concentration of plasma vitamin A was considerably higher than in untreated females. Other studies[19-33] confirmed these results, regardless of socio-economic status or other demographic characteristics. The increase was of the order of 50%, yet platelet vitamin A levels did not alter.[20] Some of this earlier work has been reviewed.[4,34] It was found that RBP also existed in higher concentrations.[33,35]

According to one investigation,[36] the change in plasma vitamin A was mediated by the estrogen component, since a progestogen (350 μg norethindrone) administered as a contraceptive pill, had no influence. Furthermore, during the course of oral contraception with a sequential type of pill, mestranol induced an increase in the level of serum vitamin A. This value was enhanced during the remainder of the cycle when the pill also contained 2 mg norethindrone, and it was felt that the progestogen component could be influential at higher concentrations. When a progestogen was administered parenterally, as with medroxyprogesterone acetate, vitamin A status was normal,[19,20,37] even after 1 year.[37]

Since the advent of synthetic steroids as a method of fertility control, there has been a tendency to reduce the dose, particularly of the estrogen component. Nevertheless, the normal vitamin A levels were exceeded for a wide range of formulations.[20,21,23,25-30] However, an ultra low-dose combined oral contraceptive had minimal effect on retinol and RBP.[38] The magnitude of the change appeared to be proportional to the concentration of the estrogen.[27] The change in vitamin A status was evident after one cycle of treatment.[19,21,26,36] In a longitudinal study[26] there was little difference between the values from the 1rst to the 6th month. High values were sustained for long periods[22,23,29] and took about 3 months to return to normal on cessation of treatment.[21]

In contrast to vitamin A, opinion has been divided as to the impression, if any, exerted by contraceptive steroids on the carotenes. They were associated with a reduction in some studies[18,19,27,28] but had no effect in others.[19,21,22,25,31,36-38] A link has been proposed between the opposite trends of vitamin A and carotene.[27]

It has been suggested that the change in plasma vitamin A is secondary to the increase in RBP.[25,35,39] Retinol normally exists in the circulation bound predominantly to this protein[40] and any increase in RBP may be accounted for in terms of the general stimulatory action of estrogens on hepatic synthesis of a number of plasma proteins.[41] The suggestion that the retinol carrier could mediate the change in vitamin A status by the pill was supported by Yeung.[25] In view of his data, he argues against a general lipemia, since in treated and untreated groups plasma vitamin A was mainly retinol and retinyl ester did not increase with oral contraception.

The question of whether altered vitamin A status extended to liver stores was pursued in animals.[25,32,39,42] However, in order to obtain the same plasma response of vitamin A and RBP in rats as in humans, an excessive dose of the oral contraceptive was required.[39] Although the equivalent human dose failed to alter the plasma level,[25,39] there was evidence for hepatic depletion of vitamin A.[25] Whether this occurs in the human remains to be clarified. It was postulated[25] that under the influence of oral contraceptives, the metabolic need for vitamin A is higher. Enhanced liver depletion without an accompanying rise in plasma vitamin A could be explained if oral contraceptives increased the turnover rate of RBP rather than increasing the amount in plasma. At very high doses, there was some evidence for liver depletion, but other aspects of vitamin A metabolism were largely unaffected.[39] In mice, liver stores were increased, but the plasma concentration of vitamin A was not reported,[32] and in monkeys there was no effect on plasma vitamin A but the liver content tended to decline.[42]

The clinical significance of altered vitamin A status has been conjectural. Some authors[18,21,43] have expressed concern that the significant increase in plasma vitamin A, although well below toxic levels, could affect the fetus of those who conceived shortly after discontinuing the pill, and a treatment-free interval was recommended to allow a return to normal. Although the problem remains unresolved, there has been little support for these fears.[19,24,35,39] This was based partly on the evidence of Wild et al.[24] on the relationship between oral contraceptives, vitamin A, and the outcome of pregnancy, and also from a consideration of the condition in which retinol exists in the circulation.[35] A functional excess of vitamin A is related to that proportion of the total concentration which is present as the free alcohol. From an analysis of the distribution of retinol and retinyl ester,[25] it was concluded that plasma vitamin A existed mainly as retinol in both control and oral contraceptive treated groups. According to experiments in rats,[39] most of the retinol was associated with a protein with the properties of RBP. Furthermore, the ratio of RBP-bound retinol to lipoprotein-bound retinyl ester was not changed by oral contraceptive administration. A more detailed analysis of the distribution of retinol in human plasma of those taking oral contraceptive steroids would appear to merit further study. The ingestion of vitamin A, which was often contained in multivitamin tablets, did not appear to exacerbate the oral contraceptive-induced plasma levels,[22,30] and the suggestion was made that it may even protect liver stores from becoming depleted.[30] Thus, in comparison with the potential risks of elevated vitamin A levels associated with excessive intake, the consensus of opinion appears to be that it is doubtful that contraceptive steroids, as a cause of the increase, pose a potential teratogenic risk.[11,12,34]

The long-term effects of raised vitamin A levels on the individul are not known, but an association between hepatic depletion of the vitamin and hepatomas has been postulated.[44] Another potential clinical consequence is that the use of oral contraceptives might lead to vitamin A deficiency in malnourished women. According to the results of animal experiments, the probability of this developing was felt to be slight,[34] but could become important in areas where vitamin A deficiency is endemic. Thromboembolism has been linked with the use of estrogen contraceptives. Altered platelet behavior could result from action of vitamin A on lysozomal membranes within the cell. However, despite a raised plasma level, the vitamin was not concentrated in platelets.[20]

III. VITAMIN B_1

Several biochemical procedures have been developed to aid the assessment of vitamin B_1 status. These include measurement of the vitamin in different forms, in body tissues and fluids, and also functional tests. This subject was comprehensively reviewed by Sauberlich and co-authors,[16] and some of the current methods have been compared.[45]

In the present context, the application of the transketolase test is of particular relevance, since most of the studies concerning the consequences of hormonal contraception on thiamin status have relied on this procedure. The data has been expressed in

Table 1
VITAMIN B_1

Vitamin status test	Experimental group	Conclusion	Ref.
	Untreated controls		
ETK activity	Combined OCA	Reduced in 15%	46
LTK activity	Combined OCA	Reduced in 30%	
	Untreated controls		
ETK activity	Combined OCA (50 μg EE + 500 μg dl-NG or 1 mg EDA)	Decreased	26
ETK activation	As above	No significant difference	
	Pretreatment		
ETK activity	Combined OCA (as above)	Decreased	
ETK activation	As above	No significant difference	
	Pretreatment		
ETK activation	Combined OCA (50 μg EE + 250 μg d-NG or 1 mg NETA)	Increased	48
	Pretreatment		
ETK activity	Combined OCA (50 μg EE + 500 μg dl-NG)	No significant difference	47
ETK activation	As above	No significant difference	
ETK activity	Progestogen OCA (50—75 μg dl-NG)	No significant difference	
ETK activation	As above	No significant difference	
	Untreated controls	(a) Upper s/e group (b) Lower s/e group	
Urinary thiamin	Combined OCA	(a) Decreased	28
	(50 μg MEE + 1 mg NED)	(b) No significant difference	
	(50 μg EE + 500 μg dl-NG)	(a) Decreased (b) No significant difference	
	Pretreatment		
ETK activity	DMPA	No significant difference	37
ETK activation	DMPA	No significant difference	
ETK test	Combined OCA (30 or 50 μg EE + 150 μg d-NG)	Unchanged	33

Note: OCA, Oral contraceptive;
EE, Ethinyloestradiol;
MEE, Mestranol;
NET, Norethisterone;
NED, Norethindrone;
d-NG, D-norgestrel;
DL-NG, DL-norgestrel;
EDA, Ethynodiol diacetate;
MGA, Megestrol acetate;
NETA, Norethisterone acetate;
DMPA, Depot medroxy progesterone acetate;
ETK, Erythrocyte transketolase;
LTK, Leukocyte transketolase;
EGR, Erythrocyte glutathione reductase;
S/E Socioeconomic

terms of transketolase activity[26,37,46,47] and activation.[26,37,47,48] Erythrocytes were usually the source of the enzyme, but leukocytes have been used.[46] One group has measured urinary thiamin.[28] The conclusions reached in these surveys differed (Table 1), but the fact that a mild thiamin hypovitaminosis could be induced in adequately nourished women who took oral contraceptives[48] raised concern that the nutritional status in undernourished populations may be aggravated. These groups were the subjects of several reports.[26,28,47] Changes in thiamin status were mild with a small, steady reduction in transketolase activity.[26] Variable individual responses to oral contraception were not significant when viewed collectively.[47]

In general, clinical signs and symptoms were not remarked on, although attention was drawn to the higher frequency of complaints indicative of malnutrition in the lower socio-economic group.[28] The decline in nutritional status during oral contraceptive treatment could be prevented with a daily supplement of 3 mg vitamin B_1.[30] The implications of changes in thiamin levels are not clear, but unless deficiency is a common condition in the community, supplementation may not be necessary. Normalization of leukocyte transketolase activity occurred after discontinuation of the steroids, but this was not as evident in erythrocytes.[46]

The extent to which oral contraceptives may induce thiamin deficiency was not clear from the preceding studies. However, the prospect of this developing, or the pre-existence of a deficiency, should be considered since it has been demonstrated in rats that thiamin deficiency may be a factor influencing the metabolism of mestranol.[49] Liver microsomes of rats which had been fed a thiamin-deficient diet metabolized mestranol faster, and incremental addition of B_1 to the diet depressed mestranol O-demethylation.

Several ways in which contraceptive steroids could affect thiamin status have been suggested, including altered absorption, metabolic clearance, phosphorylation of thiamin, and interference with the binding of thiamin pyrophosphate to the apoenzyme. Phosphorylation of thiamin may be affected by prednisolone.[50] Evidence for altered absorption came from studies in rats[51] where it was shown that active transport of thiamin by the duodenum and jejunum was decreased during folate deprivation. Deficiency of this latter vitamin is known to occur in some women during oral contraception, and in two accounts of vitamin B_1 status, erythrocyte folate levels were decreased.[26,28] However, proof of a direct relationship in individuals is required. The proposal that the coenzyme-apoenzyme interaction was altered arose from an extrapolation of data indicating displacement of pyridoxal phosphate from transaminases by steroid hormones or their metabolites.[52] Failure of thiamin pyrophosphate to bind to the apoenzyme would lead to both low enzyme activity and activation. The availability of a specific method of quantifying thiamin pyrophosphate in blood and other tissues[45] may help to clarify some of the conjecture surrounding the subject of contraceptive steroids and thiamin status.

IV. VITAMIN B_2

There have been a number of studies which have evaluated oral contraceptives and their effect on vitamin B_2 status. The indices usually employed were erythrocyte glutathione reductase (EGR) activity[26,37,53] and its activation by flavin adenine dinucleotide (FAD) in vitro,[26,37,53-56] but urinary[22,28,55,57] or erythrocyte[26] riboflavin concentration have also been measured. Table 2 summarizes the results of these studies and the general conclusion reached was that vitamin B_2 status was adversely affected by oral contraceptives, even at very low doses.[33] Parenterally administered progestogens were without effect.[37,57] Failure to demonstrate changes in women of low socio-economic status or with severe malnutrition[28,33] suggested that oral contraceptive effects were masked by a pre-existing deficiency. The duration of oral contraception was taken into account in some studies.[26,30,56] Subjects followed up regularly after commencement of therapy underwent a gradual change. This was reflected in concomitant decreasing erythrocyte riboflavin concentration and EGR activity with increasing enzyme activa-

Table 2
VITAMIN B_2

Vitamin status test	Experimental group	Conclusion	Ref.
	Untreated controls		
Urinary riboflavin	Combined OCA	Decreased	57
	DMPA	Unchanged	
	Untreated controls		
Urinary riboflavin	Combined OCA	Decreased	22
	Untreated controls		
EGR activity	Combined OCA	Decreased	53
EGR activation	Combined OCA	Increased	
	Untreated controls		
EGR activity	Combined OCA	Decreased	26
EGR activation	As above	Increased	
Erythrocyte riboflavin	As above	Decreased	
	Pre-treatment controls		
EGR activity	Combined OCA	Decreased	
EGR activation	As above	Increased	
Erythrocyte riboflavin	As above	Decreased	
	Untreated controls		
EGR activation	Combined OCA	Increased	54
	Untreated controls	(a) Upper s/e group	
		(b) Lower s/e group	
Urinary riboflavin	Combined OCA	(a) Decreased	28
		(b) Unchanged	
	Untreated controls		
EGR activation	Combined OCA	Unchanged	55
Urinary riboflavin	Combined OCA	Unchanged	
EGR activation	Untreated controls	Increased in 10%	56
	Combined OCA	Increased in 40%	
	Pretreatment controls		
EGR activity	DMPA	Unchanged	37
EGR activation	DMPA	Unchanged	
EGR test		(a) Adequate or marginal nutrition group	
		(b) Severely malnourished group	
	Combined OCA	(a) Deterioration	33
		(b) No change	

tion.[26] The frequency of riboflavin depletion, according to enzyme activation, was considerably higher in those who had taken the pill over a period of at least 3 years.[56]

Clinical riboflavin deficiency, developing as a result of oral contraception, was not diagnosed in the preceding studies covering a number of ethnic types, although early symptoms existed in some women.[26] Certain conditions, such as glucose-6-phosphate dehydrogenase deficiency, may render a person more susceptible.[11,53] Vitamin B_2 deficiency is prevalent in areas where malnutrition is common, and in these groups the problem would be exacerbated. This raises the question of whether prophylactic administration of B_2 is indicated. The decline could be prevented with the administration of 2 mg riboflavin daily,[30] although it was proposed that higher amounts may be required to correct a pre-existing deficiency common in some populations. In one study of a population which was not expected to be undernourished, a supplement of 10 mg was used.[54]

Factors which could account for the observed changes in vitamin B_2 status include absorption and metabolic breakdown of the vitamin, conversion to the active coenzyme forms, or the binding of coenzyme to apoenzyme. The manner in which steroids or their metabolites interact with the pyridoxal phosphate dependent enzymes[52] may also apply to the flavins. Prednisolone has been reported to produce a riboflavin deficiency by interfering with the conversion of the vitamin to its coenzymes.[50] In a detailed investigation an attempt was made to understand which mode of action is operating.[58] Rats were used to monitor the effects of oral contraceptive treatment and the following observations were made:

1. Hepatic FAD and vitamin B_2 concentrations and GR activity increased with oral contraceptive treatment, although the ratio of vitamin B_2 to FAD did not alter and there was no change in activity coefficient. This suggested that the increase in hepatic flavin enzymes may increase the cellular requirement of the vitamin, while also bringing about a redistribution between its enzyme systems.
2. The decrease in concentration of plasma and erythrocyte riboflavin may be the result of greater sequesteration of the vitamin from the blood by tissues such as liver, where its concentration increases as a result of higher flavin binding capacity.
3. The lower urinary excretion of B_2 could reflect its higher retention in the body.

The data from these experiments led to the conclusion that oral contraceptives do not affect either the absorption of vitamin B_2 or the synthesis of flavin coenzymes, and it was suggested that they may act at the level of flavin enzymes, bringing about a higher requirement for riboflavin. Further evidence was given[58] to suggest that the in vivo increase in activity of hepatic GR was caused by an increase in enzyme concentration, rather than modification of the existing enzyme by contraceptive steroids.

V. VITAMIN B_6

The consequences of steroid contraception on tryptophan metabolism and vitamin B_6 have been the subject of many investigations. Most of the early work concerned the analysis of the complex alterations to tryptophan metabolism, and preceded the application of vitamin B_6 status tests to assess tissue depletion. The activity and activation of the transaminases, tissue levels of pyridoxal phosphate and other compounds with vitamin B_6 activity, urinary excretion of 4-pyridoxic acid, and latterly methionine metabolism were the criteria used in the assessment.

The suspicion that contraceptive steroid hormones might affect vitamin B_6 status arose in part from the analogy drawn with the state of pregnancy, and evidence came initially from studies of the tryptophan pathway. The participation of vitamin B_6 in several of these steps is the basis of a test for detecting vitamin B_6 deficiency. Administration of an oral loading dose of tryptophan amplifies altered patterns of excretion of certain intermediates when vitamin B_6 is limiting. The diagnosis of a deficiency could be confirmed, in association with clinical signs and symptoms, if the test became normal after a course of the vitamin. An abnormality in this pathway was demonstrated in women taking a variety of combined estrogen-progestogen contraceptive preparations.[59] Xanthurenic acid excretion was increased, and the correction with pyridoxine hydrochloride implied that the effect involved pyridoxal phosphate. These studies were confirmed and extended[60] with the measurement of xanthurenic acid, and other metabolites such as 3-hydroxyanthranilic acid and 3-hydroxykynurenine. The data suggested that the modification of the tryptophan pathway was primarily caused by the estrogen. It was proposed that the estrogen induced an increase in the capacity for tryptophan to be converted to nicotinic acid. As a result, tryptophan loading produced a relative shortage of pyridoxal phosphate coenzyme. Thus, the results were taken to indicate a deficiency of vitamin B_6, although at that time, the mode of action by which estrogen modified tryptophan metabolism was not clear. Several hypotheses were proposed,[60] including competition of estrogen conjugates and pyridoxal phosphate for the apoenzyme. This line of argument stemmed from in vitro observations using kynurenine transaminase.[61,62] Alternatively, evidence for a hormonal influence on tryptophan metabolism existed, particularly on tryptophan pyrrolase, the first enzyme in the conversion of tryptophan to nicotinic acid. These points and subsequent data were discussed further in a review covering ovarian hormones, oral contraceptives, and tryptophan metabolism.[63]

Confirmation of the effect of oral contraceptives on tryptophan metabolism and its correction by vitamin B_6 administration, came from many sources.[26,27,33,47,64-88] This disturbance could be demonstrated after a tryptophan load and also without loading. That estrogen appeared to be the responsible agent was also confirmed[60,66,89] in both men and women. Not only was this the case when taking estrogen in oral contraceptives, but also as replacement therapy during the post menopause.[90] In many of the studies quoted, a relatively low dose of estrogen (50 μg) was used, and there was no improvement in the metabolic response of subjects to oral contraceptives containing 30 μg compared with 50 or 75 μg estrogen.[33,88] Oral progestogens had no effect[47,66,72,89] but in combination pills, the influence of this component was reflected in the degree of disturbance.[79,82] In subjects taking depot progestogen abnormal tryptophan metabolism was discerned in one third of the women in one study,[82] but not in any in another.[37] A summary has been compiled of the major results of some of the early work on urinary tryptophan metabolites after oral contraception.[4]

Another means by which vitamin B_6 status has been evaluated is by measurement of the activity of erythrocyte enzymes dependent on the coenzyme form of the vitamin, such as the transaminases. This has included stimulation of activity by pyridoxal phosphate in vitro. In one of the first applications of this test to the study of hormone effects,[91] the mean basal activity of erythrocyte alanine aminotransferase (EALT) was significantly lower, and the degree of stimulation of activity with cofactor was higher in oral contraceptive users than in control subjects. Thus, it appeared that vitamin B_6 depletion was elicited by oral contraceptives. Extensive data exists for both EALT and EAST (erythrocyte aspartate aminotransferase) (Tables 3 and 4). The response of these two enzymes to oral contraceptive ingestion was not necessarily similar, nor have the results for a particular enzyme been uniform.[26,29,33,37,47,54,57,69,73,74,85,87,92-101] When a significant change in enzyme activity or activation was reported, both were generally increased. An earlier review of the extensive literature pertaining to the serum transaminases as an index of liver function during oral contraception also showed disparate results.[1,4] Alternative approaches to the assessment of vitamin B_6 status are direct measurement of the vitamin and its metabolites.[16,17]

In addition to using other criteria, several groups have sought trends in the urinary excretion of 4-pyridoxic acid, a metabolite of vitamin B_6. In one study,[73] a minority of the oral contraceptive group excreted subnormal amounts in comparison with the high incidence of altered tryptophan metabolism. A significant decrease in 4-pyridoxic acid excretion has been recorded.[74,92,95] Generally, however, oral contraceptives did not have an adverse effect on the levels of this metabolite, despite the excretion of abnormal amounts of tryptophan metabolites.[79,84,85,87,96]

The earliest evaluation of serum pyridoxal phosphate in subjects on the pill indicated that there was no significant difference when these levels were compared with controls.[102,291] This was true for any age group, despite a consistent trend for vitamin B_6 levels to decrease with greater age. Subsequent attempts to define the status of vitamin B_6 by direct measurement, particularly in blood, but also in urine, have led to a range of conclusions. Both enzymatic and microbiological assays have been used. The latter

Table 3
ASPARTATE AMINOTRANSFERASE

Experimental group	Activity	Activity coefficient	Ref.
Untreated controls			
Combined OCA	Increased	No significant difference	69
Untreated controls			
Combined OCA			
3—6 months	No significant difference	No significant difference	73
6—36 months	Increased	No significant difference	
Untreated controls			
Combined OCA	Increased	No significant difference	93
Untreated controls			
Combined OCA	No significant difference	Increased	94
Untreated controls			
Combined OCA		Increased	57
DMPA	—	No significant difference	
Untreated controls			
Combined OCA	Decreased	Increased	95
Untreated controls			
Combined OCA (50 μg EE + 500 μg dℓ NG or 1 mg EDA)	Increased	Increased	26
Pretreatment controls			
Combined OCA (as above)	Increased	Increased	
Untreated controls			
Combined OCA	—	Increased	54
Untreated controls			
Combined OCA	No significant difference	No significant difference	96
Pretreatment controls			
50 μg EE + 500 μg dℓ-NG	No significant difference	No significant difference	47
50 or 75 μg dℓ-NG	No significant difference	No significant difference	
Untreated controls			
Combined OCA	Increased	No significant difference	97
Untreated controls			
Combined OCA	No significant difference	No significant difference	98
Untreated controls			
Combined OCA (80 μg MEE + 1 mg NED)	—	No significant difference	29
Pretreatment controls			
DMPA	No significant difference	No significant difference	37

Table 4
ALANINE AMINOTRANSFERASE

Experimental group	Activity	Activity coefficient	Ref.
Untreated controls			
Combined OCA	No significant difference	—	69
Untreated controls			
Combined OCA	Decreased	Increased	91
Untreated controls			
Combined OCA			73
3—6 months	No significant difference	No significant difference	
6—36 months	No significant difference	No significant difference	
Untreated controls			
Combined OCA	No significant difference	Increased	93
Untreated controls			
Combined OCA	No significant difference	No significant difference	96
Pre-treatment controls			
50 μg EE + 500 μg dl-NG	No significant difference	No significant difference	47
50 or 75 μg dl-NG	No significant difference	No significant difference	
Untreated controls			
Combined OCA	No significant difference	No significant difference	97
Untreated controls			
Combined OCA	Increased	Increased	99
Untreated controls			
Combined OCA	No significant difference	No significant difference	101
DMPA or IUD	No significant difference	No significant difference	

often lacked specificity, which may account for some conflicting results. Chemical and enzymatic methods have proved to be more reliable and reproducible.[98] In one study, pyridoxal phosphate levels in the plasma of women who chose the pill as a method of contraception were significantly lower than in other women, and 20% were considered to be below normal.[83] This prospective assessment gave the impression that these would return to normal values within approximately 6 months of commencement of treatment, implying an adaptation to altered B_6 utilization. A decrease has also been found by others,[28,84,95,98,103] but this was not a consistent finding.[69,82,85,87,96] One group[84] reported that urinary vitamin B_6 was lower for those taking the pill, whereas another[69] had considered it to be unaffected.

Urinary excretion of cystathionine is another biochemical parameter which has recently been used to assess vitamin B_6 status. The catabolism of methionine attracted interest becase pyridoxal phosphate acts as cofactor for a number of enzymes in the pathway. Cystathionine excretion, following a methionine load, was reported to be similar irrespective of steroid contraceptive use.[85,87] This was confirmed in a later report[104] which also communicated the results concerning additional metabolites such as methionine, homocysteine, cystathionine, L-cysteine sulphinic acid, and taurine, and it was again concluded that there was no difference between the two groups. Independent confirmation of this work has been published.[105] Results from the former study, part of a series, were in agreement with those obtained when excretion of 4-pyridoxic acid, plasma pyridoxal phosphate, and erythrocyte aminotransferase activity were used to assess vitamin B_6 status in the same group of women, indicating that oral contraceptives had little if any effect on the vitamin B_6 requirements of most. However, they were not compatible with the conclusion that tryptophan metabolism differed in those who took steroid contraceptives. The results from this series of papers have been summarized.[14,85,87] Further results from these studies showed that the changes in tryptophan and methionine metabolism induced by experimentally controlled vitamin B_6 depletion were accentuated by oral contraceptives, but this was not true for the other indices of vitamin B_6 status.

Although a high percentage of women who take the oral contraceptive pill exhibit abnormal tryptophan metabolism, only a minority show biochemical evidence of vitamin B_6 deficiency, the number depending on the criterion used. Significant improvements in B_6 status occurred when pyridoxine hydrochloride was administered, usually at dose levels of 20 to 40 mg daily.

The means by which oral contraceptives disturb tryptophan metabolism and vitamin B_6 status has been discussed at length in the literature.[1,3-7,9-12,14,52,63,87,106-117] Although the basis of the changes has not been established with certainty, it is widely believed that there are a number of actions, both direct and indirect, which may be attributable. The first enzyme in the tryptophan-nicotinic acid pathway, tryptophan pyrrolase, appears to be particularly susceptible to the secondary effects arising from the administration of oral contraceptives. Induction of hepatic tryptophan pyrrolase directs more tryptophan into the pathway than usual, with the result that greater utilization increases the requirement for B_6. Since a number of the steps are vitamin B_6 dependent, a relative deficiency is provoked. Simultaneous induction of aminotransferases may lead to a redistribution of the vitamin among the enzymes, partial removal from the tryptophan-nicotinic acid route, and possibly decreased circulating levels. Interference with the binding of cofactor to the enzyme may also be a contributory factor. Considered together, the effect is a relative deficiency of the vitamin and the accumulation of intermediates of tryptophan metabolism.

The first premise, referring to changes in tryptophan pyrrolase activity, has been recently challenged.[88] After reviewing the relevant literature, it was deduced that this conclusion was based on circumstantial evidence. More direct meaurements of pyrrolase activity were therefore conducted in human subjects. One method employed was the assessment of tryptophan and kynurenine appearance and disappearance in plasma following an oral tryptophan load. These experiments demonstrated that the rate of tryptophan metabolism by liver tryptophan pyrrolase was the same, irrespective of dependence on oral contraception, taking into account the use of low dose estrogen pills. Although the production of kynurenine was similar in both groups, additional data suggested that its subsequent metabolism differed in the oral contraceptive group, implying more rapid conversion to other metabolites such as acetylkynurenine. Support was given to the hypothesis that esters may interfere with the activity of some pyridoxal phosphate enzymes and the consequences therein. It was concluded that the changes in tryptophan metabolism may be caused by a relative vitamin B_6 deficiency, but an increase in hepatic tryptophan pyrrolase activity was not a contributory factor.

The consequences of altered vitamin B_6 status are diverse, and several clinical manifestations which have been described in oral contraceptive users may be related to a vitamin B_6 deficiency. The more prominent of these are psychiatric symptoms, impaired glucose tolerance, cancer of the urinary tract, and perioral dermatosis.[92,95]

The pill has frequently been blamed for mental disturbances, and a large body of evidence exists to support the claim, despite the obvious difficulties experienced in conducting objective investigations. The pertinent results and general conclusions have been amply documented,[6-8,10-12,14,106,109,111,115,117-122,123] although the underlying factors behind the development of depressive symptoms are undoubtedly complex. Psychological factors may play a role, but plausible biochemical evidence exists for a link between depressive mood changes and tryptophan metabolism. The disturbance of tryptophan and B_6 metabolism during oral contraception may cause depression through changes in the levels of brain 5-hydroxytryptamine (5-HT). Decreased levels of this may arise from a paucity of tryptophan, which is judged to be one of the consequences of increased activity of tryptophan pyrrolase whereby less tryptophan enters the alternative routes. A diversion from the indole pathway would lead to decreased synthesis of 5-HT. Direct inhibition of cerebral uptake of tryptophan would also reduce the amount available for 5-HT synthesis. The functional B_6 deficiency could be a factor regulating this pathway, by virtue of the dependence of 5-hydroxytryptophan decarboxylase on pyridoxal phosphate. This enzyme may also be susceptible to competition for pyridoxal phosphate by estrogen conjugates. A more direct interaction between these two pathways of tryptophan metabolism has also been implied.[124] The symptoms of depression could be alleviated by the administration of 20 mg pyridoxine daily in only a proportion of women, those who exhibited true vitamin B_6 deficiency.[74,125] These results suggested that the metabolic basis for the condition varies.

The effect of contraceptive steroids on carbohydrate metabolism has been extensively studied, and a relationship between these and impaired glucose tolerance has been recognized. The topic has been reviewed on several occasions.[14,126-129] Not only are the age and predisposition of the subject risk factors, but the duration, type, and dose of contraceptive dictate the extent of the problem.[129-137] The main implications of the changes in carbohydrate metabolism, seen in those who are taking steroid contraceptives, lies in the subsequent development of diabetes, although vascular problems may occur in some subjects. A number of factors which include excess free cortisol and/or a pyridoxine deficiency has been recognized as being capable of effecting the changes.[129] The view that altered tryptophan metabolism could mediate the secondary change in carbohydrate metabolism stemmed, in part, from the knowledge that impaired glucose tolerance could be ameliorated with the administration of vitamin B_6.[138-142] In a more detailed study, it appeared that this was achieved only in the women who had demonstrable vitamin B_6 deficiency.[142] Oral glucose tolerance in gestational diabetes also showed improvement after 100 mg B_6 administration,[143,144] although this was recently disputed.[145]

One way in which altered tryptophan metabolism could effect a variation in glucose tolerance is by complex formation between xanthurenic acid and insulin, thereby reducing hypoglycemic activity. Although this theory was well founded, definitive evidence was lacking[141,142] and recent views[14,142] were that this mode of action was unlikely. An alternative theory involved quinolinic acid which can inhibit gluconeogenesis by its influence on phosphoenolpyruvate carboxykinase. The extent to which the hormonal control of quinolinic acid may function in the regulation of gluconeogenesis was also discussed.[14,141,142]

Increased concentrations of tryptophan metabolites excreted have been implicated in carcinoma of the urinary tract. In addition to the structural similarity of these compounds to recognized carcinogens, their carcinogenicity has been demonstrated in the bladder of laboratory animals. Furthermore, many patients with this type of cancer excrete higher amounts of these metabolites. The evidence on which the link is based has been critically examined.[111] However, the relationship is unclear,[146-149] with the result that any role of contraceptive steroids remains speculative. The long latent period between exposure to the carcinogen and tumor formation, precludes a reliable estimation of the risk, as it is possibly too early at this stage to look for a significant increase in the incidence of this condition in the population exposed to oral contraceptives.

Not only can vitamin B_6 prevent or correct the change in status, but several clinical conditions are also responsive. Thus, many have suggested prophylactic administration.[67,68,70,82,110] However, the wisdom of this has been questioned,[14,74] since fears have been expressed concerning other adverse consequences such as alterations in amino acid metabolism. This would be particularly undesirable in populations where malnutrition is endemic. If, in fact, vitamin B_6 supplements are considered desirable, then the question of dose level has to be resolved. Doses which have been prescribed depended on the criteria used, and the optimum level remains open to debate.[30,100]

VI. VITAMIN B_{12}

The initial investigation into the relationship between hormonal contraceptive use and vitamin B_{12} status was focused on the vitamin B_{12} binding proteins.[150] Data indicating that serum vitamin B_{12} binding capacity was increased aroused interest in serum vitamin B_{12} levels, and during the next few years a number of comparative studies were published.[23,29,54,150-156,291] The observation by Wertalik et al.[152] that oral contraceptive use resulted in a significant decrease, was confirmed.[23,29,54,151-156,291] Erythrocyte concentration was unaffected[154] and there was no significant difference between low and high dose products with respect to serum B_{12}.[38] Long-term depot progestogens were also without effect.[37,153] In comparison with these earlier results, the changes seen more recently have been modest.[157,158] The salient features of the earlier work concerning vitamin B_{12} have been covered in several reviews of the literature pertaining to vitamins and oral contraceptives.[4,10-12]

The opinion that oral contraceptive use induced an increase in unsaturated vitamin B_{12} binding capacity (UBBC)[150] was not confirmed.[152,154,156-158] This was in contrast to the change in vitamin B_{12} status in pregnancy, where reduced serum levels and concomitant increased UBBC were a common finding. Thus, the change was apparently not associated with B_{12} binding proteins. However, more detailed analyses uncovered changes in the UBBC ratio of the individual transcobalamins,[157,158] despite early indications to the contrary.[152,154] When these two studies were compared, the individual changes exhibited by transcobalamins I, II, and III were dissimilar, however, both showed an increase in UBBC of transcobalamin I. Vitamin B_{12} levels could not be correlated with UBBC or individual vitamin B_{12} binding proteins.[158] In normal serum, transcobalamin II was mostly unsaturated for binding exogenous vitamin B_{12}.[157,158]

Despite the fact that many subjects had levels of vitamin B_{12} near the lower limit of normal with some below the normal range, clinical manifestations of a deficiency were not described. However, it was predicted[154] that long-term use of oral contraceptives could lead to megaloblastic anemia and the development of neurological symptoms if folic acid was used indiscriminantly. The analysis of red blood cell parameters included in some studies did not reveal significant differences, but hematological data from a large number of women was consistent with a folate and/or vitamin B_{12} deficiency.[159] It was considered that the hazard of developing clinically significant macroscopic anemia in healthy, well-nourished women was probably minimal. However, these changes may become significant if other factors such as malnutrition exist.[10]

A study of vitamin B_{12} status in an African population with inadequate standards of nutrition showed that they had low levels of the vitamin and further reduction induced by the use of oral contraceptives was considered undesirable.[153] Prophylactic supplements which included 4 μg cyanocobalamin taken daily for 1 month reversed the change in vitamin B_{12} status.[54] It is of interest that pyridoxine can also elevate oral contraceptive induced low vitamin B_{12} levels.[160] Alternatively, the use of a depot progestogen could be considered since no effect on serum vitamin B_{12} was detected using this type of contraceptive.[153]

Several ways by which oral contraceptives may reduce serum vitamin B_{12} have been referred to[154] and enhanced tissue avidity, resulting in a redistribution of the vitamin, was considered more likely. The opinion that malabsorption could be the cause of the low concentration of vitamin B_{12} in certain subjects has been excluded on the basis of normal Schilling tests.[151,154,156,158] Furthermore, administration of a small oral dose of cyanocobalamin (1 μg/day) increased the vitamin level to a suboptimal value.[154] In another study,[161] vitamin B_{12} absorption was monitored in a large number of females, half of whom were taking a type of oral contraceptive pill. Although there was no significant difference between the means of the two groups, in a minority of the treated subjects the percentage absorption was below the normal value. The possibility that vitamin B_{12} deficiency was secondary to folate depletion was also excluded, since folate had no influence on serum B_{12}.[154]

It has been speculated[11] that the pill may inhibit the production of transcobalamin I by leukocytes, and the resultant low serum vitamin B_{12} level would not cause macrocytic anemia. Recent data[157,158] may help to clarify the significance of the changes in vitamin B_{12} status with oral contraceptive use. This included comparisons of serum vitamin B_{12}, UBBC, and B_{12} binding fractions with blood counts and iron status applied to mildly anemic women.[158] One proposal was that iron status may be a factor in determining the levels of vitamin B_{12}. Another view, based on statistical correlations of the data, was that white blood cells may be a controlling factor in any observed increase in UBBC or transcobalamin I UBBC levels. Alternatively, since most of the endogenous vitamin B_{12} is carried by transcobalamin I, higher binding capacity may simply indicate increased unsaturation, reflecting the lower B_{12} content. The total vitamin B_{12} binding capacity of transcobalamin I was the same, irrespective of oral contraceptive use.[157,158] However, it has not been established whether the distribution of endogenous B_{12} is similar in the two groups.

VII. FOLIC ACID

A potentially serious problem of steroid contraceptive use is the side effect concerned with hematopoiesis, mediated by interference with vitamin B_{12} and folic acid metabolism. The first indication that serum folic acid levels were reduced in oral contraceptive users was in 1968, from the data of Shojania et al.[162] However, according to results published shortly afterwards[163] this was not confirmed, and consequently, the subject of folate status and steroid contraceptive use attracted considerable interest. During the next few years, the scientific literature contained many accounts of current research into the topic. Nevertheless, opinion was divided as to whether serum folate levels were adversely affected. According to some studies, these drugs resulted in a depression of the normal folate level.[23,29,152,154,155,164-169,291] Others claimed that there was no change.[28,37,157,170-181] It was thought that the stage of the menstrual cycle could be disregarded in these analyses,[175] but in a recent study a plasma reduction of folate was found in oral contraceptive users on day 5 but not on day 20.[182] Additional criteria of folate status have also been used, such as erythrocyte folate and formiminoglutamate (FIGLU) excretion following an oral histidine load. Decreased erythrocyte folate was found in a number of studies,[23,26,28,29,155,157,164-167,291] whereas according to others it was unchanged.[37,163,180-182] FIGLU excretion increased.[164-167] In a brief report of several vitamins in undernourished women on low-dose pills, it was claimed that red cell folate tended to be higher.[33]

The salient features of some of the earlier experimental results have been presented, and several reasons have been proposed to account for the different effects of oral contraception on folate levels.[4,7,12,13,166,167,183-185] These included the duration and type of contraception used, characteristics of the subjects, and the effect of recent food intake, including the dietary supply of folate. If this was satisfactory, the effect on blood folate was probably mild, and by itself unlikely to lead to megaloblastic anemia.[166] However, if folate intake was reduced, or other conditions prevailed which interfered with its metabolism, oral contraceptive use may result in a more severe deficiency and megaloblastic anemia.[166] In a study conducted on Canadians,[164] the lowest blood folate levels appeared to correlate with a diet containing lesser amounts of the vitamin, with the consequence that impairment of folate metabolism would be demonstrated sooner than in other investigations. Another Canadian study[182] did not resolve this point. A seasonal fluctuation of erythrocyte folate levels has been noticed.[185] Since the values were lower in winter, this was judged to be the result of the different diet adopted between the summer and winter months.

The adverse effect of oral contraceptives on folate metabolism did not appear to be reflected in a functional deficiency, resulting in megaloblastic anemia. The peripheral blood was similar in untreated and treated groups.[26,166,167,172,176,178,182,184,291] However, when the blood of a large number of women from both categories was examined for morphological signs of folate deficiency, there were modest changes in red cell parameters which could point towards deranged B_{12} and folate hematopoiesis, but would not present a clinical problem for oral contraceptive users.[159]

Since clinical consequences of altered folate status were not obvious, the problem was approached from another point of view. Women who had developed megaloblastic anemia, and who also depended on the contraceptive pill, were investigated to determine the biochemical lesion, particularly to find out whether the condition was a direct consequence of the medication. The number of women who have developed folate responsive megaloblastic anemia in association with steroid contraception has not been large.[155,164,165,183,186-200] This agent was etiologically implicated in the functional folate deficiency, unless extensive investigations of the condition revealed other disorders. Yet, the occurrence of folate deficiency in women on the pill is a rare complication compared with other causes such as malnutrition, pregnancy, and malabsorption. An extensive number of factors could be responsible for malabsorption alone.[201] Attention has been drawn to the potential effect of drugs which interfere with B_{12} or folate absorption and utilization.[13,202-204] In vitro, some commonly used drugs can markedly affect folate conjugase,[205] this being the enzyme which transforms natural polyglutamyl folate to the more readily absorbed monoglutamyl folate. Other factors such as alcohol and dietary amino acids can also be influential.[13,202] Several underlying disorders considered likely to disturb folate metabolism were, in fact, found in many megaloblastic subjects who were dependent on the pill.[184] These included celiac disease,[193,196,197,199] folate deficient diet,[196,198] and regional enteritis,[197] to name a few. Thus, it has been recommended at the patient be thoroughly investigated before oral contraceptives are incriminated as the sole cause of folate deficiency. Although there appears to be little evidence to support the view that oral contraceptives alone will cause depletion of folate or lead to megaloblastic anemia, megaloblastic changes have been described in cervical epithelial cells in a minority of women who took the pill.[184,206] Folic acid levels were similar to those of the control group. Thus, normal surveys would not have detected the difference. Ten milligrams folic acid per day, taken for 3 weeks, reversed the changes,[206] but on re-examination 3 years later the morphological changes had reoccurred.[184] The possible basis was explored and it was surmised that contraceptive hormone therapy led to a local depletion of folate coenzymes resulting in abnormal DNA synthesis and the characteristic cytology. Rarely does this become systemic. Although the changes were not malignant, views were expressed on the relationship to cervical malignancy. Fluctuations have been noted in the activity of rat endometrial conjugase in response to hormonal influences, thus altering the proportion of folates.[207] This data[184,206] shows the benefit of folate supplements for all women receiving oral contraceptives. However, indiscriminate use of folate has been discouraged on the basis of reduced vitamin B_{12} status and potential neurological symptoms.[154] One recommendation[13] was to identify subjects at risk. Thirty-five micrograms daily was considered to be an adequate prophylactic dose, but up to 100 μg may be required in some cases. Less conservative doses have been suggested by others, based on the hematological response of megaloblastic anemia to folate therapy.[165,183,187,189]

A warning has been issued with respect to the discontinuation of oral contraceptives and the onset of pregnancy.[7] Normal confinement alters folate status and this could be exacerbated if lowered levels of folate persisted from previous pill use. Residual effects on folate status have been demonstrated.[185] Women who used an oral contraceptive within 6 months of conception exhibited lower erythrocyte and plasma folate levels in the first trimester of pregnancy, but the effect was not sustained. Folate supplements were recommended for those who conceive within 6 months of discontinuing the pill.

Two further adverse effects of steroid contraception which may be associated with folic acid have been documented. Depressive illness has been frequently attributed to pyridoxine deficiency, but according to a review of clinical studies pertaining to this problem, changes in the status of other vitamins such as folate and B_{12} may also be relevant.[121] Reference has been made to a possible correlation between thromboembolic disease and folic acid deficiency induced by oral contraceptives, mediated by homocysteine.[6,208]

The absorption of folic acid has been the subject of several comprehensive reviews,[12,184,209] and the fact that in food the vitamin is present largely as the polyglutamate raised the question of malabsorption. This line of investigation was pursued by several groups[166,175,183,187,189,210,211] and applied to subjects taking contraceptive medication. The early work[183,187,189,211] pointed towards an absorption defect. One report[189] implied that oral contraceptive agents may inhibit absorption of the polyglutamate form, but not "free" folate, with the result that prolonged use could lead to the development of folate responsive megaloblastic anemia. However, it was mentioned that the American diet should be sufficiently rich in folate to prevent this development.[183,189] It was judged[183] that partial inhibition of absorption should not affect most women, but if this was moderately decreased due to some unrecognized condition, a further reduction could bring about a deficient state. This reasoning was used to explain the rare occurrence of folate deficiency among oral contraceptive users. It was also suggested that on the days when the pill was not taken, enough folate may be absorbed to prevent deficiency in some women, while having little effect on others.

Subsequent work has thrown doubt on this theory.[175,210] One group[175] deduced that there was no true inability to absorb polyglutamyl folate in those who used oral contraceptives, and they proposed increased clearance of absorbed vitamin. Another group[210] concluded that decreased absorption of polyglutamyl folate was not a uniform effect, but was impaired only in a subgroup, who may be women with an otherwise mild and asymptomatic malabsorption syndrome. The effect appeared to be correlated with serum folate.[166,210] From this data folate malabsorption was excluded as an explanation of impaired folate metabolism in oral contraceptive users. Limited in vitro experiments have been equivocal with respect to the effect of estrogens on the intestinal conjugase.[205,210] The current information, with respect to folate absorption, has been critically evaluated, and the consensus of opinion was that the evidence for an absorption defect of polyglutamyl folate during oral contraception was insufficient.[12,182,184] Other explanations for the experimental results have been discussed.[12,160,167,177,182,184] Aspects considered were plasma clearance, increased urinary excretion, absorption of N^5-formyl tetrahydrofolate, increased physiological demand for folate, binding of folate to plasma proteins, and tissue metabolism of endogenous folate. A significant abnormality in folate metabolism in a large proportion of women taking oral contraceptives has not been demonstrated conclusively. The information relating to the means by which this could occur cover a wide range of hypotheses, but these require further clarification.

Reference has been made to the uncertain status of the vitamin in oral contraceptive users, so that data on the binding protein could be relevant. A specific folic acid-binding protein (FABP) has been described in folate deficient serum,[212] measured by the uptake of exogenous tritiated pteroyl glutamic acid (^{3}H-PGA). Saturation with endogenous folate in normal serum prevented ready detection with ^{3}H-PGA. The inverse relationship between serum folate and the level of unsaturated-binding protein[212,213] suggested that these are relatively saturated with endogenous folate, but can be unsaturated in a deficiency.[214] A granulocytic source of serum FABP seemed possible on the basis of available evidence,[212] and a macromolecular factor which binds folic acid was, in fact, demonstrated in leukocyte lysates and sera of a significant number of women who were pregnant or receiving oral contraceptives.[215] Inability to detect the binder in the control subjects was thought to be a reflection of the assay, in which it was complexed with endogenous ^{3}H-PGA, since in these subjects the binder would normally be already saturated with folic acid. In this study,[215] a significant correlation could not be established between the identification of the folate binder in serum or leukocytes and the serum folate concentration. However, data from pregnant women indicated that low serum folate and concomitant high folate binder was likely. The evidence for a relationship between oral contraceptive use and increased FABP levels was not clearly supported in another study,[216] but higher serum levels of FABP with lowering of the concentration of serum folate have been recorded.[157,178] Recent data[157] showed that the capacity for FABP to bind exogenous ^{3}H-PGA was greater in the treated group. Erythrocyte folate was significantly lower and serum folate slightly lower. The changes were thought to be the result of increased synthesis of FABP. It has also been suggested[11] that the folate binder might sequester folate and hence contribute to hormone induced folate deficiency. In another context, it has been surmised[212] that FABP may be important in the understanding of folate malabsorption associated with a number of conditions, including oral contraceptive use, and should be sought in intestinal mucous and secretions.

A recent paper, pertinent to the problem, has described the absorption of mono and polyglutamyl folate in zinc-depleted man.[217] Zinc status was first reported[218] to be significantly lower in oral contraceptive users, although opinion is divided regarding the effect of these agents on the zinc level,[219,220] with recent data suggesting that it is unchanged.[221] Zinc depletion affected the absorption of polyglutamyl but not monoglutamyl folate, pointing to decreased intestinal hydrolysis of pteroyl polyglutamate.[217] It was proposed that the human conjugase may be a zinc-dependent enzyme. The possibility was raised that poor zinc status could be a factor contributing to folate deficiency. Correlations of this nature could be attempted in subjects receiving steroids for contraceptive purposes.

VIII. VITAMIN C

The influence of steroid contraception on vitamin C status in humans had not been studied when, in 1968, Clemetson[222] predicted an adverse effect. This hypothesis was based on the fact that the plasma level of ceruloplasmin, a copper transport protein with ascorbate oxidase activity, was increased by estrogens[223,224] and oral contraceptives.[225] *A priori*, the oxidation of ascorbic acid by ceruloplasmin in vitro[224] could also take place in vivo, and therefore contraceptive steroids could indirectly effect a reduction in blood ascorbate levels. Animal studies supported these ideas. Experiments with estrogen-treated guinea pigs suggested that ascorbic acid metabolism was adversely affected.[222,226] Additional investigations in women revealed that the platelet concentration of ascorbate was lower in oral contraceptive users than in controls during a depletion study.[227] Further evidence in support of changes in vitamin C status was provided in several independent reports.[7,20,22,54,228-233] Among the variety of steroid contraceptives used, depot medroxyprogesterone acetate or oral progestogens proved to be without effect.[20,228,230] The salient features of some of this work have been described by several reviewers of the subject.[4,10-12,234,235]

Rivers[234] critically examined the data of contributors to the literature in a dissertation on oral contraceptives and ascorbic acid. It was concluded that the concentration of plasma, leukocyte, and platelet ascorbic acid was decreased in women ingesting oral contraceptives. Although ascorbic acid levels were very often reduced, the absolute values were well above the range which is found in clinical scorbutics. Evidence was given to show that the estrogen component appeared to be responsible for the observed changes. It had been reported that urinary excretion of ascorbic acid was also decreased,[231] but this change was disputed.[229,234] The question of whether ascorbic acid supplements could restore blood levels in oral contraceptive users was also addressed.[234] While admitting that the data was limited, it was suggesed that ascorbic acid supplementation would increase blood levels, but not to the same extent as in the controls. It appeared that the quantity of dietary vitamin C required to maintain a normal plasma, leukocyte, or platelet concentration in women on oral contraceptives may be greater than in control subjects. Although preceding studies[54,229,230,232,233] indicated that up to 500 mg/day may be required, another group[236] concluded that single massive doses had no permanent effect. They advocated remedial action by the inclusion of vitamin C-rich foods in the diet.

The results of earlier surveys concerning ascorbic acid status have not been supported by recent work.[27-29,31,237,238] In some of the studies[27-29,31] comprehensive evaluation of a wide spectrum of nutritional factors revealed changes in other aspects of nutrient metabolism, but not in ascorbic acid. Lack of control of vitamin C intake, particularly in the diet, might account for some results.[27,31] In order to eliminate the dietary factors which may have influenced the earlier results, Weininger and King initiated controlled studies in humans[237] and monkeys.[238] They evaluated ascorbic acid status and also serum ceruloplasmin levels in two groups of well-nourished females[237] during consumption of a constant formula diet for 8 to 12 days. In contrast to a striking twofold increase in ceruloplasmin, no significant difference was noticed in plasma, leukocyte, and urinary ascorbic acid. Tissue ascorbic acid saturation was also unchanged, according to a loading test. Despite the constant diet, considerable individual variation existed for each index. It was concluded that the ascorbic acid status of well-nourished women with daily dietary intakes of at least 100 mg vitamin C was not adversely affected by oral contraceptives. The difference between the results of various groups could not be reconciled. Further data,[238] using rhesus monkeys, confirmed the results which had been found in women. Urinary ascorbic acid, however, decreased significantly during oral contraceptive use. When this was also monitored with the use of ^{14}C-ascorbate it was found that the half-life was shorter. In this work, a daily tablet containing 100 mg ascorbic acid replaced the dietary content of the vitamin. The conclusion was that the dietary vitamin C requirement was increased by oral contraceptive use.[4,237,238]

The basis of the changes found in the earlier studies have been explored and several reasons were proposed[4,234,235] for reduced ascorbic acid status, particularly in association with estrogen contraceptive use. These included increased urinary excretion of the vitamin, increased metabolism of ascorbic acid, decreased concentration of reducing compounds such as reduced glutathione, decreased absorption, or changes in tissue distribution. There is little evidence to support these proposals. An increased rate of catabolism of ascorbic acid was thought to be secondary to the increase in serum ceruloplasmin concentration.[226,228] An inverse relationship between these had been previously suspected, although the evidence was circumstantial since measurements had not been performed on the same sample. Recent studies do not support this view.[27,233,237,238] In comparison with higher dose products, a very low dose induced a more modest increase in ceruloplasmin, but not in leukocyte ascorbic acid level.[233] It was found that some of the women with the highest concentration of ceruloplasmin had the lowest levels of ascorbate.[27] However, analysis of all the data did not give statistical significance to the relationship and more recently other studies[237,238] have confirmed a lack of correlation. Rivers[234,235] argued against increased urinary excretion as a contributing factor on the basis that steroids had no effect on the rate of ascorbic acid breakdown, however, tissue uptake patterns were apparently altered. He suggested that the observed changes in blood levels of ascorbate in oral contraceptive users may be caused by changes in tissue distribution.

The role of vitamin C in the body has been researched extensively, causing considerable conjecture, and the subject continues to be widely debated. Therefore, the functional significance of any changes in vitamin C status induced by oral contraceptives remains doubtful. However, one important clinical consequence which has been ascribed to oral contraceptives, particularly the estrogen component, is the increased risk of developing cardiovascular disease. The extent to which this is related to vitamin C status is not clear, although the involvement of the latter in thromboembolic disorders has been referred to in the context of platelet disturbances.[20,227] The merits of ascorbate supplementation under normal circumstances also remain unresolved, and in view of the conflicting data on vitamin C status and oral contraceptives it is doubtful whether supplementation is advisable or necessary at this time.

IX. VITAMIN D

The biological effects of vitamin D are mediated by a metabolite, 1α, 25-dihydroxycholecalciferol which, together with its immediate precursor, 25-hydroxycholecalciferol, can be measured directly by protein-binding techniques.[17] However, evaluation of vitamin D status on a routine basis has relied mainly on indirect measurements, such as serum alkaline phosphatase activity, and calcium and phosphorous levels. The data from alkaline phosphatase, in particular, requires cautious interpretation as it is not a specific index of vitamin D status, and can be altered under many conditions. The synthesis and role of the metabolites of vitamin D have been reviewed[239] together with the hormonal regulation of calcium and phosphorous by 1α, 25-dihydroxycholecalciferol, parathyroid hormone, and calcitonin.

The relationship between hormonal contraceptives and vitamin D has aroused less interest than with other nutritional factors. An evaluation of vitamin D status during steroid contraception relies largely on changes in calcium and phosphorous levels. Reviewers of the literature pertaining to the metabolic effects of oral contraceptives[1,4] concluded that urinary calcium and serum phosphorous were decreased,[1] but referred to conflicting results for serum calcium.[4] It would appear from additional data that the latter was unchanged.[20,220,240] Depot medroxyprogesterone acetate did not affect serum calcium,[20,37,241] but caused a variable response in phosphorous.[20,241] It has been suggested that bone mineral concentrations may be higher in users of certain estrogen-containing oral contraceptives.[242]

Investigations concerned specifically with vitamin D status and contraceptive steroids were carried out by Carter el al.[243,244] They reported[243] that in three women who had taken combined oral contraceptives for 1 month, conversion of vitamin D_3 to 25-hydroxycholecalciferol was inhibited. This effect was not evident in two women treated with the estrogen component alone. Further work[244] led to the conclusion that the steroid oral contraceptive effect on the conversion of vitamin D_3 to its metabolites may be a combination of distribution and metabolism changes resulting in an increased half-life.

Results of measurement of the human transport protein for vitamin D and its metabolites, and its relationship with 25-hydroxycholecalciferol have been recently published.[245] It was found that the plasma concentration of the vitamin D-binding protein greatly exceeded the normal concentration of 25-hydroxycholecalciferol and there appeared to be no correlation between the two. The binding protein was not altered by diseases of calcium metabolism, but women on combined oral contraceptives had higher levels.

On the basis of the available evidence, it would appear that there is no change in the vitamin D requirement of women receiving contraceptive steroids.

X. VITAMIN E

The use of oral contraceptives and their effect on vitamin E status has been the subject of a number of investigations over the past few years. Initial studies[246] carried out in rats appeared to indicate that plasma vitamin E levels were lowered when oral contraceptives were administered. When vitamin E deficient rats were used the effects of the steroids were not as obvious. It was pointed out that the contraceptives may have a primary effect on lipoprotein distribution and any reduction of this type would be reflected in lower α-tocopherol levels due to the induced shortage of a suitable lipoprotein carrier.

In an extension of this early work to human subjects[247] it was found that plasma tocopherol levels were lowered by about 20%. This reduction was most evident in those subjects who were also taking vitamin E supplements at the time of the study. Other results found were that plasma lipoproteins decreased whereas cholesterol increased. Confirmation of this trend in vitamin E levels was obtained from two separate investigations,[248] one involving African women and the other Caucasians. A difference was noted in the response of the two ethnic types to oral contraceptive use. The only significant change shown was in the latter group, where a decrease was noted. No explanation could be found to account for this apparent ethnic difference in the effects of oral contraceptives. It was recommended, however, that supplementary α-tocopherol be given at dose levels of 10 mg daily in order to restore plasma levels to the pretreatment values.

Later studies[27,29,31] on the effects of a variety of oral contraceptive types have failed to confirm the earlier reported decrease in plasma vitamin E levels. In some cases an increase has been found.[29] When a depot progestogen was used, there was no change in vitamin E status.[37,57] It would appear from most of the evidence published so far that any observed changes in vitamin E status, with respect to oral contraceptive use, is closely linked by a number of related factors. The type of response found in normal healthy female subjects will be dependent on, *inter alia*, the type and dose level of oral contraceptive used, the duration of treatment, the physical characteristics of the subjects studied, and their dietary habits. Perhaps some of the conflicting results found for vitamin E levels arise, at least in part, from the lack of control over some of these parameters.

A more recent study[249] has attempted to define some of the above-mentioned variables, and in doing so, the methods of assay used to determine vitamin E were also critically evaluated. Individual tocopherols (α, β, and γ) were measured using chromatographic techniques and a comparison was carried out between this and other methods of analysis which included spectrophotometric and biochemical assays for the determination of total tocopherol. The identification and measurement of the individual tocopherols was considered of interest since attention has been drawn to the fact that types other than α- may make a significant contribution to the overall vitamin E activity in plasma.[250] The results derived from this study[249] showed that the biochemical method, using erythrocyte hemolysis, was the least satisfactory in evaluating vitamin E status and no significant differences were found among the groups studied using this method. However, the other methods of assay indicated a lowering of total plasma tocopherol. This was particularly evident in the groups where a higher dose of estrogen had been administered. In contrast to this result, plasma triglyceride levels increased in the same subjects.

Since vitamin E is known to be transported in blood principally in the β-lipoprotein fraction,[251] and since oral contraceptive use is generally associated with a state of hyperlipemia,[252-261] then a thorough investigation linking the changes in blood lipoprotein fractions with tocopherol levels during contraceptive medication would appear to warrant further study.

Vitamin E deficiency in adults is a comparatively rare condition because of the considerable tissue storage capacity of the vitamin, and it is only after prolonged dietary deprivation that clinical signs may become manifest.[250,262] For this reason the use of oral contraceptives in otherwise healthy individuals is unlikely to give rise to any noticeable pathological condition. Present evidence would therefore indicate that supplementation is not required. Nevertheless, much speculation has been directed at possible interactions in which a change in vitamin E status may be involved.

XI. VITAMIN K

The potential for steroid contraceptives to alter vitamin K status should be considered in view of its only established function in the hepatic synthesis of plasma coagulation factors, principally prothrombin, VII, IX, and X, in the post-translational conversion of existing biologically inactive precursor proteins to active zymogens by carboxylation of glutamic acid residues.

Direct measurement of the vitamin in blood is not normally carried out, and an assessment of vitamin K status is made indirectly by measuring plasma prothrombin or factors VII, IX, and X.[16,17] There have been many comparative investigations of blood coagulation in treated and untreated subjects as a result of the epidemiological link between oral contraceptives and thromboembolic disorders. These included the vitamin K dependent factors, but specific accounts of the effects of steroid contraceptives on vitamin K status have been sparse.

An early report showed striking increases in the levels of prothrombin, factors VII, IX, and X.[263] Subsequent work has confirmed these trends.[1,2,264-267] Several parameters were studied in a comparison of the effects of different oral contraceptives.[38,268] Those containing higher amounts of estrogen with various progestogens caused significant increases in factor VII levels. This was not found in subjects taking an oral contraceptive containing lower doses of estrogen and progestogen. The level of factor VII could be modified by changing the progestogen component of the combined oral contraceptives[269] and the effect of those containing only a progestogen appeared to be less.[270,271] While monitoring the effect of sequential oral contraceptives on serum clotting factors, it was observed that vitamin K dependent activity was elevated after 3 months and remained raised for the duration of the study.[272] When estrogenic hormones were administered to rats which had been fed a vitamin K deficient diet, the induced hypoprothrombinemia was reversed.[273] This suggested that estrogen reduced the need for vitamin K. The inhibitory effect of coumarin anticoagulants on hepatic synthesis of vitamin K dependent clotting factors and the concomitant hypoprothrombinemic response was also reported to be altered in women taking oral contraceptives.[274] These subjects may, therefore, require higher doses of the anticoagulant to achieve the same effect.

The realization that oral contraceptive use incurs a small, but significant risk of developing cardiovascular disease,[266,275,276] particularly thrombosis, makes an evaluation of vitamin K status important. Although the estrogen component has been implicated,[277] the risk also depends on a number of other factors, including personal characteristics such as the age of the woman and whether she is a cigarette smoker.[278,279] According to the measurement of proteins dependent on vitamin K for synthesis, and the early work relating the vitamin with estrogens, the requirement for vitamin K appears to be reduced by oral contraceptives.

XII. CONCLUSIONS

When consideration is given to the plasma protein changes attending the use of synthetic steroids involved in fertility control,[1,4,280-286] and the association of many vitamins with particular protein fractions,[287] it is, perhaps, not unexpected that altered vitamin status has been found. Since the majority of plasma proteins are synthesized in the liver, this points to a major influence of sex steroids on hepatic protein synthesis or catabolism of these moieties. Not only have changes been recognized in the concentration of specific transport proteins, but the effects may also extend to liver enzymes responsible for metabolic activation and deactivation.

The demonstration of altered status of many vitamins has come largely from independent investigations of a single vitamin, although in a number of studies, some were assessed concurrently.[22,26-31,37,54,291] One may assume that many changes in vitamin status coexist within the individual. Thus, a consideration of vitamin and mineral interactions may be relevant.[202,208] Notable among these are the relationship between vitamins B_2 and B_6,[7,26,288] B_1 and B_2,[289] B_6 and B_{12},[160] folate and vitamin C,[290] folate, B_6 and B_{12},[121,202,208] folate and B_1,[51] zinc,[217] or iron.[158] These considerations would also apply to the application of multivitamin supplements to improve vitamin status.

The efficacy of giving vitamin supplements during oral contraception has been open to conjecture. Although biochemical changes are frequently demonstrated, most women in developed countries appear healthy, lacking distinct manifestations of nutritional deficiency, but a relationship between the changes in status of many of the vitamins and some serious conditions cannot be excluded. Long-term effects need to be clarified. Lower socioeconomic groups exhibited a higher frequency of clinical signs and symptoms attributable to nutritional deficiencies,[28] so that concern must be expressed for women who inhabit areas where malnutrition is endemic. The coexistence of disease and the attendant use of drugs should also be taken into consideration. Resolution of the question of vitamin supplementation should depend on the outcome of clinical trials, currently proceeding under the direction of the World Health Organization.[292] If a positive recommendation emerges, then intermittent doses, taken during the nonhormone days, could be adopted since this was as effective and more convenient than a daily dose.[33]

REFERENCES

1. **Briggs, M. H., Pitchford, A. G., Staniford, M., Barker, H. M., and Taylor, D.,** Metabolic effects of steroid contraceptives, in *Advances in Steroid Biochemistry and Pharmacology,* Vol. 2, Briggs, M. H., Ed., Academic Press, New York, 1970, 112.
2. **Miale, J. B. and Kent, Jessie, W.,** The effects of oral contraceptives on the results of laboratory tests, *Am. J. Obstet. Gynecol.,* 120(2), 264, 1974.
3. **Weindling, H. and Henry, J. B.,** Laboratory test results altered by "the pill", *JAMA,* 229(13), 1762, 1974.
4. **Briggs, M.,** Biochemical effects of oral contraceptives, in *Advances in Steroid Biochemistry and Pharmacology,* Vol. 5, Briggs, M. H. and Christie, G. A., Eds., Academic Press, London, 1976, 66.
5. **Hauschildt, S.,** Stoffwechseluntersuchungen unter Einnahme oraler Kontrazeptiva, *Z. Ernährungswiss,* 17(1), 511, 1978.
6. **Anon.,** Oral contraceptive agents and vitamins, *Nutr. Rev.,* 30, 229, 1972.
7. **Theuer, R. C.,** Effect of oral contraceptive agents on vitamin and mineral needs: a review, *J. Reprod. Med.,* 8, 13, 1972.
8. **Butterworth, C. E.,** Interactions of nutrients with oral contraceptives and other drugs, *J. Am. Diet. Assoc.,* 62, 510, 1973.
9. **Reinken, L. and Dapunt, O.,** Vitaminstoffwechsel und Kontrazeptiva, *Geburtshilfe frauenheilkd,* 33, 503, 1973.
10. **Larsson-Cohn, U.,** Oral contraceptives and vitamins: a review, *Am. J. Obstet. Gynecol.,* 121(1), 84, 1975.
11. **Wynn, V.,** Vitamins and oral contraceptive use, *Lancet,* 1, 561, 1975.
12. **Anderson, K. E., Bodansky, O., and Kappas, A.,** Effects of oral contraceptives on vitamin metabolism, in *Advances in Clinical Chemistry,* Vol. 18, Bodansky, O. and Latner, A. L., Eds., Academic Press, New York, 1976, 247.
13. **Roe, D. A.,** Nutrition and the contraceptive pill, *Curr. Concepts. Nutr.,* 5, 37, 1977.
14. **Rose, D. P.,** Effects of oral contraceptives on nutrient utilization, in *Nutrition and Drug Interrelations,* Hathcock, J. N. and Coon, J., Eds., Academic Press, New York, 1978, 151.
15. **Benagiano, G.,** Long-acting systemic contraceptives, in *Regulation of Human Fertility,* Proc. Symp. Advances Fertility Regulation, World Health Organization, Diczfalusy, E., Ed., Scriptor, Copenhagen, 1977, 323.
16. **Sauberlich, H. E., Skala, J. H., and Dowdy, R. P.,** *Laboratory Tests for the Assessment of Nutritional Status,* CRC Press, Boca Raton, Fla., 1974.
17. **Bamji, M. S.,** Laboratory tests for the assessment of vitamin nutritional status, in press.
18. **Gal, I., Parkinson, C., and Craft, I.,** Effects of oral contraceptives on human plasma vitamin-A levels, *Br. Med. J.,* 2, 436, 1971.
19. **Briggs, M., Briggs, M., and Bennum, M.,** Steroid contraceptives and plasma carotenoids, *Contraception,* 6(4), 275, 1972.
20. **Briggs, M.,** Blood platelet biochemistry in women receiving steroid contraceptives, *Haematologia,* 7(3-4), 347, 1973.
21. **Gal, I. and Parkinson, C. E.,** Changes in serum vitamin A levels during and after oral contraceptive therapy, *Contraception,* 8(1), 13, 1973.
22. **Clements, A. S.,** The effects of long-term usage of oral contraceptive steroid drugs on vitamin A, beta carotene, ascorbic acid, riboflavin and niacin in women, *Diss. Abstr. Int. B,* 35, 915, 1974.
23. **Lawrence, J. D. and Smith, J. L.,** Effects of oral contraceptive steroids on nutritional status of low income women, *Fed. Proc. Abstr.,* 33, 698, 1974.
24. **Wild, J., Schorah, C. J., and Smithells, R. W.,** Vitamin A, pregnancy, and oral contraceptives, *Br. Med. J.,* 1, 57, 1974.
25. **Yeung, D. L.,** Effects of oral contraceptives on vitamin A metabolism in the human and the rat, *Am. J. Clin. Nutr.,* 27, 125, 1974.
26. **Ahmed, F., Bamji, M. S., and Iyengar, L.,** Effect of oral contraceptive agents on vitamin nutrition status, *Am. J. Clin. Nutr.,* 28, 606, 1975.
27. **Horwitt, M. K., Harvey, C. C., and Dahm, C. H.,** Relationship between levels of blood lipids, vitamins C, A, and E, serum copper compounds, and urinary excretions of tryptophan metabolites in women taking oral contraceptive therapy, *Am. J. Clin. Nutr.,* 28, 403, 1975.
28. **Prasad, A. S., Lei, K. Y., Oberleas, D., Moghissi, K. S., and Stryker, J. C.,** Effect on oral contraceptive agents on nutrients. II. Vitamins, *Am. J. Clin. Nutr.,* 28, 385, 1975.
29. **Smith, J. L., Goldsmith, G. A., and Lawrence, J. D.,** Effects of oral contraceptive steroids on vitamin and lipid levels in serum, *Am. J. Clin. Nutr.,* 28, 371, 1975.
30. **Ahmed, F. and Bamji, M. S.,** Vitamin supplements to women using oral contraceptives (Studies of vitamins B_1, B_2, B_6 and A), *Contraception,* 14(3), 309, 1976.

31. **Yeung, D. L.**, Relationships between cigarette smoking, oral contraceptives, and plasma vitamins A, E, C, and plasma triglycerides and cholesterol, *Am. J. Clin. Nutr.*, 29, 1216, 1976.
32. **Allred, J. P.**, Influence of oral contraceptives on vitamin A metabolism and on serum immunoglobulins in mice and women, *Diss. Abstr. Int. B*, 37, 6055, 1977.
33. **Bamji, M. S. and Prema, K.**, Effects of oral contraceptives on vitamin nutrition status of malnourished women, in *Abstr. 5th Int. Congr. Horm. Steroids*, New Delhi, India, 1978, 62.
34. **Anon.**, The effect of oral contraceptive agents on plasma vitamin A in the human and the rat, *Nutr. Rev.*, 35, 245, 1977.
35. **Briggs, M. H.**, Vitamin A and the teratogenic risks of oral contraceptives, *Br. Med. J.*, 2, 170, 1974.
36. **Yeung, D. L. and Chan, P. L.**, Effects of a progestogen and a sequential type oral contraceptive on plasma vitamin A, vitamin E, cholesterol and triglycerides, *Am. J. Clin. Nutr.*, 28, 686, 1975.
37. **Amatayakul, K., Sivasomboon, B., and Thanangkul, O.**, Vitamin and trace mineral metabolism in medroxyprogesterone acetate users, *Contraception*, 18(3), 253, 1978.
38. **Briggs, M. H. and Briggs, M.**, Clinical and biochemical investigations of an ultra low-dose combined type oral contraceptive, *Curr. Med. Res. Opin.*, 3(9), 618, 1976.
39. **Supopark, W. and Olson, J. A.**, Effect of *Ovral*, a combination type oral contraceptive agent, on vitamin A metabolism in rats, *Int. J. Vitam. Nutr. Res.*, 45, 113, 1975.
40. **Kanai, M., Raz, A., and Goodman, D. S.**, Retinol-binding protein: the transport protein for vitamin A in human plasma, *J. Clin. Invest.*, 47, 2025, 1968.
41. **Laurell, C. B., Kullander, S., and Thorell, J.**, Effect of administration of a combined estrogen-progestin contraceptive on the level of individual plasma proteins, *Scand. J. Clin. Lab. Invest.*, 21, 337, 1968.
42. **Belavady, B., Krishnamurthi, D., Mohiuddin, S. M., and Rao, P. U.**, Metabolic effects of oral contraceptives in monkeys fed adequate protein and low protein diets, *Ind. J. Exp. Biol.*, 11, 15, 1973.
43. **Gal, I.**, Vitamin A, pregnancy, and oral contraceptives, *Br. Med. J.*, 2, 560, 1974.
44. **Gal, I.**, Steroidal contraceptive therapy: hepatomas and vitamin A, *Lancet*, 1, 684, 1975.
45. **Warnock, L. G., Prudhomme, C. R., and Wagner, C.**, The determination of thiamin pyrophosphate in blood and other tissues, and its correlation with erythrocyte transketolase activity, *J. Nutr.*, 108, 421, 1978.
46. **Markkanen, T., Himanen, P., and Peltola, O.**, Pentose phosphate pathway inhibition in blood cells during anticonceptive hormone therapy, *Pharmacology*, 11, 108, 1974.
47. **Joshi, U. M., Lahiri, A., Kora, S., Dikshit, S. S., and Virkar, K.**, Short-term effect of Ovral and Norgestrel on the vitamin B_6 and B_1 status of women, *Contraception*, 12(4), 425, 1975.
48. **Briggs, M. H. and Briggs, M.**, Thiamine status and oral contraceptives, *Contraception*, 11(2), 151, 1975.
49. **Wade, A. E. and Evans, J. S.**, The influence of thiamin deficiency on the metabolism of the oral contraceptive mestranol (3-methoxy-17-ethynyl-1, 3,5(10)-estratrien-17β-ol) by female rat liver enzymes, *Steroids*, 30(2), 275, 1977.
50. **Nakamura, T., Nomoto, N., Yagi, R., and Oya, N.**, Metabolic vitamin B complex deficiency due to the administration of glucocorticoid, *J. Vitaminol. (Japan)*, 16, 89, 1970.
51. **Howard, L., Wagner, C., and Schenker, S.**, Malabsorption of thiamine in folate deficient rats, *Am. J. Clin. Nutr.*, 27 (Abstr.), 440, 1974.
52. **Mason, M., Ford, J., and Wu, H. L. C.**, Effects of steroid and nonsteroid metabolites on enzyme conformation and pyridoxal phosphate binding, *Ann. N.Y. Acad. Sci.*, 166(1), 170, 1969.
53. **Sanpitak, N. and Chayutimonkul, L.**, Oral contraceptives and riboflavin nutrition, *Lancet*, 1, 836, 1974.
54. **Briggs, M. and Briggs, M. H.**, Changes in biochemical indices of vitamin nutrition in women using oral contraceptives during treatment with "Surbex 500", *Curr. Med. Res. Opin.*, 2, 626, 1975.
55. **Guggenheim, K. and Segal, S.**, Oral contraceptives and riboflavin nutriture, *Int. J. Vitam. Nutr. Res.*, 47(3), 234, 1977.
56. **Newman, L. J., Lopez, R., Cole, H. S., Boria, M. C., and Cooperman, J. M.**, Riboflavin deficiency in women taking oral contraceptive agents, *Am. J. Clin.Nutr.*, 31, 247, 1978.
57. **Briggs, M. and Briggs, M.**, Oral contraceptives and vitamin nutrition, *Lancet*, 1, 1234, 1974.
58. **Ahmed, F. and Bamji, M. S.**, Biochemical basis for the "riboflavin defect" associated with the use of oral contraceptives. A study in female rats, *Contraception*, 14(3), 297, 1976.
59. **Rose, D. P.**, Excretion of xanthurenic acid in the urine of women taking progestogen-oestrogen preparations, *Nature (London)*, 210, 196, 1966.
60. **Rose, D. P.**, The influence of oestrogens on tryptophan metabolism in man, *Clin. Sci.*, 31, 265, 1966.
61. **Mason, M. and Gullekson, E. H.**, Estrogen-enzyme interactions: inhibition and protection of kynurenine transaminase by the sulfate esters of diethylstilbestrol, estradiol, and estrone, *J. Biol. Chem.*, 235(5), 1312, 1960.

62. **Scardi, V., Iaccarino, M., and Scarano, E.,** The action of sulphate and phosphate esters of oestrogens on the reconstitution of two pyridoxal 5-phosphate-dependent enzymes, *Biochem. J.*, 83, 413, 1962.
63. **Rose, D. P.,** Effect of ovarian hormones and oral contraceptives on tryptophan metabolism, *J. Clin. Pathol.*, 3 (Suppl. 23), 37, 1970.
64. **Price, J. M., Thornton, M. J., and Mueller, L. M.,** Tryptophan metabolism in women using steroid hormones for ovulation control, *Am. J. Clin. Nutr.*, 20(5), 452, 1967.
65. **Rose, D. P. and Toseland, P. A.,** The determination of 3-hydroxy anthranilic acid in urine by gas-liquid chromatography, *Clin. Chim. Acta*, 17, 235, 1967.
66. **Brown, R. R., Rose, D. P., Price, J. M., and Wolf, H.,** Tryptophan metabolism as affected by anovulatory agents, *Ann. N.Y. Acad. Sci.*, 166, 44, 1969.
67. **Price, S. A. and Toseland, P. A.,** Oral contraceptives and depression, *Lancet*, 2, 158, 1969.
68. **Luhby, A. L., Davis, P., Murphy, M., Gordon, M., Brin, M., and Spiegel, H.,** Pyridoxine and oral contraceptives, *Lancet*, 2, 1083, 1970.
69. **Aly, H. E., Donald, E. A., and Simpson, M. H. W.,** Oral contraceptives and vitamin B_6 metabolism, *Am. J. Clin. Nutr.*, 24, 297, 1971.
70. **Luhby, A. L., Brin, M., Gordon, M., Davis, P., Murphy, M., and Spiegel, H.,** Vitamin B_6 metabolism in users of oral contraceptive agents. I. Abnormal urinary xanthurenic acid excretion and its correction by pyridoxine, *Am. J. Clin. Nutr.*, 24, 684, 1971.
71. **Price, S. A., Rose, D. P., and Toseland, P. A.,** Effects of dietary vitamin B_6 deficiency and oral contraceptives on the spontaneous urinary excretion of 3-hydroxyanthranilic acid, *Am. J. Clin. Nutr.*, 25, 494, 1972.
72. **Rose, D. P. and Adams, P. W.,** Oral contraceptives and tryptophan metabolism: effects of oestrogen in low dose combined with a progestogen and of a low-dose progestogen (megestrol acetate) given alone, *J. Clin. Pathol.*, 25, 252, 1972.
73. **Rose, D. P., Strong, R., Adams, P. W., and Harding, P. E.,** Experimental vitamin B_6 deficiency and the effect of oestrogen-containing oral contraceptives on trytophan metabolism and vitamin B_6 requirements, *Clin. Sci.*, 42, 465, 1972.
74. **Adams, P. W., Wynn, V., Rose, D. P., Seed, M., Folkard, J., and Strong, R.,** Effect of pyridoxine hydrochloride (vitamin B_6) upon depression associated with oral contraception, *Lancet*, 1, 897, 1973.
75. **Leklem, J. E., Rose, D. P., and Brown, R. R.,** Effect of oral contraceptives on urinary metabolite excretions after aministration of L-tryptophan or L-kynurenine sulfate, *Metabolism*, 22(12), 1499, 1973.
76. **Leklem, J. E., Brown, R. R., Rose, D. P., Arend, R. A., and Linkswiler, H. M.,** Tryptophan metabolism in oral contraceptive users and controls on defined intakes of vitamin B_6, *J. Nutr.*, 103(Abstr.), 18, 1973.
77. **Luhby, A. L., Brin, H., Spiegel, H., Brown, R. R., and Leklem, J. E.,** Reliability of urinary xanthurenic acid excretion after tryptophan load as index of vitamin B_6 deficiency and problem of its Fe^{+++} assay in oral contraceptive users, *Fed. Proc. Abstr.*, 32, 891, 1973.
78. **Lumeng, L., Cleary, R. E., and Li, T. K.,** Plasma pyridoxal-5′-phosphate in oral contraceptive users and in chronic alcohol abuse, *Fed. Proc. Abstr.*, 32, 891, 1973.
79. **Rose, D. P., Adams, P. W., and Strong, R.,** Influence of the progestogenic component of oral contraceptives on tryptophan metabolism, *J. Obstet. Gynaecol. Br. Commonw.*, 80, 82, 1973.
80. **Rose, D. P. and Toseland, P. A.,** Urinary excretion of quinolinic acid and other tryptophan metabolites after deoxypyridoxine or oral contraceptive administration, *Metabolism*, 22(2), 165, 1973.
81. **Rose, D. P., Brown, R. R., Arend, R. A., and Leklem, J. E.,** Effect of oral contraceptives on tryptophan and kynurenine metabolism, *Fed. Proc. Abstr.*, 32, 891, 1973.
82. **Coelingh Bennink, H. J. T. and Schreurs, W. H. P.,** Disturbance of tryptophan metabolism and its correction during hormonal contraception, *Contraception*, 9(4), 347, 1974.
83. **Lumeng, L., Cleary, R. E., and Li, T. K.,** Effect of oral contraceptives on the plasma concentration of pyridoxal phosphate, *Am. J. Clin. Nutr.*, 27, 326, 1974.
84. **Miller, L. T., Benson, E. M., Edwards, M. A., and Young, J.,** Vitamin B_6 metabolism in women using oral contraceptives, *Am. J. Clin. Nutr.*, 27, 797, 1974.
85. **Brown, R. R., Rose, D. P., Leklem, J. E., and Linkswiler, H. M,** Effects of oral contraceptives on tryptophan metabolism and vitamin B_6 requirements in women, *Acta Vitaminol. Enzymol. (Milano)*, 29, 151, 1975.
86. **Leklem, J. E., Brown, R. R., Rose, D. P., Linkswiler, H., and Arend, R. A.,** Metabolism of tryptophan and niacin in oral contraceptive users receiving controlled intakes of vitamin B_6, *Am. J. Clin. Nutr.*, 28, 146, 1975.
87. **Leklem, J. E., Brown, R. R., Rose, D. P., and Linkswiler, H. M.,** Vitamin B_6 requirements of women using oral contraceptives, *Am. J. Clin. Nutr.*, 28, 535, 1975.

88. **Green, A. R., Bloomfield, M. R., Woods, H. F., and Seed, M.**, Metabolism of an oral tryptophan load by women and evidence against the induction of tryptophan pyrrolase by oral contraceptives, *Br. J. Clin. Pharmacol.*, 5, 233, 1978.
89. **Wolf, H., Brown, R. R., Price, J. M., and Madsen, P. O.**, Studies on tryptophan metabolism in male subjects treated with female sex hormones, *J. Clin. Endocrinol.*, 31, 397, 1970.
90. **Haspels, A. A., Coelingh Bennink, H. J. T., and Schreurs, W. H. P.**, Disturbance of tryptophan metabolism and its correction during oestrogen treatment in postmenopausal women, *Maturitas*, 1, 15, 1978.
91. **Doberenz, A. R., Van Miller, J. P., Green, J. R., and Beaton, J. R.**, Vitamin B_6 depletion in women using oral contraceptives as determined by erythrocyte glutamic-pyruvic transaminase activities, *Proc. Soc. Exp. Biol. Med.*, 137, 1100, 1971.
92. **Reinken, L., Dapunt, O., and Kammerlander, H.**, Vitamin-B_6-Verarmung bei Einnahme oraler Contraceptiva, *Int. Z. Vit. Ern. Forsch.*, 43, 20, 1973.
93. **Rose, D. P., Strong, R., Folkard, J., and Adams, P. W.**, Erythrocyte aminotransferase activities in women using oral contraceptives and the effect of vitamin B_6 supplementation, *Am. J. Clin. Nutr.*, 26, 48, 1973.
94. **Salkeld, R. M., Knörr, K., and Körner, W. F.**, The effect of oral contraceptives on vitamin B_6 status, *Clin. Chim. Acta*, 49, 195, 1973.
95. **Reinken, L. and Dapunt, O.**, Vitamin-B_6-Mangel als Ursache der perioralen Dermatose nach Einnahme oraler Contraceptiva, *Int. Z. Vitam. Ern. Forschung.*, 44, 75, 1974.
96. **Brown, R. R., Rose, D. P., Leklem, J. E., Linkswiler, H., and Anand, R.**, Urinary 4-pyridoxic acid, plasma pyridoxal phosphate, and erythrocyte aminotransferase levels in oral contraceptive users receiving controlled intakes of vitamin B_6, *Am. J. Clin. Nutr.*, 28, 10, 1975.
97. **Miller, L. T., Johnson, A., Benson, E. M., and Woodring, M. J.**, Effect of oral contraceptives and pyridoxine on the metabolism of vitamin B_6 and on plasma tryptophan and α-amino nitrogen, *Am. J. Clin. Nutr.*, 28, 846, 1975.
98. **Shane, B. and Contractor, S. F.**, Assessment of vitamin B_6 status. Studies on pregnant women and oral contraceptive users, *Am. J. Clin. Nutr.*, 28, 739, 1975.
99. **Driskell, J. A., Geders, J. M., and Urban, M. C.**, Vitamin B_6 status of young men, women, and women using oral contraceptives, *J. Lab. Clin. Med.*, 87(5), 813, 1976.
100. **Kishi, H., Kishi, T., Williams, R. H., Watanabe, T., and Folkers, K.**, Deficiency of vitamin B_6 in women taking contraceptive formulations, *Res. Commun. Chem. Pathol. Pharmacol.*, 17(2), 283, 1977.
101. **Wien, E. M.**, Vitamin B_6 status of Nigerian women using various methods of contraception, *Am. J. Clin. Nutr.*, 31, 1392, 1978.
102. **Davis, R. E. and Smith, B. K.**, Pyridoxine and depression associated with oral contraception, *Lancet*, 1, 1245, 1973.
103. **Hontz, A. C., György, P., Balin, H., Rose, C. S., and Shaw, D. L.**, Interaction of contraceptive steroids with metabolic functions of vitamin B_6, *Am. J. Clin. Nutr.*, 27 (Abstr.), 440, 1974.
104. **Leklem, J. E., Linkswiler, H. M., Brown, R. R., Rose, D. P., and Anand, C. R.**, Metabolism of methionine in oral contraceptive users and control women receiving controlled intakes of vitamin B_6, *Am. J. Clin. Nutr.*, 30, 1122, 1977.
105. **Miller, L. T., Dow, M. J., and Kokkeler, S. C.**, Methionine metabolism and vitamin B_6 status in women using oral contraceptives, *Am. J. Clin. Nutr.*, 31, 619, 1978.
106. **Curzon, G.**, Tryptophan pyrrolase — a biochemical factor in depressive illness?, *Br. J. Psychiatry*, 115, 1367, 1969.
107. **Mason, M.**, Effects of conjugated steroids on enzymes, in *Metabolic Conjunction and Metabolic Hydrolysis*, Vol. 1, Fishman, W. H., Ed., Academic Press, New York, 1970, 121.
108. **Moursi, G. E., Abdel-Daim, M. H., Kelada, N. L., Abdel-Tawab, G. A., and Girgis, L. H.**, The influence of sex, age, synthetic oestrogens, progestogens and oral contraceptives on the excretion of urinary tryptophan metabolites, *Bull. W.H.O.*, 43, 651, 1970.
109. **Rose, D. P. and Braidman, I. P.**, Excretion of tryptophan metabolites as affected by pregnancy, contraceptive steroids, and steroid hormones, *Am. J. Clin. Nutr.*, 24, 673, 1971.
110. **Brown, R. R.**, Normal and pathological conditions which may alter the human requirement for vitamin B_6, *J. Agric. Food. Chem.*, 20(3), 498, 1972.
111. **Rose, D. P.**, Aspects of tryptophan metabolism in health and disease: a review, *J. Clin. Pathol.*, 25, 17, 1972.
112. **Anon.**, Oral contraceptives and vitamin B_6, *Nutr. Rev.*, 31(2), 49, 1973.
113. **Rose, D. P.**, Assessment of tryptophan metabolism and vitamin B_6 nutrition in pregnancy and oral contraceptive users, in *Biochemistry of Women: Methods for Clinical Investigations*, Curry, A. S. and Hewitt, J. V., Eds., CRC Press, Boca Raton, Fla., 1974, 317.
114. **Wolf, H.**, Studies on tryptophan metabolism in man, *Scand. J. Clin. Lab. Invest. Suppl.*, 136, 1974.

115. **Wynn, V., Adams, P. W., Folkard, J., and Seed, M.**, Tryptophan, depression and steroidal contraception, *J. Steroid. Biochem.*, 6, 965, 1975.
116. **Anon.**, Tryptophan catabolism by liver tryptophan pyrrolase, *Nutr. Rev.*, 34(1), 27, 1976.
117. **Larsson-Cohn, U.**, Some effects of oral contraceptives on vitamins and on carbohydrate and lipid metabolism, *Acta Obstet. Gynecol. Scand. Suppl.*, 54, 5, 1976.
118. **Bucci, L.**, Drug-induced depression and tryptophan metabolism, *Dis. Nerv. Syst.*, 33, 105, 1972.
119. **Winston, F.**, Oral contraceptives, pyridoxine, and depression, *Am. J. Psychiatry*, 130(11), 1217, 1973.
120. **Leeton, J.**, Depression induced by oral contraception and the role of vitamin B_6 in its management, *Aust. N.Z. J. Psychiatry*, 8, 85, 1974.
121. **Kane, F. J.**, Evaluation of emotional reactions to oral contraceptive use, *Am. J. Obstet. Gynecol.*, 126, 968, 1976.
122. **Malek-Ahmadi, P. and Behrmann, P. J.**, Depressive syndrome induced by oral contraceptives, *Dis. Nerv. Syst.*, 37, 406, 1976.
123. **Anon.**, Drug-induced depression, *Lancet*, 2, 1333, 1977.
124. **Joseph, M. H.**, Tryptophan metabolism, oral contraceptives, and pyridoxine, *Lancet*, 1, 661, 1978.
125. **Adams, P. W., Wynn, V., Seed, M., and Folkard, J.**, Vitamin B_6, depression and oral contraception, *Lancet*, 2, 516, 1974.
126. **Spellacy, W. N.**, A review of carbohydrate metabolism and the oral contraceptives, *Am. J. Obstet. Gynecol.*, 104(3), 448, 1969.
127. **Bingel, A. S. and Benoit, P. S.**, Oral contraceptives: therapeutics versus adverse reactions, with an outlook for the future. II, *J. Pharm. Sci.*, 62, 349, 1973.
128. **Beck, P.**, Contraceptive steroids: modifications of carbohydrate and lipid metabolism, *Metabolism*, 22(6), 841, 1973.
129. **Spellacy, W. N.**, Carbohydrate metabolism in male infertility and female fertility-control patients, *Fertil. Steril.*, 27(10), 1132, 1976.
130. **Goldman, J. A.**, Effect of ethynodiol diacetate and a combination-type oral contraceptive compound on carbohydrate metabolism, *Diabetologia*, 13, 89, 1977.
131. **Lageder, H., Irsigler, K., Schneider, W. H. F., Spona, J., Bauer, P., and Wohlzogen, F. X.**, Changes in glucose tolerance, serum insulin and blood lipids during contraceptive medication in women with normal metabolism, *Klin. Wochenschr.*, 8, 276, 1977.
132. **Leis, D., Bottermann, P., Ermler, R., and Maurer, U.**, Comparison of ethinylestradiol and mestranol in sequential type oral contraceptives in their effects on blood glucose and serum insulin in oral glucose tolerance tests, *Fertil. Steril.*, 28(7), 737, 1977.
133. **Spellacy, W. N., Newton, R. E., Buhi, W. C., and Birk, S. A.**, The effects of a "low-estrogen" oral contraceptive on carbohydrate metabolism during six months of treatment: a preliminary report of blood glucose and plasma insulin values, *Fertil. Steril.*, 28(8), 885, 1977.
134. **Goldman, J. A.**, Intravenous glucose tolerance after 18 months on progestogen or combination-type oral contraceptive, *Isr. J. Med. Sci.*, 14(3), 324, 1978.
135. **Spellacy, W. N., Buhi, W. C., and Birk, S. A.**, Effect of estrogen treatment for one year on carbohydrate and lipid metabolism in women with normal and abnormal glucose tolerance test results, *Am. J. Obstet. Gynecol.*, 131(1), 87, 1978.
136. **Spellacy, W. N., Buhi, W. C., Dumbaugh, V. A., and Birk, S. A.**, The effects of a once-a-week steroid contraceptive (R2323) on lipid and carbohydrate metabolism in women during three months of use, *Fertil. Steril.*, 30(3), 289, 1978.
137. **Spellacy, W. N., Mahan, C. S., Buhi, W. C., Dumbaugh, V. A., and Birk, S. A.**, Blood glucose, insulin, cholesterol and triglyceride levels in women treated for six months with the weekly oral contraceptive R2323, *Contraception*, 18(2), 121, 1978.
138. **Spellacy, W. N., Buhi, W. C., and Birk, S. A.**, The effects of vitamin B_6 on carbohydrate metabolism in women taking steroid contraceptives: preliminary report, *Contraception*, 6(4), 265, 1972.
139. **Adams, P. W., Cramp, D. G., Rose, D. P., and Wynn, V.**, The effect on carbohydrate metabolism of correcting pyridoxine (B6) deficiency in women taking oral contraceptive steroids (O.C.), *8th Congr. Int. Diabetes Federation*, Excerpta Medica, Amsterdam, Int. Congr. Ser. No. 280, 1973, 69.
140. **Rose, D. P., Leklem, J. E., Brown, R. R., and Linkswiler, H. M.**, Impairment of glucose tolerance by dietary vitamin B_6 deficiency in oral contraceptive users, *J. Nutr.*, 103(7), 18, 1973.
141. **Rose, D. P., Leklem, J. E., Brown, R. R., and Linkswiler, H. M.**, Effect of oral contraceptives and vitamin B_6 deficiency on carbohydrate metabolism, *Am. J. Clin. Nutr.*, 28, 872, 1975.
142. **Adams, P. W., Wynn, V., Folkard, J., and Seed, M.**, Influence of oral contraceptives, pyridoxine (vitamin B_6), and tryptophan on carbohydrate metabolism, *Lancet*, 1, 759, 1976.
143. **Coelingh Bennink, H. J. T. and Schreurs, W. H. P.**, Improvement of oral glucose tolerance in gestational diabetes by pyridoxine, *Br. Med. J.*, 3, 13, 1975.
144. **Spellacy, W. N., Buhi, W. C., and Birk, S. A.**, Vitamin B_6 treatment of gestational diabetes mellitus. Studies of blood glucose and plasma insulin, *Am. J. Obstet. Gynecol.*, 127(6), 599, 1977.

145. **Perkins, R. P.**, Failure of pyridoxine to improve glucose tolerance in gestational diabetes mellitus, *Obstet. Gynecol.*, 50(3), 370, 1977.
146. **Bryan, G. T.**, The role of urinary tryptophan metabolites in the etiology of bladder cancer, *Am. J. Clin. Nutr.*, 24, 841, 1971.
147. **Schievelbein, H., Löschenkohl, K., and Kuntze, I.**, Tryptophane metabolism, cancer of the urinary bladder and smoking habits, *Z. Klin. Chem. Klin. Biochem.*, 10, 445, 1972.
148. **Gailani, S., Murphy, G., Kenny, G., Nussbaum, A., and Silvernail, P.**, Studies on tryptophan metabolism in patients with bladder cancer, *Cancer Res.*, 33, 1071, 1973.
149. **Teulings, F. A. G., Fokkens, W., Kaalen, J. G., and van der Werf-Messing, B.**, The concentration of free and conjugated 3-hydroxyanthranilic acid in the urine of bladder tumour patients before and after therapy, measured with an enzymatic method, *Br. J. Cancer*, 27, 316, 1973.
150. **Bianchine, J. R., Bonnlander, B., Macaraeg, P. V. J., Hersey, R., Bianchine, J. W., and McIntyre, P. A.**, Serum vitamin B_{12} binding capacity and oral contraceptive hormones, *J. Clin. Endocrinol. Metab.*, 29, 1425, 1969.
151. **Shojania, A. M.**, Effect of oral contraceptives on vitamin-B_{12} metabolism, *Lancet*, 2, 932, 1971.
152. **Wertalik, L. F., Metz, E. N., LoBuglio, A. F., and Balcerzak, S. P.**, Decreased serum B_{12} levels secondary to oral contraceptive agents, *Am. J. Clin. Nutr.*, 24 (Abstr.), 603, 1971.
153. **Briggs, M. H. and Briggs, M.**, Endocrine effects on serum vitamin B_{12}, *Lancet*, 2, 1037, 1972.
154. **Wertalik, L. F., Metz, E.N., LoBuglio, A. F., and Balcerzak, S. P.**, Decreased serum B_{12} levels with oral contraceptive use, *JAMA*, 221(12), 1371, 1972.
155. **Alperin, J. B.**, Folate metabolism in women using oral contraceptive agents (OCA), *Am. J. Clin. Nutr.*, 26(Abstr.), 19, 1973.
156. **Rosenthal, H. L. and Wilbois, R. P.**, Influence of oral contraceptive agents (OCA) on vitamin B_{12} absorption and plasma level, *Fed. Proc. Abstr.*, 34, 905, 1975.
157. **Areekul, S., Panatampon, P., Doungbarn, J., Yamarat, P., and Vongyuthithum, M.**, Serum vitamin B_{12}, serum and red cell folates, vitamin B_{12} and folic acid binding proteins in women taking oral contraceptives, *Southeast Asian J. Trop. Med. Public Health*, 8(4), 480, 1977.
158. **Costanzi, J. J., Young, B. K., and Carmel, R.**, Serum vitamin B_{12} and B_{12}-binding protein levels associated with oral contraceptives, *Tex. Rep. Biol. Med.*, 36, 69, 1978.
159. **Fisch, I. R. and Freedman, S. H.**, Oral contraceptives and the red blood cell, *Clin. Pharm. Ther.*, 14(2), 245, 1973.
160. **Boots, L., Cornwell, P. E., and Beck, L. R.**, Effect of ethynodiol diacetate and mestranol on serum folic acid and vitamin B_{12} levels and on tryptophan metabolism in baboons, *Am. J. Clin. Nutr.*, 28, 354, 1975.
161. **Béla, K., Árpád, T., Sarolta, N., István, K., Tibor, P., and Tibor, J.**, Oralis contraceptivumok és B_{12} vitamin felszívódás, *Orv. Hetil.*, 117(7), 404, 1976.
162. **Shojania, A. M., Hornady, G., and Barnes, P. H.**, Oral contraceptives and serum-folate level, *Lancet*, 1, 1376, 1968.
163. **Spray, G. H.**, Oral contraceptives and serum-folate levels, *Lancet*, 2, 110, 1968.
164. **Shojania, A. M., Hornady, G., and Barnes, P. H.**, Oral contraceptives and folate metabolism, *Lancet*, 1, 886, 1969.
165. **Luhby, A. L., Shimizu, N., Davis, P., and Cooperman, J. M.**, Folic acid deficiency in users of oral contraceptive agents, *Fed. Proc. Abstr.*, 30, 239, 1971.
166. **Shojania, A. M., Hornady, G. J., and Barnes, P. H.**, The effect of oral contraceptives on folate metabolism, *Am. J. Obstet. Gynecol.*, 111(6), 782, 1971.
167. **Gaafar, A., Toppozada, H. K., Hozayen, A., Abdel-Malek, A. T., Moghazy, M., and Youssef, M.**, Study of folate status in long-term Egyptian users of oral contraceptive pills, *Contraception*, 8(1), 43, 1973.
168. **Roetz, R. and Nevinny-Stickel, J.**, Serumfolat, serumeisen und totale Eisenbindungskapazitat des Serums unter hormonaler Kontrazeption. Ergebnisse einer prospektiven Untersuchung, *Geburtshilfe Frauenheilkd.*, 33, 629, 1973.
169. **Shojania, A. M., Hornady, G. J., and Scaletta, D.**, The effect of oral contraceptives on folate metabolism. III. Plasma clearance and urinary folate excretion, *J. Lab. Clin. Med.*, 85, 185, 1975.
170. **Maniego-Bautista, L. P. and Bazzano, G.**, Effects of oral contraceptives on serum lipid and folate levels, *J. Lab. Clin. Med.*, 74 (Abstr.), 988, 1969.
171. **McLean, F. W., Heine, M. W., Held, B., and Streiff, R. R.**, Relationship between the oral contraceptive and folic acid metabolism, *Am. J. Obstet. Gynecol.*, 104(5), 745, 1969.
172. **Castrén, O. M. and Rossi, R. R.**, Effect of oral contraceptives on serum folic acid content, *J. Obstet. Gynaecol. Br. Commonw.*, 77, 548, 1970.
173. **Kahn, S. B., Fein, S., Rigberg, S., and Brodsky, I.**, Correlation of folate metabolism and socioeconomic status in pregnancy and in patients taking oral contraceptives, *Am. J. Obstet. Gynecol.*, 108(6), 931, 1970.

174. **Pritchard, J. A., Scott, D. E., and Whalley, P. J.**, Maternal folate deficiency and pregnancy wastage, *Am. J. Obstet. Gynecol.*, 109(3), 341, 1971.
175. **Stephens, M. E. M., Craft, I., Peters, T. J., and Hoffbrand, A. V.**, Oral contraceptives and folate metabolism, *Clin. Sci.*, 42, 405, 1972.
176. **Paine, C. J., Dickson, V. L., and Eichner, E. R.**, Effect of oral contraceptives on serum folate level and on hematologic status, *Clin. Res.*, 22 (Abstr.), 25A, 1974.
177. **Butterworth, C. E., Krumdieck, C. L., Stinson, H. N., and Cornwell, P. E.**, A study of the effect of oral contraceptive agents on the absorption, metabolic conversion and urinary excretion of a naturally-occurring folate (citrovorum factor), *Ala. J. Med. Sci.*, 12, 330, 1975.
178. **Paine, C. J., Grafton, W. D., Dickson, V. L., and Eichner, E. R.**, Oral contraceptives, serum folate, and hematologic status, *JAMA*, 231(7), 731, 1975.
179. **Heilmann, E. and Bönninghoff, E.**, Beeinflussen orale Kontrazeptiva den Folsäure-Spiegel im Plasma, *Med. Welt.*, 27(47), 2291, 1976.
180. **Karlin, R. and Bourgeay, M.**, Étude de l'action de contraceptifs oraux sur les taux sanguins des folates, *Pathol. Biol.*, 24(4), 251, 1976.
181. **Karlin, R., Dumont, M., and Long, B.**, Étude des taux sanguins d'acide folique au cours des traitements oestro-progestatifs, *J. Gynecol. Obstet. Biol. Reprod.*, 6, 489, 1977.
182. **Pietarinen, G. J., Leichter, J., and Pratt, R. F.**, Dietary folate intake and concentration of folate in serum and erythrocytes in women using oral contraceptives, *Am. J. Clin. Nutr.*, 30, 375, 1977.
183. **Streiff, R. R.**, Folate deficiency and oral contraceptives, *JAMA*, 214(1), 105, 1970.
184. **Lindenbaum, J., Whitehead, N., and Reyner, F.**, Oral contraceptive hormones, folate metabolism, and the cervical epithelium, *Am. J. Clin. Nutr.*, 28, 346, 1975.
185. **Martinez, O. and Roe, D. A.**, Effect of oral contraceptives on blood folate levels in pregnancy, *Am. J. Obstet. Gynecol.*, 128(3), 255, 1977.
186. **Paton, A.**, Oral contraceptives and folate deficiency, *Lancet*, 1, 418, 1969.
187. **Streiff, R. R.**, Malabsorption of polyglutamic folic acid secondary to oral contraceptives, *Clin. Res.*, 17 (Abstr.), 345, 1969.
188. **Holmes, R. P.**, Megaloblastik anemia precipitated by the use of oral contraceptives, *N.C. Med. J.*, 31, 17, 1970.
189. **Necheles, T. F. and Snyder, L. M.**, Malabsorption of folate polyglutamates associated with oral contraceptive therapy, *N. Engl. J. Med.*, 282(15), 858, 1970.
190. **Buhac, I. and Finn, J. W.**, Folsäuremangelanämie als Folge des langfristigen Gebrauchs peroraler kontrazeptiver Mittel, *Schweiz. Med. Wochenschr.*, 101, 1879, 1971.
191. **Palva, P.**, Megaloblastisk anemi orsakad av orala kontraseptiva, *Nord. Med.*, 86 (Abstr.), 1491, 1971.
192. **Ryser, J. E., Farquet, J. J., and Petite, J.**, Megaloblastic anemia due to folic acid deficiency in a young woman on oral contraceptives, *Acta Haematol.*, 45, 319, 1971.
193. **Toghill, P. J. and Smith, P. G.**, Folate deficiency and the pill, *Br. Med. J.*, 1, 608, 1971.
194. **Flury, R. and Angehrn, W.**, Folsäuremangelanämie infolge Einnahme oraler Kontrazeptiva, *Schweiz. Med. Wochenschr.*, 102, 1628, 1972.
195. **Salter, W. M.**, Megaloblastic anemia and oral contraceptives, *Minn. Med.*, 55, 554, 1972.
196. **Wood, J. K., Goldstone, A. H., and Allan, N. C.**, Folic acid and the pill, *Scand. J. Haematol.*, 9, 539, 1972.
197. **Johnson, G. K., Geenen, J. E., Hensley, G. T., and Soergel, K. H.**, Small intestinal disease, folate deficiency anemia, and oral contraceptive agents, *Dig. Dis.*, 18(3), 185, 1973.
198. **Meguid, M. M. and Loebl, W. Y.**, Megaloblastic anaemia associated with the oral contraceptive pill, *Postgrad. Med. J.*, 50, 470, 1974.
199. **Veyssier, P., Philippe, J. M., and Dusehu, E.**, Dysérythropoièse aiguë par carence folique révélant une maladie coeliaque, *Ann. Méd. Interne*, 128(10), 789, 1977.
200. **Mendes de Leon, D. E.**, Foliumzuurdeficiëntie, "de pil" en de verzwegen anamnese, *Ned. Tijdschr. Geneeskd.*, 122(5), 146, 1978.
201. **Hoffbrand, A. V.**, The role of malabsorption in the development of folate deficiency, *Clin. Med.*, 79(1), 19, 1972.
202. **Waxman, S., Corcino, J. J., and Herbert, V.**, Drugs, toxins and dietary amino acids affecting vitamin B_{12} or folic acid absorption or utilization, *Am. J. Med.*, 48, 599, 1970.
203. **Anon.**, Drugs and folic acid utilization, *Nutr. Rev.*, 29(2), 34, 1971.
204. **Rosenberg, I. H.**, Drugs and folic acid absorption, *Gastroenterology*, 63, 353, 1972.
205. **Streiff, R. R. and Greene, B.**, Drug inhibition of folate conjugase, *Clin. Res.*, 18 (Abstr.), 418, 1970.
206. **Whitehead, N., Reyner, F., and Lindenbaum, J.**, Megaloblastic changes in the cervical epithelium. Association with oral contraceptive therapy and reversal with folic acid, *JAMA*, 226(12), 1421, 1973.
207. **Krumdieck, C. L., Boots, L. R., Cornwell, P. E., and Butterworth, C. E.**, Estrogen stimulation of conjugase activity in the uterus of ovariectomized rats, *Am. J. Clin. Nutr.*, 28, 530, 1975.
208. **Butterworth, C. E.**, Interaction of drugs and nutrients, *JAMA*, 214(1), 137, 1970.

209. **Bernstein, L. H., Gutstein, S., Weiner, S., and Efron, G.,** The absorption and malabsorption of folic acid and its polyglutamates, *Am. J. Med.*, 48, 570, 1970.
210. **Shojania, A. M. and Hornady, G. J.,** Oral contraceptives and folate absorption, *J. Lab. Clin. Med.*, 82(6), 869, 1973.
211. **Snyder, L. M. and Necheles, T. F.,** Malabsorption of folate polyglutamates associated with oral contraceptive therapy, *Clin. Res.*, 17 (Abstr.), 602, 1969.
212. **Waxman, S. and Schreiber, C.,** Characteristics of folic acid-binding protein in folate-deficient serum, *Blood*, 42(2), 291, 1973.
213. **Waxman, S. and Schreiber, C.,** Measurement of serum folate levels and serum folic acid-binding protein by ^{3}H-PGA radioassay, *Blood*, 42(2), 281, 1973.
214. **Waxman, S.,** Folate binding proteins, *Br. J. Haematol.*, 29, 23, 1975.
215. **Da Costa, M. and Rothenberg, S. P.,** Appearance of a folate binder in leukocytes and serum of women who are pregnant or taking oral contraceptives, *J. Lab. Clin. Med.*, 83(2), 207, 1974.
216. **Eichner, E. R., Paine, C. J., Dickson, V. L., and Hargrove, M. D.,** Clinical and laboratory observations on serum folate-binding protein, *Blood*, 46(4), 599, 1975.
217. **Tamura, T., Shane, B., Baer, M. T., King, J. C., Margen, S., and Stokstad, E. L. R.,** Absorption of mono- and polyglutamyl folates in zinc-depleted man, *Am. J. Clin. Nutr.*, 31, 1984, 1978.
218. **Halsted, J. A., Hackley, B. M., and Smith, J. C.,** Plasma-zinc and copper in pregnancy and after oral contraceptives, *Lancet*, 2, 278, 1968.
219. **Margen, S. and King, J. C.,** Effect of oral contraceptive agents on the metabolism of some trace minerals, *Am. J. Clin. Nutr.*, 28, 392, 1975.
220. **Prasad, A. S., Oberleas, D., Lei, K. Y., Moghissi, K. S., and Stryker, J. C.,** Effect of oral contraceptive agents on nutrients. I. Minerals, *Am. J. Clin. Nutr.*, 28, 377, 1975.
221. **Sing, E. J., Baccarini, I. M., O'Neill, H. J., and Olwin, J. H.,** Effects of oral contraceptives on zinc and copper levels in human plasma and endometrium during the menstrual cycle, *Arch. Gynecol.*, 226, 303, 1978.
222. **Clemetson, C. A. B.,** Caeruloplasmin and green plasma, *Lancet*, 2, 1037, 1968.
223. **Russ, E. M. and Raymunt, J.,** Influence of estrogens on total serum copper and caeruloplasmin, *Proc. Soc. Exp. Biol. Med.*, 92, 465, 1956.
224. **Humoller, F. L., Mockler, M. P., Holthaus, J. M., and Mahler, D. J.,** Enzymatic properties of caeruloplasmin, *J. Lab. Clin. Med.*, 56(2), 222, 1960.
225. **Schenker, J. G., Jungreis, E., and Polishuk, W. Z.,** Oral contraceptives and correlation between serum copper and ceruloplasmin levels, *Int. J. Fertil.*, 17, 28, 1972.
226. **Saroja, N., Mallikarjuneswara, V. R., and Clemetson, C. A. B.,** Effect of estrogens on ascorbic acid in the plasma and blood vessels of guinea pigs, *Contraception*, 3(4), 269, 1971.
227. **Kalesh, D. G., Mallikarjuneswara, V. R., and Clemetson, C. A. B.,** Effects of estrogen-containing oral contraceptives on platelet and plasma ascorbic acid concentrations, *Contraception*, 4(3), 183, 1971.
228. **Briggs, M. and Briggs, M.,** Vitamin C requirements and oral contraceptives, *Nature (London)*, 238, 277, 1972.
229. **Rivers, J. M. and Devine, M. M.,** Plasma ascorbic acid concentrations and oral contraceptives, *Am. J. Clin. Nutr.*, 25, 684, 1972.
230. **Briggs, M. and Briggs, M.,** Vitamin C and colds, *Lancet*, 1, 998, 1973.
231. **Harris, A. B., Hartley, J., and Moor, A.,** Reduced ascorbic-acid excretion and oral contraceptives, *Lancet*, 2, 201, 1973.
232. **McLeroy, V. J. and Schendel, H. E.,** Influence of oral contraceptives on ascorbic acid concentrations in healthy, sexually mature women, *Am. J. Clin. Nutr.*, 26, 191, 1973.
233. **Briggs, M. H. and Briggs, M.,** Ascorbic acid and the pill, *Chem. Br.*, 11, 74, 1975.
234. **Rivers, J. M.,** Oral contraceptives and ascorbic acid, *Am. J. Clin. Nutr.*, 28, 550, 1975.
235. **Rivers, J. M. and Devine, M. M.,** Relationships of ascorbic acid to pregnancy, and oral contraceptive steroids, *Ann. N.Y. Acad. Sci.*, 258, 465, 1975.
236. **Harris, A. B., Pilley, M., and Hussein, S.,** Vitamins and oral contraceptives, *Lancet*, 2, 82, 1975.
237. **Weininger, J. and King, J. C.,** Effect of oral contraceptives on ascorbic acid status of young women consuming a constant diet, *Nutr. Rep. Int.*, 15(3), 255, 1977.
238. **Weininger, J. and King, J. C.,** Ascorbic acid metabolism in rhesus monkeys before and during use of oral contraceptive agents, *Fed. Proc. Abstr.*, 37, 359, 1978.
239. **Briggs, M. H. and Briggs, M.,** Vitamin D hormones. I, *Med. J. Aust.*, 1, 838, 1974; II, *Med. J. Aust.*, 1, 891, 1974.
240. **Goulding, A. and McChesney, R.,** Oestrogen-progestogen oral contraceptives and urinary calcium excretion, *Clin. Endocrinol.*, 6, 449, 1977.
241. **Simpson, G. R. and Dale, E.,** Serum levels of phosphorous, magnesium, and calcium in women utilizing combination oral or long-acting injectable progestational contraceptives, *Fertil. Steril.*, 23(5), 326, 1972.

242. **Goldsmith, N. F. and Johnston, J. O.**, Bone mineral: effects of oral contraceptives, pregnancy, and lactation, *J. Bone Jt. Surg., Am. Vol.*, 57(5), 657, 1975.
243. **Carter, D., Bressler, R., Haussler, M., Hughes, M., Christian, C. D., and Heine, M. W.**, Effect of oral contraceptives on drug and vitamin D_3 metabolism, *Clin. Pharm. Ther.*, 15 (2, Abstr.), 202, 1974.
244. **Carter, D. E., Bressler, R., Hughes, M. R., Haussler, M. R., Christian, C. D., and Heine, M. W.**, Effect of oral contraceptives on plasma clearance, *Clin. Pharm. Ther.*, 18(6), 700, 1975.
245. **Bouillon, R., van Baelen, H., and de Moor, P.**, The measurement of the vitamin D-binding protein in human serum, *J. Clin. Endocrinol. Metab.*, 45(2), 225, 1977.
246. **Aftergood, L. and Alfin-Slater, R. B.**, Oral contraceptive-α-tocopherol interrelationships, *Lipids*, 9(2), 91, 1974.
247. **Aftergood, L., Alexander, A. R., and Alfin-Slater, R. B.**, Effect of oral contraceptives on plasma lipoproteins, cholesterol and α-tocopherol levels in young women, *Nutr. Rep. Int.*, 11(4), 295, 1975.
248. **Briggs, M. and Briggs, M.**, Vitamin E status and oral contraceptives, *Am. J. Clin. Nutr.*, 28, 436, 1975.
249. **Tangney, C. C. and Driskell, J. A.**, Vitamin E status of young women on combined-type oral contraceptives, *Contraception*, 17(6), 499, 1978.
250. **Bieri, J. G. and Farrell, P. M.**, Vitamin E, *Vitam. Horm.*, 34, 31, 1976.
251. **Horwitt, M. K., Harvey, C. C., Dahm, C. H., and Searcy, M. T.**, Relationship between tocopherol and serum lipid levels for determination of nutritional adequacy, *Ann. N.Y. Acad. Sci.*, 203, 223, 1972.
252. **Harman, D.**, Vitamin E. Effect on serum cholesterol and lipoproteins, *Circulation*, 22, 151, 1960.
253. **McCormick, E. C., Cornwell, D. G., and Brown, J. B.**, Studies on the distribution of tocopherol in human serum lipoproteins, *J. Lipid Res.*, 1(3), 221, 1960.
254. **Davies, T., Kelleher, J., and Losowsky, M. S.**, Interrelation of serum lipoprotein and tocopherol levels, *Clin. Chim. Acta*, 24, 431, 1969.
255. **Rubinstein, H. M., Dietz, A. A., and Srinavasan, R.**, Relation of vitamin E and serum lipids, *Clin. Chim. Acta*, 23, 1, 1969.
256. **Larsson-Cohn, U., Berlin, R., and Vikrot, O.**, Effects of combined and low-dose gestagen oral contraceptives on plasma lipids; including individual phospholipids, *Acta Endocrinol.*, 63, 717, 1970.
257. **Kekki, M. and Nikkilä, E. A.**, Plasma triglyceride turnover during use of oral contraceptives, *Metabolism*, 20(9), 878, 1971.
258. **Larsson-Cohn U.**, Metabolic effects of contraceptive steroids with special reference to the lipid metabolism, *Acta Endocrinol.*, Suppl. 155 (Abstr.), 173, 1971.
259. **Van der Steeg, H. J. and Pronk, J. C.**, The effect of an oral contraceptive on serum lipoproteins and skinfold thickness in young women, *Contraception*, 16(1), 29, 1977.
260. **Bostofte, E., Hemmingsen, L., Alling Møller, K. J., Serup, J., and Weber, T.**, Serum lipids and lipoproteins during treatment with oral contraceptives containing natural and synthetic oestrogens, *Acta Endocrinol.*, 87, 855, 1978.
261. **Brockerhoff, P., Rathgen, G. H., and Weis, H.**, The effects of hormonal steroids on lipid metabolism — a prospective double blind study, *J. Steroid Biochem.*, 9(9), 858, 1978.
262. **Bieri, J. G.**, Vitamin E, *Nutr. Rev.*, 33(6), 161, 1975.
263. **Rutherford, R. N., Hougie, C., Banks, A. L., and Coburn, W. A.**, The effects of sex steroids and pregnancy on blood coagulation factors, *Obstet. Gynecol.*, 24(6), 886, 1964.
264. **Poller, L.**, Relation between oral contraceptive hormones and blood clotting, *J. Clin. Pathol.*, 23 (Suppl. 3), 67, 1970.
265. **Carey, H. M.**, Principles of oral contraception. II. Side effects of oral contraceptives, *Med. J. Aust.*, 2, 1242, 1971.
266. **McQueen, E. G.**, Hormonal steroid contraceptives. III. Adverse reactions, *Drugs*, 2, 20, 1971.
267. **Ambrus, J. L., Mink, I. B., Courey, N. G., Niswander, K., Moore, R. H., Ambrus, C. M., and Lillie, M. E.**, Progestational agents and blood coagulation. VII. Thromboembolic and other complications of oral contraceptive therapy in relationship to pretreatment levels of blood coagulation factors: summary report of a ten-year study, *Am. J. Obstet. Gynecol.*, 125(8), 1057, 1976.
268. **Briggs, M.**, Effect of a low-oestrogen combined oral contraceptive ("Nordette") on metabolic parameters especially related to the regulation of coagulation, blood pressure, and intermediary metabolism, in *Proc. 1st Int. Congr. Asian Fed. Obstet. Gynecol.*, 2, 175, 1976.
269. **Briggs, M. H.**, Thromboembolism and oral contraceptives, *Br. Med. J.*, 2, 503, 1974.
270. **Poller, L., Thomson, J. M., Thomas, W., and Wray, C.**, Blood clotting and platelet aggregation during oral progestogen contraception: a follow-up study, *Br. Med. J.*, 1, 705, 1971.
271. **Lira, P., Rivera, L., Diáz, S., and Croxatto, H. B.**, Study of blood coagulation in women treated with megestrol acetate implants, *Contraception*, 12(6), 639, 1975.

272. **Mink, I. B., Courey, N. G., Niswander, K. R., Moore, R. H., Lillie, M. A., and Ambrus, J. L.,** Progestational agents and blood coagulation. V. Changes induced by sequential oral contraceptive therapy, *Am. J. Obstet. Gynecol.*, 119, 401, 1974.
273. **Mellette, S. J.,** Interrelationships between vitamin K and estrogenic hormones, *Am. J. Clin. Nutr.*, 9(4, Suppl.), 109, 1961.
274. **Schrogie, J. J., Solomon, H. M., and Zieve, P. D.,** Effect of oral contraceptives on vitamin K-dependent clotting activity, *Clin. Pharm. Ther.*, 8(5), 670, 1967.
275. **Vessey, M. P. and Doll, R.,** Investigation of relation between use of oral contraceptives and thromboembolic disease, *Br. Med. J.*, 2, 199, 1968.
276. **Vessey, M. P. and Doll, R.,** Investigation of relation between use of oral contraceptives and thromboembolic disease: a further report, *Br. Med. J.*, 2, 651, 1969.
277. **Inman, W. H. W., Vessey, M. P., Westerholm, B., and Engelhund, A.,** Thromboembolic disease and the steroidal content of oral contraceptives. A report to the committee on safety of drugs, *Br. Med. J.*, 2, 203, 1970.
278. Royal College of General Practitioners, Mortality among oral-contraceptive users, *Lancet*, 2, 727, 1977.
279. **Vessey, M. P., McPherson, K., and Johnson, B.,** Mortality among women participating in the Oxford/Family Planning Association contraceptive study, *Lancet*, 2, 731, 1977.
280. **Laurell, C. B., Kullander, S., and Thorell, J.,** Plasma proteins after continuous, oral use of a progestogen-chloromadionone acetate- as a contraceptive, *Scand. J. Clin. Lab. Invest.*, 24, 387, 1969.
281. **Mendenhall, H. W.,** Effect of oral contraceptives on serum protein concentrations, *Am. J. Obstet. Gynecol.*, 106(5), 750, 1970.
282. **Dale, E. and Spivey, S. H.,** Serum proteins of women utilizing combination oral or long-acting injectable progestational contraceptives, *Contraception*, 4(4), 241, 1971.
283. **Briggs, M. and Briggs, M.,** Effects of some contraceptive steroids on serum proteins of women, *Biochem. Pharmacol.*, 22, 1, 1973.
284. **Gleichmann, W., Bachmann, G. W., Dengler, H. J., and Dudeck, J.,** Effects of hormonal contraceptives and pregnancy on serum protein pattern, *Eur. J. Clin. Pharmacol.*, 5, 218, 1973.
285. **Briggs, M.,** Effect of oral progestogens on estrogen-induced changes in serum protein, in *Proc. 2nd Int. Norgestrel Symp. London*, Excerpta Medica, Amsterdam, Int. Congr. Ser. No. 344, 1974, 35.
286. **Ramcharan, S., Sponzilli, E. E., and Wingerd, J. C.,** Serum protein fractions. Effects of oral contraceptives and pregnancy, *Obstet. Gynecol.*, 48(2), 211, 1976.
287. **Baker, H., Frank, O., Feingold, S., and Leevy, C. M.,** Vitamin distribution in human plasma proteins, *Nature (London)*, 215, 84, 1967.
288. **Sharada, D. and Bamji, M. S.,** Erythrocyte glutathione reductase activity and riboflavin concentration in experimental deficiency of some water soluble vitamins, *Int. J. Vitam. Nutr. Res.*, 42, 43, 1972.
289. **Vo-Khactu, K. P., Sims, R. L., Clayburgh, R. H., and Sandstead, H. H.,** Effect of simultaneous thiamin and riboflavin deficiencies on the determination of transketolase and glutathione reductase, *J. Lab. Clin. Med.*, 87(5), 741, 1976.
290. **Blair, J., Ratanasthien, K., Leeming, R., Melikian, V., and Cooke, W. T.,** Human serum folates in health and disease, *Am. J. Clin. Nutr.*, 27 (Abstr.), 441, 1974.
291. **Davis, R. E. and Smith, B. K.,** Pyridoxal, vitamin B_{12} and folate metabolism in women taking oral contraceptive agents, *S. Afr. Med. J.*, 48, 1937, 1974.
292. World Health Organization, Special Programme of Research, Development and Research Training in Human Reproduction, 7th Ann. Rep., 1978.
293. **Ram, M. M. and Bamji, M. S.,** Serum vitamin A and retinol-binding protein in malnourished women treated with oral contraceptives: Effects of estrogen dose and duration of treatment, *Am. J. Obstet. Gynecol.*, 135, 470, 1979.
294. **Bohner, J.,** Vitamin A und orale kontrazeptiva, *Deutche Med. Wochenschr.*, 104(13), 480, 1979.
295. **Anon.,** The effect of oral contraceptives on blood vitamin A levels and the role of sex hormones, *Nutr. Rev.* 37(11), 346, 1979.
296. **Vahlquist, A., Johnsson, A., and Nygren, K.,** Vitamin A transporting plasma proteins and female sex hormones, *Am. J. Clin. Nutr.*, 32(7), 1433, 1979.
297. **Bamji, M. S. and Ahmed, F.,** Effect of oral contraceptive steroids on vitamin A status of women and female rats, *World Rev. Nutr. Diet,*. 31, 135, 1978.
298. **Vir, S. C. and Love, A. H. G.,** Effect of oral contraceptive agents on thiamin status, *Int. J. Vitam. Nutr. Res.*, 49, 291, 1979.
299. **Vir, S. C. and Love, A. H. G.,** Riboflavin nutriture of oral contraceptive users, *Int. J. Vitam. Nutr. Res.*, 49, 286, 1979.
300. **Carrigan, P. J., Machinist, J., and Kershner, R. P.,** Riboflavin nutritional status and absorption in oral contraceptive users and nonusers, *Am. J. Clin. Nutr.*, 32, 2047, 1979.

301. **Wolf, H., Walter, S., Brown, R. R., and Arend, R. A.,** Effect of natural oestrogens on tryptophan metabolism: evidence for interference of oestrogens with kynureninase, *Scand. J. Clin. Lab. Invest.*, 40, 15, 1980.
302. **Williams, S. A. and Boots, L. R.,** Uterine and kidney aspartate aminotransferase activity in ethinylestradiol and norgestrel treated rates, *Contraception,* 21(6), 659, 1980.
303. **Tant, D.,** Megaloblastic anaemia due to pyridoxine deficiency associated with prolonged ingestion of an oestrogen-containing oral contraceptive, *Br. Med. J.*, 2, 979, 1976.
304. **Anon,** The vitamin B_6 requirement in oral contraceptive users, *Nutr. Rev.*, 37, (11), 344, 1979.
305. **Bender, D. A., Coulssn, W. F., Papadaki, L. and Pugh, M.,** Effects of oestrone on apparent vitamin B_6 status in peri- and post-menopausal women, *Proc. Nutr. Soc.*, 40, 00A, 1981(abstr.), in press.
306. **Wynick, D. and Bender, D. A.,** The effect of oestrone sulphate on the activity of kynureninase in vitro, *Proc. Nutr. Soc.*, 40, 00A, 1981(abstr.), in press.
307. **Calabrese, E. J.,** Does the use of oral contraceptives enhance the toxicity of carbon disulphide through interactions with pyridoxine and tryptophan metabolism, *Med. Hypotheses,* 6(1), 21, 1980.
308. **Bosse, T. R. and Donald, E. A.,** The vitamin B_6 requirement in oral contraceptive users. I. Assessment by pyridoxal level and transferase activity in erythrocytes, *Am. J. Clin. Nutr.*, 32(5), 1015, 1979.
309. **Donald, E. A. and Bosse, T. R.,** The vitamin B_6 requirement in oral contraceptive users, Ii. Assessment by tryptophan metabolites, vitamin B_6 and pyridoxic acid levels in urine, *Am. J. Clin. Nutr.*, 32(5), 1024, 1979.
310. **Parry, B. L. and Rush, A. J.,** Oral contraceptives and depressive symptomatology: biologic mechanisms, *Compr. Psychiatry,* 20(4), 347, 1979.
311. **Anon.,** Depresseion and oral contraceptives: the role of pyridoxine, *Drug. Ther. Bull.*, 16(22), 86, 1978.
312. **Applegate, W. V., Forsythe, A. and Bauernfeind, J. B.,** Physiological and psychological effects of vitamins E and B_6 on women taking oral contraceptives, *Int. J. Vitam. Nutr. Res.*, 49((1), 43, 1979.
313. **Grobe, H.,** Homocystinuria and oral contraceptives (letter), *Lancet,* 1, 158, 1978.
314. **Moller, S. E.,** Tryptophan and tyrosine availability and oral contraceptives (letter), *Lancet,* 2, 472, 1979.
315. **Warnes, H. and Fitzpatrick, C.,** Oral contraceptives and depression, *Psychosomatics,* 20(3), 187, 1979.
316. **Korzon, T. and Korzon, M.,** Significance of tryptophan metabolism disorders in obstetrics and gynecology, *Ginekol. Pol.*, 49(10), 917, 1978.
317. **Rose, D. P.,** The interactions between vitamin B_6 and hormones, *Vitam. Horm.*, 36, 53, 1978.
318. **Vir, S. C. and Love, A. H. G.,** Effect of oral contraceptives on vitamin B_6 nutriture of young women, *Int. J. Vitam. Nutr. Res.*, 50, 29, 1980.
319. **Adams, P. W.,** Pyridoxine, the pill and depression, *J. Pharmacotherapy,* 3(1), 20, 1980.
320. **Briggs, M. H.,** Effects of steroid therapy: oral contraceptives and depression, *J. Steroid Biochem.*, 9, 858, 1978.
321. **El-Zoghby, S. M., El-Kholy, Z. A., Saad, A. A., Mostafa, M. H., Abdel-Tawab, G. A., El-Dardiri, N., El-Kabariti, H. and Abdel-Rafea, A.,** The effect of environment on tryptophan metabolism via kynurenine in oral contraceptive users, *Acta Vitaminol. Enzymol. (Miland),* 32, 167, 1978.
322. **Oliveira, H. S., Oliveira, C., and Abraul, E.,** Influence of the vitamin B_6 on the alterations of the glucose and insulin levels in normal women receiving oral contraceptive drugs, Abstracts 11th World Congress of Gynecology and Obstetrics, Tokyo, October, 1979, 151.
323. **Bamji, M. S.,** Implications of oral contraceptive use on vitamin nutritional status, *Indian J. Med. Res.*, Suppl. 68, 80, 1978.
324. **Joshi, U. M.,** The effects of oral contraceptives on carbohydrate, lipid and protein metabolism in subjects with altered nutritional status and in association with lactation, *J. Steroid Biochem.*, 11, 483, 1979.
325. **Shojania, A. M. and Wylie, B.,** The effect of oral contraceptives on vitamin B_{12} metabolism, *Am. J. Obstet. Gynecol.*, 135, 129, 1979.
326. **Bianco, G., Accatino, G., Ciotta Vasino, A. and Neretto, G.,** Anemia megaloblastica da carenza di acido folico conseguente a somministrazione di contraccettivi orali, *Minerva Med.*, 69, 1513, 1978.
327. **Barone, C., Bartoloni, C., Ghirlanda, G. and Gentiloni, N.,** Megaloblastic anemia due to folic acid deficiency after oral contraceptives, *Haematologica,* 64(2), 190, 1979.
328. **Bamji, M. S. and Lakshmaiah, N.,** Half-life and Metabolism of Folic Acid in Oral Contraceptive Treated Rats, paper presented at Proc. Int. Congr. Nutrition, Rio de Janiero, August 1978.
329. **Lakshmaiah, N. and Bamji, M. S.,** Effect of oral contraceptives on folate economy — a study in female rats, *Horm. Metab. Res.*, 11(1), 64, 1979.
330. **Heilmank, E.,** Drug-induced megaloblastic anemia, *Med. Klin.*, 74(34), 1207, 1979.

331. **Hudiburgh, Neva K. and Milner, Alicen N.**, Influence of oral contraceptives on ascorbic acid and triglyceride status, *J. Am. Diet. Assoc.*, 75, 19, 1979.
332. **Clemetson, C. A. B.**, Some thoughts on the epidemiology of cardiovascular disease, (with special reference to women "on the pill"), role of ascorbic acid, *Med. Hypotheses*, 5, 825, 1979.
333. **Prasad, A. S., Lei, K. Y., Moghisse, K. S., Stryker, J. C., and Oberleas, D.**, Effect of oral contraceptives on nutrients. III. Vitamins B_6 B_{12} and folic acid, *Am. J. Obstet. Gynecol.*, 125(8), 1065, 1976.
334. **Bamji, M. S., Prema, B. A., Lakshmi, R., Ahmed, F., and Jacob, C. M.**, Oral contraceptive use and vitamin nutrition status of malnourished womem — effects of continuous and intermittent vitamin supplements, *J. Steroid Biochem.*, 11, 487, 1979.
335. **Massey, L. K. and Davison, M. A.**, Effects of oral contraceptives on nutritional status, *Am. Fam. Physician*, 19(1), 119, 1979.
336. **Lewis, C. M. and King, J. C.**, Effect of oral contraceptive agents on thamin, riboflavin and pantothenic acid status in young women, *Am. J. Clin. Nutr.*, 33, 832, 1980.
337. **Cervantes, L. F., Lopez, B. J. and Calderon, M.**, Changes in laboratory tests after treatment with a new contraceptive agent, *Ginecol. Obstet. Mex.*, 43, 285, 1978.
338. **Prasad, A. S., Lei, K. Y. and Moghissi, K. S.**, The effect of oral contraceptives on micronutrients, in *Nutrition and Human Reproduction*, Mosley, W. H., Ed., Plenum Press, New York, 1977, 61.
339. **Gaspard, U.**, Metabolic repercussions of oral contraception, *J. Pharm. Belg.*, 33(5), 312, 1978.
340. **Heilmann, E.**, Orale contrazeptiva und vitamine, *Dtsch. Med. Wochenschr.*, 104(4), 144, 1979.
341. **Worthington, B. S.**, Nutrition during pregnancy, lactation and oral contraception, *Nurs. Clin. North Am.*, 14(2), 269, 1979.
342. **Anon.**, Effects of oral contraceptives on laboratory test results, *Med. Lett. Drugs Ther.*, 21(13), 54, 1979.
343. **Nash, A. L., Cornish, E. J. and Hain, R.**, Metabolic effects of oral contraceptives containing 30 micrograms and 50 micrograms of oestrogen, *Med. J. Aust.*, 2(6), 277, 1979.

Chapter 3

VITAMIN STATUS OF THE ELDERLY

Delia M. Flint and D. M. Prinsley

TABLE OF CONTENTS

I. Introduction 66

II. Causes of Vitamin Deficiency 66

III. Vitamin D 67
A. Clinical Aspects of Vitamin D Deficiency in the Elderly 67
B. Vitamin D Status of the Elderly 68
C. Causes of Vitamin D Deficiency 69

IV. Ascorbic Acid 69
A. Clinical Aspects of Vitamin C Deficiency in the Elderly 69
B. Vitamin C Status of the Elderly 70
C. Causes of Vitamin C Deficiency in the Elderly 70

V. Folacin 70
A. Clinical Aspects of Folacin Deficiency in the Elderly 70
B. Causes of Folacin Deficiency 71

VI. Vitamin B_{12} 72
A. Clinical Aspects of Vitamin B_{12} Deficiency in the Elderly 73
B. Causes of Vitamin B_{12} Deficiency 74

VII. Thiamin 74
A. Clinical Manifestations of Thiamin Deficiency in the Elderly 74
B. Thiamin Deficiency in the Elderly 75
C. Causes of Thiamin Deficiency 75

VIII. Niacin 75

IX. Conclusions 75

References 76

I. INTRODUCTION

For the first time in the history of survival old age has become accepted as normal for the majority of the population. This phenomenon is taking place not only in western developed society, but also in less developed parts of the world. Medical advances, especially in the conquest of infectious diseases and improved health care, have increased the life span markedly in the western world. Not very much is known about the nutritional status of elderly members of primitive societies. Because of closer family ties it may be that their meals more closely resemble those of the other members of the family. They are not isolated and left to fend for themselves, which is so often the fate of the aged in our allegedly more advanced western society. As more of our citizens live beyond 65 years, the health of this group will assume more importance. Nutritional considerations are likely to be relevant.

Nutrient requirements vary little beyond adolescence. However, life style, psychological and economic changes, or chronic disease in the later years of life may alter dietary habits and therefore nutrient intake.

It is important to exclude underlying disease when a nutrient deficiency is discovered.[1] Vitamin deficiencies with a clear-cut clinical presentation are, nevertheless, not common in the elderly in spite of the fact that adequate vitamin intake frequently appears to be restricted because of cost of food stuffs, problems with cooking, and sheer lack of interest or initiative. Deficiency states in the elderly are most likely to occur from inadequate intake of vitamins of the B group, vitamin C, and vitamin D.[2]

II. CAUSES OF VITAMIN DEFICIENCY

Decreased mobility of the elderly reduces the distance which can be covered outside the home. Trips to markets and other sources of fresh fruit and vegetables become difficult or impossible because of lack of transport or ability to carry parcels. Consequently, convenience food and tinned food becomes increasingly part of the diet. Such diets may lack adequate vitamin content.

Lack of mobility within the house due to unsteady gait, arthritic arms and legs, or degenerative neurological diseases leads to neglect in preparation of proper meals.

Mental frailty in the form of poor memory, poor attention span, and lack of initiative is one of the common failings of old age. If the old person lives alone, this compounds the inability to plan and prepare balanced meals.

Adequate income in old age, ideal but unusual, would enable pensioners to purchase foods containing the necessary vitamins, if they were aware of the correct items to buy.

The tongue remains powerful and active throughout life, but teeth decay and are removed. Dentures, if supplied, tend to become loose with gum shrinkage and consequently ineffective and sometimes are removed for eating. The result is a restricted choice of foods that can be masticated.

There are a number of relatively common pathological conditions in old people which cause change in dietary habit or difficulty in absorption with consequent malnutrition.[3,4] Narrowing of the lower end of the esophagus due to regurgitation of gastric juice with inflammation and later scarring, "heart burn" of so many obese multiparous women is relatively common. It does not reduce life span. It may exist for years and be accepted that it no longer is possible to eat hard or bulky food. Many sufferers from this condition have been advised always to eat a soft diet which they could manage for many years without further thought, and the potential for malnutrition is obvious.

Loss of function of the parietal cells of the stomach with gastric atrophy in some aged people makes absorption of vitamin B_{12} impossible because of lack of intrinsic factor. Incidence of gastric atrophy in old age is not well documented, but the ever increasing number of cases of megaloblastic anemia due to B_{12} deficiency which come to light in old age suggest that the condition is not uncommon.

Surgical indications for elective partial gastrectomy for peptic ulcer, curative for the ulcer, may, years later, cause complicated malabsorption patterns, especially the development of megaloblastic anemia. In western societies diverticulosis of the colon associated with chronic constipation is better understood. Management of this condition with high fiber diets is now advised. Fruits and vegetables are no longer avoided by sufferers from diverticular disease, but there are still many old people who have persisting faith in medical advice given many years ago who are still on low fiber vitamin deficient diets.

Certain vulnerable groups of elderly can be identified who are at greater risk of developing vitamin deficiency than the average old person living in the community in good health. The house-bound who never have exposure to sunshine have little chance of synthesizing adequate vitamin D. Vitamin D is unlikely to be present in adequate quantities in the diet. Many institutionalized old people in rest homes, hostels, and nursing homes have little fresh air and sunlight and have little choice of diet, which may be appetizing and adequate but, nevertheless, has been mass produced. Institutional catering, however carefully conducted, is liable to produce over-cooked vegetables which are kept hot for long periods before reaching the residents. A long-term resident under these conditions, after 3 months of institutional catering, is likely to run short of folic acid because of over-cooking of green vegetables. Visually handicapped or mentally infirm patients who require feeding by the nursing staff, have to manage with varying degrees of nursing skills in channeling adequate calories in soft diets to their stomachs. Such skill and patience and suitably prepared diet are not always available. Patients requiring special diets for prolonged periods for medical conditions, such as peptic ulcer, may receive totally inadequate vitamin content in the meals. Long-term medication in old age, for a variety of conditions, may inhibit vitamin absorption from the intestinal tract. Some sedatives, anticonvulsants, antibiotics, and slow release potassium tablets are known to interfere with vitamin utilization.

Clinical manifestations of vitamin deficiency in the elderly are not easy to identify. Some of the described stigmata of vitamin deficiencies may be seen in healthy old people whose vitamin status is not in question. Angular stomatitis is far more commonly caused by badly fitting false teeth than by vitamin B deficiency. Cheilosis, the area of denuded epithelium along the lips, and rough skin and corneal vascularization is seen, but rarely due to vitamin deficiency. Definite clinical manifestations of individual avitaminosis, however, may be obvious and the response to appropriate treatment is usually magnificent.

III. VITAMIN D

A. Clinical Aspects of Vitamin D Deficiency in the Elderly

Osteomalacia of adults occurs as a result of vitamin D deficiency.[5-8] An investigation of 200 patients admitted to a geriatric assessment unit for women found 16 cases of osteomalacia.[8] Similar findings have been reported by Chalmers et al., who believe it is due to dietary lack as well as lack of exposure to sunlight.[6]

House-bound and institutionalized aged are susceptible to vitamin D deficiency because of limited exposure to sunlight and inadequate dietary intake.

Osteomalacia is of great importance, because it is clearly related to vitamin D deficiency and is completely curable. The bones in osteomalacia contain the normal amount of bone tissue, but the mineral content is reduced. Lack of vitamin D tends to reduce absorption of calcium and calcium deposition in the bones. The active metabolites of vitamin D evolve through a series of events in the skin, in the liver, and finally in the kidneys, where dihydrocholecalciferol is produced. In the elderly, exposure of the skin to ultraviolet radiation is reduced by lack of outdoor activity and by wearing too much clothing. Southern climates with less sun and more cold weather further increases the liability to vitamin D deficiency. Chronic renal disease with heavy calcium loss in the urine and malabsorption syndromes are also potential causes of osteomalacia.

The clinical picture is not diagnostic. Presentation with lack of mobility and proximal bone pain may initially be presumed to be "arthritis". Backache, difficulty in rising from a chair, difficulty in climbing stairs, and the so-called waddling gait associated with bone tenderness are features of osteomalacia, and there may be episodes of pain associated with partial fractures of bone. Radiological investigation shows incomplete fractures of bone, without displacement, in certain characteristic sites. These are rami of the pelvis, the medial aspect of the upper femur along the lateral edge of the scapular, and the lower border of ribs. The pseudo fractures or Looser zones are often bilateral. Biochemical investigation confirms the diagnosis by revealing raised plasma alkaline phosphatase and a tendency to low calcium and phosphate levels.

One of the most gratifying medical treatments is to witness the improvement within days, with loss of pain and increased mobility in an otherwise incapacitated old person when treated with vitamin D supplementation. Fifty thousand IU daily for 2 weeks followed by a smaller maintenance dose, will cure osteomalacia and heal pseudo fractures.

Development of osteoporosis in old age results from a variety of causes. Deficiency of estrogen, lack of adequate calcium in the diet, and reduced activity are probably the most important. There is reduction in the amount of bony tissue relative to the volume of anatomical bone, but the bony tissue remaining contains the normal amount of mineral content as calcium. The reduction of hard bone leads to fractures caused by minimal trauma, common sites being neck of femur, wrist, and vertebral bodies. Osteoporosis is not caused by vitamin D deficiency, but osteomalacia may coexist with osteoporosis. Treatment is still a matter for debate. The best results seem to be obtained using estrogen and calcium. Calcium and vitamin D was less effective.[9]

B. Vitamin D Status in the Elderly

It has been shown by histological methods[10] that 20 to 30% of women and 40% of men with fracture of the proximal femur had osteomalacia. Faccini and co-workers confirmed the significance of vitamin D deficiency as an important factor in the pathogenesis of fracture of the femoral neck.[11] In 1976, Brown found significant lower levels of 25-hydroxy cholecalciferol in patients with fractures of the femoral neck, compared with those in controls of similar age, from whom blood samples were taken at the same time of the year.[12] Chalmers et al. recognized the importance of vitamin D deficiency as a cause of fractures and other orthopedic problems in the elderly, particularly in women.[6]

Various workers have measured vitamin D metabolites in the blood of elderly people.[5-8,12-21] Seasonal variation has been shown to have a significant effect on vitamin D status.[10,12,14] The radiosteroassay of 25-hydroxycholecalciferol levels have shown seasonal variations in both young and old. Contrary to the views of Lamb et al.,[15] Stamp and Round[14] concluded that summer sunlight is an important determinant of vitamin D nutrition in Britain. Pettifor et al.,[19] analyzed the serum of elderly South Africans monthly. They concluded that the seasonal decline in 25-hydroxycholecalciferol concentrations may be related to a fall in environmental temperature, resulting in less time spent out of doors and greater skin covering by clothing.

Workers have reported[18] an absence of osteomalacia in the presence of low plasma 25-hydroxycholecalciferol levels, which have been associated with vitamin D deficiency osteomalacia. Therefore, care should be taken in interpreting the low plasma 25-hydroxycholecalciferol concentrations. In a study by Vir and Love[16] of institutionalized and noninstitutionalized aged requiring medical care, biochemical deficiency of vitamin D was noted in over 40% of the subjects, while osteomalacia was diagnosed biochemically in only 5.2%. Vitamin D status in this study also demonstrated no relationship to bone density.

Biochemical procedures used to evaluate vitamin D nutriture cannot be correlated with dietary intake of the vitamin, as the requirement of vitamin D is met by ingestion of food as well as by exposure of skin to sunlight.[19]

C. Causes of Vitamin D Deficiency

Deficiency of the vitamin does not occur due to a dietary lack of vitamin D alone. Unless the diet is unusually rich in fish products, there are no important ingested sources of vitamin D.[20] But diet can be a contributory cause together with lack of exposure to ultraviolet light.[16] Institutionalized and house-bound elderly are very vulnerable to deficiency of the vitamin.[10] Recent surveys have confirmed that a low intake of vitamin D is common in the diet of elderly people today.[10,16,18,22,23]

In a study by Exton-Smith and colleagues,[24] it was found that 48% of house-bound women, 70 to 79 years, had a dietary intake of less than 30 IU/day compared with 13% of active women of similar age. Brown reported that the reduced efficiency in formation of vitamin D metabolite in the liver is responsible for low levels of serum 25-hydroxycholecalciferol and for increasing incidence of osteomalacia in the elderly.[12] A study in Scotland reported that a substantial proportion of all subjects had low intakes of vitamin D, which agrees with the high prevalence of osteomalacia in Scotland.[25]

The panel on Nutrition of the Elderly,[23] recommended a trial in which prophylactic vitamin D is given to all persons who are at risk of vitamin D deficiency. Among other causes of deficiency are malabsorption, increased physiological requirement, impaired conversion of 25-hydroxycholecalciferol to the active 1,25-hydroxycholecalciferol due to a decline in renal function in old age,[26,27] and to increased breakdown of vitamin D metabolites due to the action of certain drugs.[18]

It follows that it is usually necessary to improve the vitamin D status of the elderly, either by supplementation or by ensuring exposure to ultraviolet light.

IV. ASCORBIC ACID

A. Clinical Aspects of Vitamin C Deficiency in the Elderly

In old age, it would seem probable that scurvy would be a common event at the end of winter when fresh fruit is scarce and expensive in countries which do not have local produce available all year round. Clinical scurvy, however, is a rare condition, although subclinical cases which are not diagnosed may be less rare.

The condition is found only where a history of gross malnutrition over a prolonged period is elicited. Elderly men cooking for themselves are the most likely candidates, occurring in spring after a winter when no fresh fruit has been eaten. The usual clinical picture is one of anemia and malnutrition and general neglect of personal hygiene. Old scars of varicose ulcers on the legs may have broken down and unexplained areas of s.c. hemorrhage may be present, especially around the lower legs. Hyperkeratosis of hair follicles on the abdomen, thighs, and outer aspects of the upper arms should be sought, with plugging of the follicles which contain curled up hairs. Around the plugged follicles there may be small hemorrhages.

The combination of anemia, subnutrition, body neglect, skin changes, and hemorrhage is virtually diagnostic, without the bleeding gums described in younger cases of scurvy. A much less florid clinical picture of scurvy may be seen in patients with old, breaking down ulcers on the legs, and anemia which does not respond readily to iron, but improves rapidly when vitamin C is added.[28] This form of dyshaemopoietic anemia is almost certainly due to lack of other factors such as proteins, in addition to vitamin C, which are contained in a normal diet.

Old people with scurvy, in apparently poor condition, respond rapidly to treatment with vitamin C,[29] and the possibility of this diagnosis should always be in the mind of physicians treating patients with a poor dietary history.

B. Vitamin C Status of the Elderly

Overt clinical manifestations of vitamin C deficiency are rare, but body tissue reserves are low. Leukocyte and serum vitamin C levels are widely accepted as a sensitive index of vitamin C deficiency.[30,31] Low levels of leukocyte vitamin C have been reported in the elderly[23,32-48] and are lower than younger subjects,[32-36] lower in winter than summer,[31] and lower in men than women.[37-42] Institutionalization and smoking also decreases leukocyte vitamin C concentrations.[40,43,45] Elderly men living alone have been found to be in a poor status of vitamin C.[37,46-48] In a study conducted on elderly men living alone in New Zealand, the low dietary intakes of vitamin C significantly correlated with both plasma and leukocyte vitamin C levels.[45]

Other workers have found that elderly subjects are well nourished with respect to vitamin C.[49-54] Older women in nursing homes and private homes had good vitamin C status.[49] Serum vitamin C values reported among institutionalized aged women observed by Justice et al.[50] and women 59 years or over in the Ten State Nutrition Survey[51] indicated that vitamin C deficiency was not prevalent among these groups studied in the U.S.

A dietary intake study conducted on the elderly living in a rural communtiy in Northeast U.S.[52] compared with that of nursing home residents, who prior to entering the institution resided in the same geographic area, all had intakes of vitamin C which met the Allowances.[55]

Wilson et al.[56] determined whether higher mortality, associated with low leukocyte and platelet ascorbic acid levels, could be reduced by administration of a daily dose of 200 mg vitamin C. Evidence suggested that though poor dietary intake may be responsible for low leukocyte vitamin C levels, in some cases they are more commonly a consequence of a wide variety of stressful states and a reflection of severity of illness rather than a dietary deficiency state.

C. Causes of Vitamin C Deficiency in the Elderly

It is believed that some illnesses which occur in the elderly can depress leukocyte vitamin C levels,[2,56] and also a number of drugs.[57] The elderly in institutions are vulnerable to vitamin C deficiency caused by poor cooking techniques and inadequate intake of fresh fruit and vegetables.

V. FOLACIN

A. Clinical Aspects of Folacin Deficiency in the Elderly

Deficiency of folacin in the elderly has been reported by many workers,[58-71] and it is believed that folacin deficiency contributes to the prevalence of megaloblastic anemia which is common in this population. However, Girdwood et al.[71] failed to find any difference in serum folacin levels between young and older age groups in Edinburgh.

Megaloblastic anemia due to folacin deficiency may result from a variety of causes, including inadequate dietary intake, destruction of folacin by incorrect food preparation, and by failure of absorption and by administration of certain drugs, specially anticonvulsants. The symptoms produced are those associated with any anemia, with a possibility that dementia may be a presenting feature.

Mental changes may precede the anemia and there have been investigations which show the relationship between folacin deficiency and neurological disorders.[72-76]

In a study carried out by Herbert to delineate the sequence of events in developing folate deficiency in man, forgetfullness appeared, and gradually became progressive during the 4th month of the low folacin diet.[77]

Strachan and Henderson[78] noted a response in two patients with dementia to prolonged folacin administration. The relationship between poor folacin nutrition and dementia was confirmed in a study by Sneath et al.[79] He found that in 115 consecutive admissions to a geriatric department, 14 were diagnosed as having dementia. Their mean red cell folacin concentration (279 ng/mℓ) was significantly lower than the mean (394 ng/mℓ) for the group as a whole. Intellectual function was assessed on a 16 point mental test score[80] and there was some correlation between intellectual function and the red cell folacin concentration in those with low red cell folacin levels (below 200 ng/mℓ).

B. Causes of Folacin Deficiency

The main causes are inadequate intake, malabsorption, increased utilization, and impaired effectiveness.[2,81,82]

In a recent study of dietary folacin utilization,[81] it was suggested that the folacin deficits, so commonly in the elderly, are caused by impaired ability to obtain folacin from ingested foods. The resultant folacin deficit may induce changes in the epithelial structure and enzyme secretion of the small bowel which further exacerbate folacin malabsorption. Maintenance of food folacin utilization depends upon pteroylpolyglutamate hydrolase (conjugase) in the small bowel.[82] Therefore a nutritionally adequate diet may not suffice as a source of folacin in the elderly.[2]

Dietary inadequacy of folacin is an important cause of deficiency. The vitamin is readily destroyed by sunlight, oxidation, and cooking. Nutritional deficiency of this vitamin is common among patients admitted to geriatric units.[83] A nutritional survey of 46 elderly long-term surgical patients receiving a sufficient hospital diet revealed inadequate intakes of folacin in most patients. As part of an investigation on folacin status, the folacin intake of a group of elderly patients was measured. It was found that the hospital patients received considerably less folacin per day (101.1 μg) than young and elderly noncontrol groups, who had an intake of 225.1 μg and 145.5 μg of folacin, respectively. Food served in the hospital was low in folacin.[84,85]

The results of a study on the biochemical status of folacin in a group of community and institutionalized are summarized in Figures 1 and 2.[39] For the community based old people, the serum folacin level was 6.8 ± 0.4 ng/mℓ whereas, for the institutionalized subjects, it was 4.1 ± 0.3 ng/mℓ. ($p \sim <0.001$) The red cell folacin community value was 553 ± 63 ng/mℓ and the institutionalized 374 ± 31 ng/mℓ. ($p \sim <0.001$) There was no significant difference in serum and red cell folacin concentration between the males and females. Inadequate dietary intake by the institutionalized may account for these differences.

One must be cautious when interpreting dietary intakes of folacin as it is only recently that total and free folacin activity have been measured in foodstuffs. The values are available in the food composition tables by Hoppner[86] and Paul and Southgate.[87]

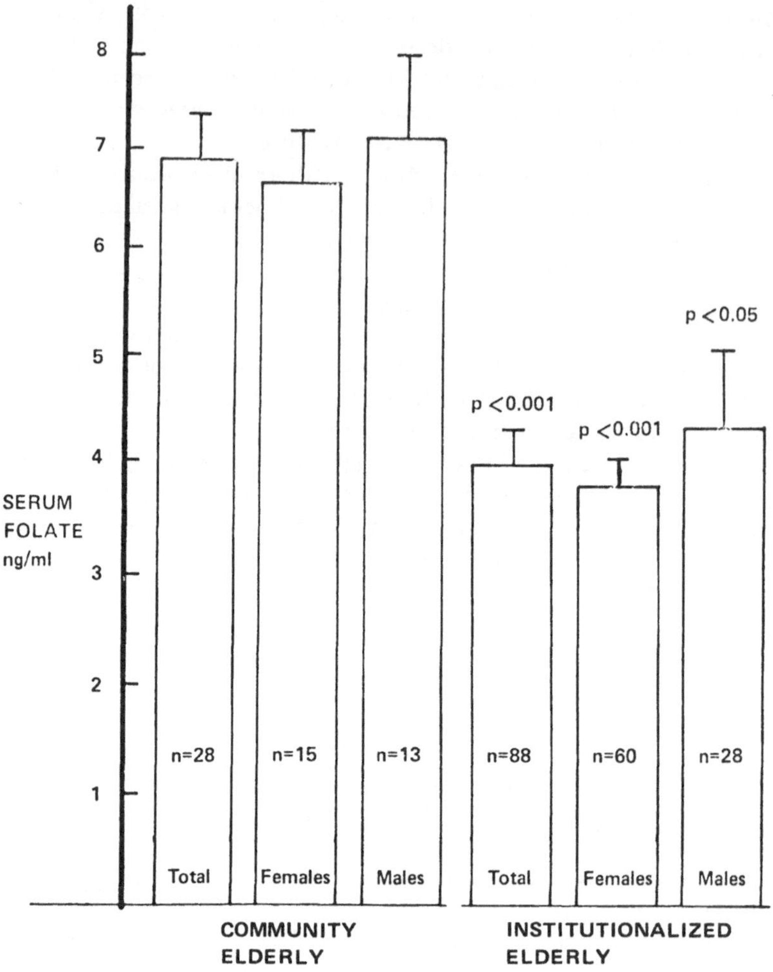

FIGURE 1. Mean (± SE) serum folacin concentrations in community and institutionalized elderly (n = number).

VI. VITAMIN B_{12}

There is a close metabolic interrelationship between folacin and vitamin B_{12},[88-90] particularly in the similarity of changes that occur in the hemopoietic system, and the mental disturbances when a deficiency of one of these vitamins is present. Therefore, vitamin B_{12} nutritional status must be evaluated in parallel with folacin status.

Serum vitamin B_{12} levels measurement has been done by many workers.[88-95] There may be a reduction with increasing age, but pathological levels are not reached unless the patient has either pernicious anemia or gastrointestinal disease. Other workers have shown that age has no effect on serum vitamin B_{12} concentration.[93-95]

A study by Meindock and Dvorsky[69] was done on groups of elderly admitted from their own homes and homes for the aged, together with a group of healthy young subjects. It was concluded that the significantly higher level of serum vitamin B_{12} found in the geriatric group was possibly due to the increased use of vitamin B_{12} supplements in the treatment of the elderly over the past few years.

Droller and Dossett[94] found that although age itself had no effect on the serum vitamin B_{12} concentration, there was a correlation between mental failure and low serum vitamin B_{12} level.

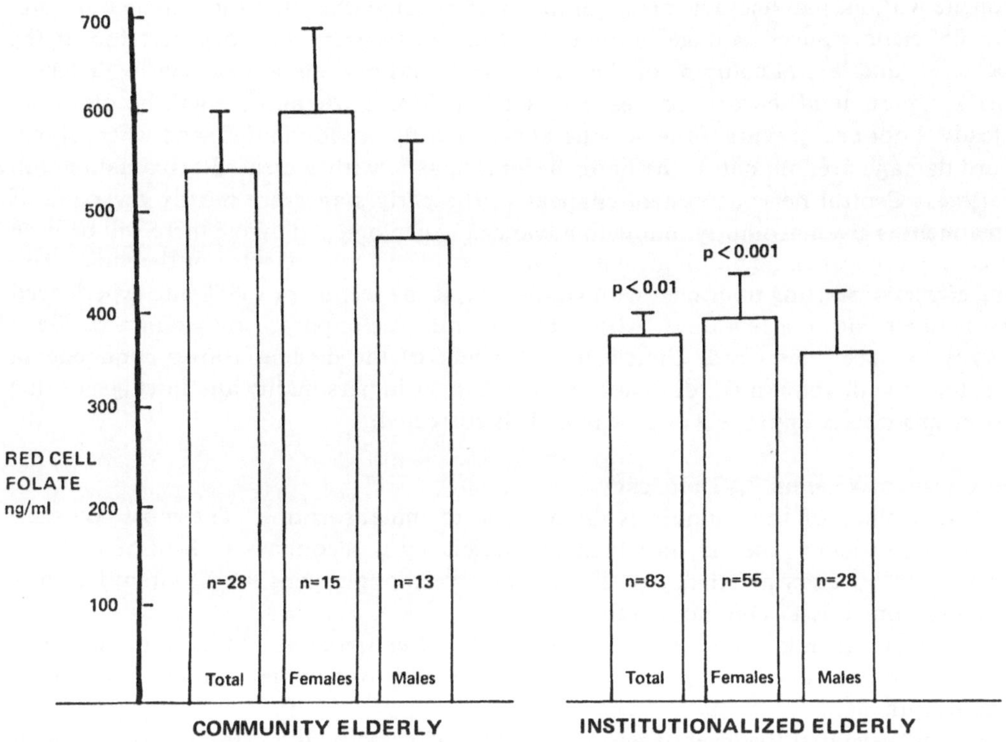

FIGURE 2. Mean (± SE) red cell folacin concentrations in community and institutionalized elderly (n = number).

A. Clinical Aspects of Vitamin B_{12} Deficiency in the Elderly

The clinical aspects of vitamin B_{12} deficiency include megaloblastic anemia,[96,97] subacute combined degeneration of the cord, peripheral neuropathy, and mental changes.[100,101] Mental changes may predominate in patients with a deficiency of this vitamin. Occurrence of mental disturbance with vitamin B_{12} deficiency is not believed by all workers. Buxton et al. suggest that vitamin B_{12} deficiency is unlikely to be a common cause of mental disturbance in the elderly.[99] A study of acute admissions to a geriatric ward reported a higher incidence of mental derangement in subjects with a low serum vitamin B_{12} level. A higher incidence of psychiatric symptoms has been found in patients with anemia due to B_{12} deficiency than in those patients whose anemia was due to other causes.[100]

The importance of recognizing early vitamin B_{12} deficiency has been emphasized.[99,100] Abramsky[100] believes that the neurological damage may become completely irreversible if treatment is delayed by failure in early diagnosis. Work by Strachan[78] has focused attention on the deficiency being presented as a nonspecific, often fluctuating mental change, occurring sometimes in the absence of anemia, or megaloblastic changes in the marrow.

Hematologists in acute general hospitals become aware of its frequency in elderly patients being treated for fractures in orthopedic services. Gastric mucosal atrophy in old age results in lowered secretion of intrinsic factor and lack of absorption of B_{12}. Immunological disturbances in old age include the development of parietal cell antibodies. Dietary deficiency of B_{12} is unlikely. Symptoms of anemia may be very slow to develop and some old patients with pernicious anemia may be still ambulant and uncomplaining with remarkably low levels of hemoglobin. Patients have a smooth tongue without papillae but rarely complain of or remember that the tongue was sore. B_{12} deficiency causes damage to the central nervous system with degeneration of the posterior and lateral columns of the spinal cord, and peripheral neuropathy. There is marked peripheral sensory loss, especially of position sense and the gait becomes unsteady. Foot drop occurs if the peripheral neuropathy predominated, and where spinal cord damage predominates, the limbs become spastic with increased proximal tendon reflexes. Central nervous system changes in the early stages are mostly reversible if treatment is given promptly, but with advanced neurological damage there will be little recovery even when hematological normality has been achieved. One of the most striking effects of starting treatment with vitamin B_{12} is the feeling of well being experienced by patients within a few days. Although dementia accompanied by vitamin B_{12} deficiency has been described, clinical improvement of the dementia after commencing treatment with vitamin B_{12} does not occur. Changes in personality and intelligence due to anemia clearly improve when the anemia is corrected.

B. Causes of Vitamin B_{12} Deficiency

A deficiency of this vitamin is mainly due to malabsorption. The most common cause is pernicious anemia, in which the deficiency is secondary to lack of intrinsic factor.[104,105] Dietary deficiency is very rare except among vegans, as the vitamin occurs in most animal tissues but not plants.

Pernicious anemia is a disease of later life. MacLennan et al.[106] found an incidence of 2.5% in two surveys of people aged 65 years and over living at home in Glasgow and Kilsyth.

It is believed that the level of serum B_{12} is also affected by the tissue status of folacin and iron.[107] Therefore, vitamin B_{12} status cannot be accurately assessed by measurement of serum levels alone.

VII. THIAMIN

A. Clinical Manifestations of Thiamin Deficiency in the Elderly

Thiamin deficiency, classically described in populations eating polished rice, can be the result of any inadequate diet. In geriatric practice, the group particularly at risk are elderly alcoholics whose calorie intake is mainly in fluid form. Wet beriberi with edema, effusions, and cardiac enlargement is occasionally seen. The dry form of beriberi manifests itself as polyneuritis with degeneration of peripheral nerves and damage to the central nervous system, especially in the area of the third ventricle. Degeneration of peripheral nerves leads to a typical clinical picture of weakness and tenderness of leg muscles with absent tendon reflexes, foot drop, and gross ataxic gait. In an advanced state the patient is unable to walk or stand unaided. Damage to the central nervous system causes disturbance of intelligence, apathy, and a curious upset of the mechanism of eye movements with double vision, nystagmus, and drooping of the eyelids. This may be of sudden onset and usually responds rapidly with resolution of ocular symptoms, if thiamin is administered systemically. The condition is known as Wernicke's encephalopathy. More persisting effects of thiamin deficiency are seen with deterioration of intelligence and loss of memory, especially recent memory. Patients tend to cover up their loss of memory by inventing suitable answers to questions requiring memory. Confabulation may be convincing, the more so if previous intelligence and personality has been good. Memory loss in this condition tends to persist together with confabulation and the syndrome has been named Korsakoff's psychosis.

The clinical response to treatment of thiamin deficiency in old age is only moderate. Wernicke's encephalopathy usually recovers, but the peripheral neuropathy, with gait disturbance, needs prolonged rehabilitation procedures to reestablish mobility. Korsakoff psychosis tends to improve slowly, with improved orientation of the patient. Measures to prevent Korsakoff psychosis have been suggested — such as the addition of thiamin to beer which is cheap and undetectable and thiamin is not destroyed with storage in alcohol.[108]

B. Thiamin Deficiency in the Elderly

Specific deficiencies of thiamin are rare in developed countries, but the thiamin status of the elderly has been measured by some workers.[38,109-113] A study was done by Brin et al.[38] to evaluate the nutritional status on the basis of established criteria on a group of elderly subjects. According to ICNND[114] criteria, 18% of the whole group were thiamin deficient.

Vir and Love[109] concluded that though the dietary intake of thiamin was adequate, deficiency as assessed by erythrocyte enzyme function test existed at the cellular level. Dietary intake data does not satisfactorily assess thiamin nutriture.

C. Causes of Thiamin Deficiency

Thiamin deficiency normally occurs because of an inadequate intake of the vitamin, which is believed to be the reason why a thiamin deficiency occurs frequently with chronic alcoholism. Thiamin deficiency may also develop in disease states characterized by anorexia, vomiting, or diarrhea and in postoperative patients.[115]

VIII. NIACIN

Niacin deficiency, associated with protein malnutrition, causes the clinical syndrome of pellagra. The picture in old age of a patient who is demented, has a rough, dry over pigmented skin in the areas exposed to sunlight, and has fecal soiling with loose motions, could be suffering from nicotinic acid deficiency. The presence of a sore, swollen, purple tongue, with inflammatory changes in the skin at the angles of the mouth and rough sore lips would be additional evidence. The triad of dementia, diarrhea, and dermatitis should at least raise the possibility of nicotinic acid deficiency in the mind of the clinician, because treatment with nicotinaminide produces rapid improvement of skin and bowel symptoms.

A history of dietary inadequacy, poverty, and poor sanitary conditions may be blamed alone for the neglected state of some old people admitted to geriatric units as social emergencies and a therapeutic opportunity thereby missed.

IX. CONCLUSIONS

Nutrition of older people is increasingly of concern to professionals and the community because of the rapid increase in the number of older people in our population.

The older population are vulnerable to deficiency of ascorbic acid, Vitamin D, folacin, B_{12}, and to a lesser extent, thiamin. In the face of a decreased energy requirement with advancing years, the intake of all other nutrients remain the same. Whether it is advisable to advocate vitamin supplements to those at risk elderly groups is not clear. Recent studies have not confirmed the value of vitamin supplements in correcting some of the abnormal signs attributed to vitamin deficiency in the elderly.[47,116] Care must be taken when interpretating biochemical findings of low vitamin concentrations. Clinical manifestations of overt vitamin deficiency are rare, but subclinical vitamin deficiency does occur among our elderly population.

REFERENCES

1. **Exton-Smith, A. N.,** Malnutrition in the elderly, *Proc. R. Soc. Med.*, 70, 615, 1978.
2. **Exton-Smith, A. N.,** Nutrition in the elderly, in *Nutrition in the Clinical Management of Disease*, Dickerson, J. W. T. and Lee, H. A., Eds., Edward Arnold, London, 1978, chap. 4.
3. **Montgomery, R. D., Haeney, M. R., Ross, I. N., Sammons, H. G., Barford, A. V., Balakreshnan, S., Mayer, P. P., Culank, L. S., Field, J., and Gosling, P.,** The ageing gut: a study of intestinal absorption in relation to nutrition in the elderly, *Quart. J. Med. N. Ser.*, 48, 197, 1978.
4. **Balacki, J. A. and Dobbins, W. D.,** Maldigestion and malabsorption: making up for lost nutrients, *Geriatrics*, 29, 157, 1974.
5. **Preece, M. A., Tomlinson, S., Ribot, K. A., Pietrek, J., Korn, H. T., Davies, D. M., Ford, J. A., Dunnigan, M. G., and O'Riordan, J. H.,** Studies on vitamin D deficiency in man, *Quart. J. Med.*, 44, 575, 1975.
6. **Chalmers, J., Conacher, W. D. H., Gardner, D. L., and Scott, P. R.,** Osteomalacia a common disease in elderly women, *J. Bone Jt. Surg.*, 498, 403, 1967.
7. **Hodkinson, H. M.,** Sunlight, vitamin D and osteomalacia in the elderly, *Age and Ageing*, 2, 129, 1973.
8. **Anderson, I., Campbell, A. E. R., Dunn, A., and Runcan, J. B. M.,** Osteomalacia in elderly women, *Scott. Med. J.*, 11, 429, 1966.
9. **Nordin, C.,** Osteoporosis and osteomalacia, *Med. Aust.*, 1, 597, 1979.
10. **Conacher, W. D.,** Metabolic bone disease in the elderly, *Practitioner*, 210, 315, 1973.
11. **Faccini, J. M., Exton-Smith, A. N., and Boyde, A.,** Disorders of bone and fracture of the femoral neck, *Lancet*, 1, 1089, 1976.
12. **Brown, I. R. F., Bakowska, A., and Millard, P. H.,** Vitamin D status of patients with femoral neck fractures, *Age and Ageing*, 5, 127, 1976.
13. **Smith, R. W., Rizek, J., Frame, B., and Mansour, J.,** Determinants of serum antirachitic activity, *Am. J. Clin. Nutr.*, 14, 98, 1964.
14. **Stamp, T. C. B. and Round, J. M.,** Seasonal changes in human plasma levels of 25-hydroxyvitamin D, *Nature (London)*, 247, 563, 1974.
15. **Lamb, G. A., Mawer, E. B., and Stanbury, S. W.,** The apparent vitamin D resistance of chronic renal failure: a study of the physcology of vitamin D in man, *Am. J. Med.*, 50, 421, 1971.
16. **Vir, S. Love, A. H. G.,** Vitamin D status of elderly at home and institutionalized in hospital, *Int. J. Vitam. Res.*, 48, 123, 1978.
17. **McLaughlin, M., Fairney, A., Lester, E., Raggatt, P. R., Brown, D. J., and Wills, M. R.,** Seasonal variation in serum 25-hydroxycholecalciferol in healthy people, *Lancet*, 1, 536, 1974.
18. **Corless, D., Beer, M., Boucher, B. J., Gupta, S. P., and Cohen, R. D.,** Vitamin D status in long stay geriatric patients, *Lancet*, 1, 1404, 1975.
19. **Pettifor, J. M., Ross, F. P., and Solomon, L.,** Seasonal variation in serum 25-OH cholecalciferol concentrations in elderly South African patients with fractures of femoral neck, *Br. Med. J.*, 1, 826, 1978.
20. **Davie, M., Lawson, D. E. M., and Jung, R. T.,** Low plasma-25-hydrocyvitamin D without osteomalacia, *Lancet*, 1, 820, 1978.
21. **Sauberlich, H. E., Dowdy, R. P., and Skala, J. H.,** in *Laboratory Tests for the Assessment of Nutritional Status*, CRC Press, Boca Raton, Fla., 1974, 80.
22. **Anwar, M.,** Nutritional hypovitamindsis-D and the genesis of osteomalacia in the elderly, *J. Am. Geriatr. Soc.*, 26, 309, 1978.
23. Department of Health and Social Security, A Nutrition Survey of the Elderly, Report on Health and Social Subjects, No. 3, Her Majesty's Stationery Office, London, 1972.
24. **Exton-Smith, A. N., Stanton, B. R., and Windsor, A. C. M.,** *Nutrition of Housebound Old People*, King Edward's Hosp. Fund for London, 1972.
25. **Evans, E. and Stock, A.,** Dietary intakes of geriatric patients in hospital, *Nutr. Metab.*, 13, 21, 1971.
26. **Fraser, D. R. and Kodicek, E.,** Unique biosynthesis by kidney of a biologically active vitamin D metabolite, *Nature (London)*, 228, 764, 1970.
27. **Lund, B., Hjorth, L., Kjaer, I., Reimann, I., Friss, T., Anderson, R. B., and Sprenson, D. H.,** Treatment of osteoporosis of ageing with 1α-hydroxycholecalciferol, *Lancet*, 11,1168, 1975.
28. **Mitra, M. L.,** Vitamin C deficiency in the elderly and its manifestations, *Am. Geriatr. Soc. J.*, 18, 67, 1970.
29. **Scobie, B. A.,** Scurvy in the adult, *N.Z. Med. J.*, 70, 398, 1969.
30. **Denson, K. W. and Bowers, E. F.,** The determination of ascorbic acid in white blood cells, *Clin. Sci.*, 21, 157, 1961.

31. **Andrews, J. and Brook, M.,** Leucocyte-vitamin-C content and clinical signs in the elderly, *Lancet,* 1, 1350, 1966.
32. **Bowers, E. F. and Kubik, M. M.,** Vitamin C levels in old people and the response to ascorbic acid to the juice of the acerola, *Br. J. Clin. Pract.,* 19, 141, 1965.
33. **Andrews, J., Letcher, M., and Brook, M.,** Vitamin C supplementation in the elderly, *Br. Med. J.,* 2, 416, 1969.
34. **Attwood, E. C., Robey, E. D., Ross, J., Bradley, F., and Kramer, J.,** Determination of platelet and leucocyte vitamin C and the levels found in normal subjects, *Clin. Chim. Acta,* 54, 95, 1974.
35. **Morgan, A. G., Kelleher, J., Walker, B. E., Losowsky, M. S., Droller, H., and Middleton, R. S. W.,** A nutritional survey in the elderly blood and urine vitamin levels, *Int. J. Vitam. Nutr. Res.,* 45, 448, 1975.
36. **Loh, H. S. and Wilson, C. W. M.,** The relationship between leucocyte ascorbic acid and haemolobin levels at different ages, *Int. J. Vitam. Nutr. Res.,* 41, 259, 1971.
37. **Milne, J. S., Lonergan, M. E., Williamson, J., Moore, F. M. L., McMaster, R., and Percey, N.,** Leucocyte ascorbic acid levels and vitamin C intake in older people, *Br. Med. J.,* 4, 383, 1971.
38. **Brin, M., Dibble, M., Peel, A., McMullen, E., Borquin, A., and Chen, N.,et al,** Some preliminary findings on the nutritional status of the aged in Onondaga County, New York, *Am. J. Clin. Nutr.,* 17, 240, 1965.
39. **Flint, D. M., Wahlqvist, M. L., Prinsley, D. M., Parish, A. E., Fazio, V., Peters, K., and Richards, B.,** *Food and Nutrition Notes Review,* 36, 173, 1979.
40. **Burn, M. C., Elwood, P. C., Hurley, R. J., and Hughes, R. E.,** Plasma and leukocyte ascorbic acid levels in the elderly, *Am. J. Clin. Nutr.,* 27, 144, 1974.
41. **Woodhill, J. M.,** Australian dietary surveys with special reference to vitamins, *Int. J. Vitam. Res.,* 40, 520, 1970.
42. **Dodds, M. L.,** Sex as a factor in blood levels of ascorbic acid, *J. Am. Diet. Assoc.,* 54, 32, 1969.
43. **Andrews, J.,** Vitamin C status of elderly long-stay hospital patients, *Gerontol. Clin.,* 15, 221, 1973.
44. **Vir, S. C. and Love, A. H. G.,** Vitamin C status of institutionalized and non-institutionalized aged, *Int. J. Vitam. Nutr. Res.,* 48, 274, 1978.
45. **McClean, H., Weston, R., Beauen, D. W., and Riley, C. G.,** Nutrition of elderly men living alone. I. Intakes of energy and nutrients, *N. Z. Med. J.,* 84, 305, 1976.
46. **Exton-Smith, A. N.,** Nutrition surveys and the problems of detection of malnutrition in the elderly, *Nutrition,* 24, 218, 1970.
47. **Dymock, S. M. and Brocklehurst, J. C.,** Clinical effects of water soluble vitamin supplementation in geriatric patients, *Age and Ageing,* 2, 172, 1972.
48. **Griffiths, L. L.,** Biochemical findings of the farnborough survey, in *Vitamins in the Elderly,* Exton-Smith, A. N. and Scott, D. L., Eds., John Wright & Sons, Bristol, 1968, 34.
49. **Harrill, I. and Cerudiye, N.,** Vitamin status of older women, *Am. J. Clin. Nutr.,* 30, 431, 1977.
50. **Justice, C., Howe, J., and Clark, H.,** Dietary intakes and nutritional status of elderly patients, *J. Am. Diet. Assoc.,* 65, 639, 1974.
51. Ten State Nutrition Survey, 1968-1970. IV. Biochemical. Department of Health, Education and Welfare Publication No. (H.S.M.) 72-8132, U.S. Department of Health, Education and Welfare, Center for Disease Control, Atlanta, 1972.
52. **Brown, P., Bergan, J., Parsons, E., and Kool, I.,** Dietary status of elderly people, *J. Am. Diet. Assoc.,* 77, 41, 1977.
53. **Stiedemann, M., Jansen, C., and Harrill, J.,** Nutritional status of elderly men and women, *J. Am. Diet. Assoc.,* 73, 132, 1978.
54. **Morgan, A. F., Gilliem, H. L., and Williams, R. I.,** Nutritional status of the aging. III. Serum ascorbic acid and intake, *J. Nutr.,* 55, 431, 1955.
55. Food and Nutrition Board, *Recommended Dietary Allowances,* 8th ed., National Academy of Science, Washington, D.C., 1974.
56. **Wilson, T. S., Datta, S. B., Murrell, J. S., and Andrews, C. T.,** Relation of vitamin C levels to mortality in a geriatric hospital: a study of the effect of vitamin C administration, *Age and Ageing,* 2, 163.
57. **Windsor, A. C. M., Hobbs, C. B., Treby, D. A., and Cowper, R. A.,** Effects of tetracyclcise on leucocyte ascorbic acid levels, *Br. Med. J.,* 1, 732, 1972.
58. **Forshaw, J., Moorhouse, E. H., and Harwood, L.,** Megaloblastic anaemia due to dietary deficiency, *Lancet,* 1, 1004, 1964.
59. **Raper, C. G. L. and Choudhury, M.,** Early detection of folic acid deficiency in elderly patients, *J. Clin. Pathol.,* 31, 44, 1978.
60. **Read, A. E., Gough, K. R., Pardoe, J. L., and Nicholas, A.,** Nutritional studies on entrants to an old people's home, with particular reference to folic acid deficiency, *Br. Med. J.,* 2, 843, 1965.

61. **Jagerstad, M.**, et al., Folate intake and blood folate in elderly subjects, a study using the double sampling portion technique, *Nutr. Metab.*, 21(Suppl. 1), 29, 1976.
62. **Hurdle, A. D. F. and Picton Williams, T. C.**, Folic acid deficiency in elderly patients admitted to hospital, *Br. Med. J.*, 2, 208, 1966.
63. **Harant, Z. and Goldberger, J. V.**, Treatment of anaemia in the aged: a common problem and challenge, *J. Am. Geriatr. Soc.*, 23, 127, 1975.
64. **Batata, M., Spray, G. H., Bolton, F. G., Higgins, G., and Wollner, L.**, Blood and bone marrow changes in elderly patients with special reference to folic and vitamin B_{12}, iron and ascorbic acid, *Br. Med. J.*, 1, 667, 1964.
65. **Elwood, P. C., Shinton, N. K., Wilson, C. I. D., Sweetman, P., and Frazer, A. C.**, Haemoglobin, vitamin B_{12} and folate levels in the elderly, *Br. J. Haematol.*, 21, 557, 1971.
66. **Bose, S. K., Andrews, J., and Roberts, P. D.**, Haematological problems in a geriatric unit with special reference to anaemia, *Gerontol. Clin.*, 12, 339, 1970.
67. **Morgan, A. G., Kelleher, J., Walker, B. E., Losowsky, M. S., Droller, H., and Middleton, R. S. W.**, A nutritional survey in the elderly: haematological aspects, *Int. J. Vitam. Nutr. Res.*, 43, 462, 1973.
68. **Chanarin, J.**, Dietary deficiency of vitamin B_{12} and folic acid, in *Nutritional Deficiencies in Modern Society*, Howard, A. N. and Baird, I., Eds., Newman, London, 1973.
69. **Meindok, H. and Dvorsky, R.**, Serum folate and vitamin B_{12} levels in the elderly, *J. Am. Geriatr. Soc.*, 18, 317, 1970.
70. **Flint, D. M., Wahlqvist, M. L., Prinsley, D. M., Parish, A.E., and Peters, K.**, unpublished data, 1979.
71. **Girdwood, R. H., Thomson, A. D., and Williamson, J.**, Folate study in the elderly, *Br. Med. J.*, 2, 670, 1967.
72. **Reynolds, E. H.**, Folate responsive neuropathy, *Br. Med. J.*, 11, 42, 1976.
73. **Mitra, M. L.**, Confusional states in relation to vitamin deficiencies in the elderly, *J. Am. Geriatr. Soc.*, 19(b), 536, 1971.
74. **Botez, M. J., Cadotte, M., Beaulieu, R., Pichette, L. P., and Pison, C.**, Neurologic disorders responsive to folic acid therapy, *C.M.A. J.*, 115, 217, 1976.
75. **Botez, M. I., Peyronnard, J., Bachevalier, J., and Charron, L.**, Polyneuropathy and folate deficiency, *Arch. Neurol.*, 35, 581, 1978.
76. **Thornton, W. E. and Thornton, B. P.**, Geratric mental function and serum folate: a review and survey, *South. Med. J.*, 70, 919, 1977.
77. **Herbert, V.**, Biochemical and haematological lesions in folic acid deficiency, *Am. J. Clin. Nutr.*, 20, 562, 1967.
78. **Strachan, R. W. and Henderson, J. G.**, Psychiatric syndromes due to avitaminosis B_{12} with normal blood and marrow, *Quart. J. Med.*, 34, 303, 1967.
79. **Sneath, P., Chanarin, I., Hodkinson, H. M., McPherson, C. K., and Reynolds, E. H.**, Folate status in a geriatric population and its relationship to dementia, *Age and Ageing*, 2, 177, 1973.
80. **Denham, M. J. and Jeffreys, P. M.**, Routine mental testing in the elderly, *Mod. Geriatr.*, 2, 275, 1972.
81. **Baker, H., Jaslow, S., and Frank, D.**, Severe impairment of dietary folate utilization in the elderly, *Am. Geriatr. Soc. J.*, 26, 218, 1978.
82. **Rosenberg, I. H.**, Absorption and malabsorption of folates, *Clin. Haematol.*, 5, 589, 1976.
83. **Morgan, A. G., Kelleher, J., Walker, B. E., Losowsky, M. S., Droller, H., and Middleton, R. S. W.**, A Nutritional Survey in the Elderly Blood and Urine Vitamin Levels.
84. **Hessov, I. and Elsborg, L.**, Nutritional studies on long-term surgical patients with special reference to intake of vitamin B_{12} and folic acid, *Int. J. Vitam. Nutr. Res.*, 46, 1976.
85. **Hurdle, A. D. F.**, An assessment of the folate intake of elderly patients in hospital, *Med. J. Aust.*, 2, 101, 1968.
86. **Hoppner, K., Lampi, B., and Perrin, D. E.**, The free and total folate activity in foods available on the Canadian market, *J. Inst. Can. Sci. Technol. Altment*, 5, 60, 1972.
87. **Paul, A. and Southgate, D. A. T.**, in *McCance and Widdowson's, The Composition of Foods*, Her Majesty's Stationery Office, London, 1978.
88. **Vilter, R. W., Will, J. J., Wright, T., and Rullman, D.**, Inter-relationships of vitamin B_{12}, folic acid and ascorbic acid in the megaloblastic anaemias, *Am. J. Clin. Nutr.*, 12, 130, 1963.
89. **Herbert, V. and Zalusky, R.**, Inter-relations of vitamin B_{12} in folic acid metabolism: folic acid tolerance studies, *J. Clin. Invest.*, 41, 1263, 1962.
90. **Nixon, P. F. and Bertino, J. R.**, Inter-relationships of vitamin B_{12} and folate in man, *Am. J. Med.*, 48, 555, 1970.
91. **Luhby, A. L. and Cooperman, J. M.**, *Folic Acid Deficiency in Man and Its Inter-Relationship with Vitamin B_{12} Metabolism in Advances in Metabolic Disorders*, Vol. 1, Academic Press, New York, 1964, 263.

92. **Elwood, P. C., Burr, M. L., Hole, D., Harrison, A., Morris, T. K., Wilson, C. I. D., Richardson, R. W., and Shinton, N. K.**, Nutitional state of elderly Asian and English subjects in coventry, *Lancet*, 3, 1224, 1972.
93. **Badenoch, J.**, Stealorrhoea in the adult, *Br. Med. J.*, 2, 879, 1960.
94. **Droller, H. and Dossett, J. A.**, Vitamin B_{12} levels in senile dementia and confusional states, *Gerontol Clin. (Basel)*, 1, 96, 1959.
95. **Schulman, R.**, A survey of vitamin B_{12} deficiency in an elderly psychiatric population, *Br. J. Psychiatry*, 113, 241, 1967.
96. **Baker, S. J.**, Human vitamin B_{12} deficiency, *World Rev. Nutr. Diet*, 8, 62, 1967.
97. **Sullivan, L. W.**, Vitamin B_{12} metabolism and megaloblastic anaemia, *Semin. Haematol.*, 7, 6, 1970.
98. **Kahn, S. B.**, Recent advances in the nutritional anaemias, *Med. Clin. N. Am.*, 54, 641, 1970.
99. **Buxton, P. K., Davison, W., Hyams, D. E., and Irvine, W. J.**, Vitamin B_{12} status in mentally disturbed elderly patients, *Gerontol. Clin.*, 11, 22, 1969.
100. **Abramsky, A.**, Common and uncommon neurological manifestations as presenting symptoms of vitamin B_{12} deficiency, *J. Am. Geriatr. Soc.*, 20, 93, 1972.
101. **Schulman, P.**, Present status of vitamin B_{12} and folic acid deficiency in psychiatric illness, *Can. Psych. Assoc. J.*, 17, 205, 1972.
102. **Castle, W. B.**, Current concepts of pernicious anaemia, *Am. J. Med.*, 48, 541, 1970.
103. **Herbert, V.**, The five possible causes of the nutrient deficiency: illustrated by deficiencies of vitamin B_{12} and folic acid, *Am. J. Clin. Nutr.*, 26, 77, 1973.
104. **Herbert, V.**, Megaloblastic anemia as a problem in world health, *Am. J. Clin. Nutr.*, 21, 1115, 1968.
105. **Lind, D. E.**, Anaemia in the elderly, *J. Geriatr.*, 4, 19, 1973.
106. **MacLennan, W. J., Andrews, G. R., MacLeod, C., and Caird, F. I.**, Anaemia in the elderly, *Age and Ageing*, 1, 99, 1973.
107. **Mollin, D. L., Waters, A. H., and Harriss, E.**, Clinical aspects of the metabolic inter-relationships between folic acid and vitamin B_{12}, in *Vitamin B_{12} and Instrinsic Factor*, Heinrich, H. C., Ed., Erike, Stuggart, 1963.
108. **Price, J. and Theodoros, M. T.**, Prevention of Korsokoffs Psychoses, *Med. J. Aust.*, 1, 285.
109. **Vir, S. and Love, A. H. G.**, Nutritional evaluation of B group of vitamins in the aged, *Int. J. Vitam. Nutr. Res.*, 4, 211, 1977.
110. **Harrill, J. and Ervone, N.**, Vitamin status of older women, *Am. J. Clin. Nutr.*, 30, 431, 1977.
111. **Fisher, S., Henricks, D. G., and Mahoney, A. W.**, Nutritional assessment of senior rural Utahns by biochemical and physical measurements, *Am. J. Clin. Nutr.*, 31, 667, 1978.
112. **O'Hanlon, P. and Kohrs, M.**, Dietary studies of older Americans, *Am. J. Clin. Nutr.*, 31, 1257, 1978.
113. **McClean, H., Dodds, P., Stewart, A. W., Beaven, D., and Riley, C. G.**, Nutrition of elderly men living alone. II. Vitamin C and thiamine status, *N. Z. Med. J.*, 84, 345, 1976.
114. **I.C.N.N.C.**, Suggested guide for interpreting dietary findings in nutrition surveys, *Public Health Rep.*, 75, 699, 1960.
115. **Goldsmith, G. A.**, The B vitamins: thiamine, riboflavin, niacin, in *Nutrition: A Comprehensive Treatise*, Vol. 11, Beaton, G. H. and McHenry, E. W., Eds., Academic Press, New York, 1972.
116. **MacLeod, R. D. M.**, Abnormal tongue appearances and vitamin status of the elderly — a double blind trial, *Age and Ageing*, 1, 99, 1972.

Chapter 4

EFFECTS ON PLASMA CHOLESTEROL OF NICOTINIC ACID AND ITS ANALOGUES

Mark L. Wahlqvist

TABLE OF CONTENTS

I. Introduction 82

II. Rationale for Lipid Lowering 82

III. Mechanism of Action of Nicotinic Acid 82

IV. Effectiveness of Nicotinic Acid 83
- A. Plasma Lipids 83
- B. Xanthomata 83
- C. Atherosclerotic Vascular Disease 83
- D. Angina Pectoris 83
- E. Myocardial Infarction 86

V. Use in Special Situations 86
- A. Children 86
- B. Renal Disease 86
- C. Alcohol Abuse 86

VI. Side Effects 86
- A. Flushing 86
- B. Gastrointestinal 87
- C. Cutaneous 87
- D. Hyperuricemia 87
- E. Glucose Tolerance 87
- F. Hepatic Dysfunction 87
- G. Arrhythmia 88
- H. White Cell Count 88
- I. Maculopaphy 88
- J. Other 88

VII. Derivatives and Analogues 88

VIII. Combined Drug Therapy 89

IX. Conclusions 89

References 89

I. INTRODUCTION

The plasma cholesterol lowering properties of nicotinic acid or niacin were first recognized by Altschul and colleagues in 1955.[1,2] The pharmacological dosages of 3 to 9 g/day required are well in excess of the daily allowances of about 20 mg/day for niacin as a vitamin. Although low plasma and arterial wall levels of pyridine compounds have been observed in hypercholesterolemic animals,[3] overt niacin deficiency has not been recognized nor advanced as a basis for nicotinic acid therapy. The cholesterol lowering properties of nicotinic acid at megadosage appear distinct properties. They are not evident with nicotinamide[4,5] and do not appear dependent on nicotinic acid metabolites.[6-9]

II. RATIONALE FOR LIPID LOWERING

The ultimate rationale for the pharmacological use of nicotinic acid must be a reduction in total mortality through a decrease in cardiovascular events attributable to atherosclerotic vascular disease.[10] There is ample experimental and epidemiological evidence that the severity of atherosclerotic vascular disease is dependent, in part, on plasma cholesterol concentration,[11,12] plasma triglyceride concentration,[11,13] and inversely, on high density lipoprotein concentration.[14-17] There is increasing evidence that reduction of plasma cholesterol by diet[18-20] or drugs[21-23] will reduce coronary events, although not necessarily fatal events or total mortality.[21-23]

III. MECHANISM OF ACTION OF NICOTINIC ACID

An agent which reduces plasma cholesterol or triglycerides will do so through decreased synthesis and/or increased removal. It is likely, too, that the nature of the underlying defect in lipid metabolism will influence the effectiveness of the agent.

An impressive metabolic action of nicotinic acid in megadosage is antilipolysis.[24-26] The reduced flux of free fatty acids (FFA) to the liver ought to decrease very low density lipoprotein triglyceride (VLDL TG) production, and with it, cholesterol production.[27-29] In any case, VLDL is a precursor of low density lipoprotein (LDL), the major cholesterol-bearing lipoprotein.[30] Nicotinic acid does decrease the rate of synthesis of LDL.[31] In addition, there is evidence that nicotinic acid directly inhibits cholesterogenesis. Not all studies, however, support the view that cholesterogenesis is inhibited and some actually suggest it is enhanced,[32-40] but where assessment was 14 hr after the last dose of nicotinic acid, rebound effects might have been operative.[40]

Enhanced cholesterol oxidation has been reported.[41] Negative neutral steroid balance has been observed with nicotinic acid.[42,43] Although bile acid excretion does not appear enhanced,[42] the biliary cholic acid to chenodeoxycholic acid ratio is increased by nicotinic acid[44] which may suggest a differential effect of nicotinic acid on the alternative pathways for bile acid formation, 7 α-hydroxylation and 26-hydroxy cholesterol generation.[45] The interpretation of sterol balance studies in this context is complicated by, potentially, the mobilization of cholesterol from tissue deposits, and, therefore, nonsteady-state conditions. Triglyceride removal may also be enhanced[46] as lipoprotein lipase activity is stimulated by nicotinic acid.[47,48] The mechanism for antilipolysis with nicotinic acid remains incompletely understood. There does appear to be a combined effect on re-esterification and lipolysis.[49] Nicotinic acid does inhibit adenyl cyclase activity and cyclic AMP formation in adipose tissue, and this may relate to its antilipolytic effect.[51] Little or no effect on cyclic AMP is likely through phosphodiesterase inhibition.[50] Prostaglandin synthesis might be involved.[52-54]

The determination of nicotinic acid in blood[55-58] allows an examination of the relationship between pharmacokinetics[6,7,9] and action.[29,50] Plasma concentrations of nicotinic acid of greater than 2 μg/mℓ lower free fatty acids.[29] Often, plasma free fatty acids are used as an index of the duration of action of a nicotinic acid preparation.

When lipoprotein lipase activity is increased and triglyceride clearance enhanced, as it appears to be with nicotinic acid, HDL cholesterol concentration increases. On this basis, HDL elevation with nicotinic acid might be a secondary phenomenon. However, nicotinic acid does lead to an enrichment of apoproteins $apoA_I$ in the HDL_2 subfraction and $apoA_{II}$ in the HDL_3 subfraction.[60] The plasma $apoA_I$ to $apoA_{II}$ ratio is increased[61,62] as also is the HDL_2 to HDL_3 ratio.[60,63] These features appear to result from reciprocal changes in the synthetic rates of $apoA_I$ which is increased, and of $apoA_{II}$, which is decreased, as well as a redistribution of $apoA_I$ and $apoA_{II}$ between HDL_2 and HDL_3.[62] The HDL subfraction changes with nicotinic acid will undoubtedly assume more importance as their relationship to atherosclerotic vascular disease is defined.

IV. EFFECTIVENESS OF NICOTINIC ACID

A. Plasma Lipids

Modification of dietary fat intake in its own right will reduce plasma lipid concentrations[18,19,64,65] and, generally, this is done prior to the introduction of lipid-lowering drugs. Several clinical trials attest to the plasma cholesterol and triglyceride lowering properties of nicotinic acid.[1,2,21,27-29,66-72] Reductions of cholesterol concentration by 10 to 25% and triglyceride concentration by 23 to 46% have been observed and these effects are at least comparable to those of the other principal lipid-lowering drugs, clofibrate (Atromid-S), and the resin cholestyramine (Questran) (Table 1). With resins, only cholesterol is lowered while triglyceride is often elevated.[68,73] The only hypertriglyceridemia in which nicotinic acid appears not to be effective is the hyperchylomicronemia seen in Type 1 hyperlipoproteinemia.[74,75]

Of considerable potential importance is the consistent increase in high density lipoprotein (HDL) cholesterol seen with nicotinic acid therapy.[27] HDL has the capacity to decrease arterial cholesterol deposition by interference with LDL cholesterol uptake[76] and through an increase in the removal of free cholesterol.[77]

B. Xanthomata

There are reports of tuberous and tendinous Xanthomata softening, reducing in size, and even disappearing on nicotinic acid therapy.[27]

C. Atherosclerotic Vascular Disease

An early report of Öst and Stenson[78] of serial arteriography indicated that atherosclerosis might regress during nicotinic acid therapy. With computerized femoral angiography,[79] the extent of risk factor correction has been related to regression of preclinical atherosclerotic vascular disease. In rabbits[80,81] and mini-pigs[82] nicotinic acid or its derivatives, niceritrol and β-pyridyl carbinol, limit the development of atherosclerotic lesions. It has been suggested that nicotinic acid might directly and favorably affect arterial cholesterol deposition, independent of its effect on blood lipids.[82] This would be consistent with the recognized metabolic activity of atherosclerotic lesions.[83-85]

D. Angina Pectoris

Nicotinic acid markedly alters myocardial fuel supply in favor of carbohydrate and away from FFA.[86-90] Since glucose is the only fuel from which ATP can be obtained anaerobically and since the oxygen cost for ATP generation is less with glucose than

Table 1
DRUG THERAPY OF HYPERLIPOPROTEINEMIA

	Nicotinic acid and derivatives			Other agents			Combined therapy		
	Nicotinic acid	Aluminum nicotinate	Niceritrol	β Pyridyl carbinol	Clofibrate	Cholestyramine	Niceritrol & clofibrate	Nicotinic acid & cholestyramine	Ref.
Total cholesterol	−25%	−25%[a]							104
	−13%		−21%[a]						69
				−10%					135
				−18%[b]					136
	−10%				−7%				21
	−11%		−16%		−14%		−28%		66
			−12%		−17%	−23%	−23%		133
	−21%					−26%			68
	(1)					0%		−46%	43
	(2) −36%[d]							−28%	
Total triglyceride	−46%		−11%						69
				−15%					135
				−18%[b]					136
	−26%				−22%				21
			−26%		−21%		−38%		133
	−23%					−28%			68
High density lipoprotein cholesterol			+28%[d]						69
							0[c]		133

Note: Standard dosages were used alone or in combination unless otherwise indicated (nicotinic acid 3 g/day, niceritrol 3 g/day, β pyridyl carbinol 1.2 or 1.8 g/day, clofibrate 1.5 or 2 g/day, cholestyramine 16 g/day). Percentage changes from untreated values are shown.

[a] Dosage of nicotinic acid individualized rather than 3 g/day.
[b] In this study, comparison with placebo indicated that the fall in cholesterol was highly significant whereas that in triglyceride was not.
[c] Data for types IIa and IIb hyperlipoproteinemia combined.
[d] Two cases studied are shown separately.

for FFA, nicotinic acid could protect the ischemic myocardium. Angina pectoris occurs less often when sufferers have their hearts paced during an infusion of nicotinic acid.[91] In addition to its metabolic effects, nicotinic acid has a favorable effect on distribution of coronary blood flow during experimental coronary oclusion in dogs.[92]

E. Myocardial Infarction

Electrocardiographic evidence of myocardial infarction, by way of ST segment elevation, is less when a nicotinic acid analogue, 5-fluoronicotinate, is administered.[93] A 3-year prospective study of secondary prevention of myocardial infarction with a combination of nicotinic acid and clofibrate, produced a significant 50% reduction in nonfatal reinfarction.[94] Greater lipid lowering was seen in this study than in the Coronary Drug Project, also a secondary prevention study.[21,22] In the Coronary Drug Project a reduction in nonfatal coronary events with nicotinic acid was also found.

V. USE IN SPECIAL SITUATIONS

A. Children

Clofibrate is relatively ineffective in children with familial hypercholesterolemia, and cholestyramine is generally used.[95] There is probably a place for nicotinic acid in resistant cases. Effective lipid lowering in a child at risk from premature ischemic heart disease is likely to outweigh the disadvantage of prolonged therapy. It may also allow a reduction in resin therapy with its risk of interference with fat soluble vitamin availability.

B. Renal Disease

The management of hyperlipidemia in renal failure, renal transplant patients, and in the nephrotic syndrome is difficult. Dietary therapy alone is probably the least difficult management, but not always sufficiently effective.[96,97] In patients on chronic hemodialysis, nicotinic acid therapy achieves a 20% reduction in plasma cholesterol and a 35% reduction in plasma triglycerides.[97] A limiting factor to nicotinic acid use in renal disease is the presence of hyperuricemia which may be exacerbated.

C. Alcohol Abuse

Nicotinic acid is effective in alcohol-sensitive hyperlipidemia.[98] In rats it potentiates ethanol fatty liver.[99] Species differences may be important, however, since in the rat, plasma triglyceride, but not cholesterol, is lowered by nicotinic acid.

VI. SIDE EFFECTS

A. Flushing

Cutaneous vasodilation occurs within about 1 hr of ingestion of plain nicotinic acid during introduction of therapy. To minimize flushing, niacin is taken with meals and dosage is increased progressively. It is usually convenient to begin with 250 mg thrice daily with increments of 250 mg thrice daily every 1 to 3 days until a daily dose of 3 g is reached. Flushing only occurs while plasma nicotinic acid concentrations increase.[9,100] With constant i.v. infusion of nicotinic acid, the flush disappears when steady-state plasma concentrations have been achieved.[9] Cutaneous and muscle blood flows increase and total peripheral resistance falls.[9]

When an aluminum nicotinate (Nicalex®) is used, the occurrence of the flush is much less predictable and it may occur hours away from the time of injection.[138] This presumably represents altered and variable absorption of nicotinic acid with this preparation. Pentaerythritoltetranicotinate (niceritrol or Pericyt®) produces less flushing,[101] probably because of prolonged action and more constant blood nicotinic acid concentrations[9]

There is evidence that, at least in part, the vasodilation is induced by prostaglandin and that it can be prevented by indomethacin.[102]

B. Gastrointestinal

Nausea, vomiting, abdominal pain, and diarrhea are occasionally reported during nicotinic acid therapy.[103] In the Coronary Drug Project, only abdominal pain occurrence was significantly increased over placebo.[21] Activation of peptic ulcer has also been reported,[104] but it would be of interest to reexamine this question now that endoscopic facilities are more acceptable and available. Whether or not nicotinic acid can be used in conjunction with cimetidine, now commonly used in the management of peptic ulcer, needs to be studied. If it could, this may allow treatment of a patient group otherwise denied this form of lipid lowering therapy.

C. Cutaneous

Cutaneous side effects include pruritus, dryness of the skin, pigmentation in flexural creases, and scars which may resemble acanthosis nigricans.[21,105] These effects are rarely a problem and are reversible on cessation of therapy.

D. Hyperuricemia

Serum uric acid increases significantly on treatment with megadosage nicotinic acid[21] and acute gouty arthritis can occur.[21] Nicotinic acid is probably antiuricosuric by a renal tubular mechanism.[106]

E. Glucose Tolerance

Impairment of glucose tolerance is found in a proportion of healthy and diabetic subjects given nicotinic acid.[21,106-109] In the Coronary Drug Project, fasting blood glucose did not change significantly over a period of 5 years, although the 1 hr blood glucose rose significantly from 168 to 186 mg/100 mℓ.

There are diabetics, however, whose carbohydrate status improves on nicotinic acid.[110,111] In vitro, nicotinic acid stimulates insulin release from isolated islets of mouse pancreas.[112] It has been suggested that in diabetics whose carbohydrate status improves on nicotinic acid, a decrease in FFA, according to Randle's glucose-fatty acid cycle hypothesis, is responsible.[111] As far as the human heart is concerned, at lower FFA concentrations more glucose, lactate, and pyruvate are extracted.[86-89] One of the potential problems when thrice daily plain nicotinic acid therapy is used that a rebound rise in plasma FFA occurs as the evening dose wears off.[46] However, most studies have been unable to relate whole body glucose handling, as assessed by a glucose tolerance test, with FFA concentrations during nicotinic acid therapy.[46,110] The observation that there is often a lag in the development of impaired carbohydrate tolerance[69,113] has suggested that the impairment might relate to hepatic dysfunction.[114]

In acute studies with nicotinic acid infusion, nicotinic acid has been shown to reduce hepatic ketone production in relation to a decrease in splanchnic FFA flux.[115] It might be expected that, conversely, during the rebound rise in FFA flux after withdrawal of nicotinic acid, ketone production would rise. An interesting aspect of nicotinic acid and glucose tolerance, is that as a complex with chromium, the glucose tolerance factor (GTF), it facilitates insulin action.[116,117]

F. Hepatic Dysfunction

Hepatic enzyme activities in serum are often raised with nicotinic acid therapy.[21,118,119] Jaundice has been seen.[8,9,120,121] In the Coronary Drug Project, however, less nicotinic acid treated individuals had serum bilirubin outside specified limits than did those on placebo.[21] The enzyme changes are reversible on withdrawal of therapy. Ultrastructural changes of mitochondria and endoplasmic reticulum are seen in liver biopsies from persons treated with nicotinic acid.[122] A case of hepatic fibrosis has been reported.[123] Where liver disease is present, nicotinic acid therapy should be avoided.

G. Arrhythmia

There was an excess of atrial fibrillation and other arrythmias in the nicotinic acid treated group in the Coronary Drug Project.[21] It is possible this could relate to the rebound rise in FFA and be a case for a longer acting nicotinic acid derivative or analogue.[93,124] Another possibility is that lower levels of serum potassium, seen with nicotinic acid, were responsible.[21]

H. White Cell Count

Nicotinic acid lowered total white cell count (WCC) and absolute neutrophil count (NC) from means of 7470 to 6610/cmm, and 4580 to 3970/cmm over 5 years, respectively, in the Coronary Drug Project.[21] There was an excess over placebo, for total WCC, of 7% below 3500/cmm and for NC, of 2% below 1800/cmm.

I. Maculopaphy

An atypical form of cystoid macular edema and loss of central vision has been reported with high dose nicotinic acid.[125] There is no capillary leakage evident on fluorescein angiography. It is reversible on cessation of nicotinic acid therapy.

J. Other

Whereas with clofibrate in the Coronary Drug Project, cholelithiasis, including cholecystectomy, was significantly increased by comparison with placebo. This was not the case for nicotinic acid.[21] Also in the Coronary Drug Project, unexpected loss of appetite, loss of weight, and excessive sweating were seen more commonly with nicotinic acid than with placebo or clofibrate.[21] Serum CPK was significantly increased with nicotinic acid in the Coronary Drug Project.[21]

VII. DERIVATIVES AND ANALOGUES

Derivatives and analogues have been sought with two objects in mind:

1. Reduction in side effects, particularly cutaneous flushing and gastrointestinal symptoms
2. A more prolonged action so as to reduce tablet frequency and to overcome the rebound rise in plasma FFA

In general, it can be said that those pharmaceutical preparations which have been designed to release nicotinic acid slowly into the gut for more prolonged absorption and sustained plasma concentrations, have met with little success. This appears to be because nicotinic acid is a weak acid, and is, therefore, poorly absorbed from the more distal gastrointestinal tract.[59] Aluminum nicotinate (Nicalex®) was developed to reduce gastrointestinal side effects.[126,127]

Various esters have been prepared,[128,129] but of them, niceritrol (Pericyt®) or pentaerythritol tetranicotinate has been used longest and most extensively.[59,66,81,82,101,130-134] It does not lead to hyperuricemia. Nicotinyl alcohol (Ronicol®)

or β-pyridyl carbinol has also been used to lower plasma cholesterol and appears about three or fourfold more potent than nicotinic acid on a weight for weight basis.[135,136] It is not significantly effective in hypertriglyceridemia.[36] The relative effectiveness of these derivatives and analogues is shown in Table 1.

VIII. COMBINED DRUG THERAPY

As in antihypertensive and antitumor therapy, it seems rational to combine antihyperlipidemic agents.[137] In this way, additive effects or possibly synergism can be sought. For instance, a drug which reduces cholesterol synthesis could be combined with a drug which increases its removal. Smaller dosages of each drug should be possible with combined therapy and this would lessen side effects. In this regard, the combination of nicotinic acid and a resin such as cholestyramine, colestipol, or DEAE Sephadex (Secholex®) is one of the most attractive regimens. Available studies of combined therapy are shown in Table 1.

IX. CONCLUSIONS

Nicotinic acid is an agent which can contribute to the management of hypertriglyceridemia (except pure hyperchylomicronemia), hypercholesterolemia, and combinations of these. Although a number of side effects may be seen, they are usually reversible and not of a serious kind. There is experimental evidence that less atherosclerotic vascular disease may be seen with nicotinic acid therapy. Secondary prevention of ischemic heart disease with less nonfatal mycardial infarctions has occurred with its use. It has yet to be shown whether total mortality is favorably influenced. Data available should provide a stimulus for the development of derivatives and analogues and for an examination of niacin therapy in combination with other agents to improve efficacy. In this way, nicotinic acid might contribute to the management of hyperlipoproteinemia which is presently unsatisfactory.

REFERENCES

1. **Altschul, R., Hoffer, A., and Stephen, J. D.,** Influence of nicotinic acid on serum cholesterol in man, *Arch. Biochem.*, 54, 558, 1955.
2. **Altschul, R.,** Niacin (nicotinic acid) and serum cholesterol, *JAMA*, 166, 822, 1958.
3. **Harthon, L., Brattsand, R., and Lundholm, L.,** Influence of nicotinic acid derivatives on NAD levels of blood, liver, adipose tissue and aorta during hyperlipemic conditions, in *Metabolic Effects of Nicotinic Acid and its Derivatives*, Gey, K. F. and Carlson, L. A., Eds., Hans Huber, Bern, 1971, 115.
4. **Dalton, C., Van Trabert, T. C., and Dwyer, J. X.,** Hyperlipidemic effects of nicotinamide in the rat arising from its transformation to nicotinic acid, in *Metabolic Effects of Nicotinic Acid and its Derivatives*, Gey, K. F. and Carlson, L. A., Eds., Hans Huber, Bern, 1971, 65.
5. **Parsons, W. B., Jr. and Flinn, J. H.,** Reduction in elevated blood cholesterol levels by large doses of nicotinic acid. Preliminary report, *JAMA*, 164, 234, 1957.
6. **Fumagalli, R.,** Pharmacokinetics of nicotinic acid and some of its derivatives in *Metabolic Effects of Nicotinic Acid and its Derivatives*, Gey, K. F. and Carlson, L. A., Eds., Hans Huber, Bern, 1971, 33.
7. **Svedmyr, N., Harthon, L., and Lundholm, L.,** The relationship between plasma concentration of free nicotinic acid and some of its pharmacologic effects in man, *Clin. Pharmacol. Ther.*, 10, 559, 1969.
8. **Lee, K. W., Abelson, D. M., and Quon, Y. O.,** Nicotinic acid -6- ^{14}C metabolism in man, *Am. J. Clin. Nutr.*, 21, 223, 1968.

9. **Svedmyr, N., Harthon, L., and Lundholm, L.**, Dose-response relationship between concentration of free nicotinic acid in plasma and some metabolic and circulatory effects after administration of nicotinic acid and pentaaerythritol tetranicotinate, in *Metabolic Effects of Nicotinic Acid and its Derivatives*, Gey, K. F. and Carlson, L. A., Eds., Hans Huber, Bern, 1971, 1085.
10. **Fitzgerald, J. D.**, The evaluation of lipid lowering agents, in *Principles and Practice of Clinical Trials*, Harris, E. L. and Fitzgerald, J. D., Eds., E.S. Livingston, 1970, 165.
11. **Carlson, L. A. and Böttiger, L. E.**, Ischaemic heart disease in relation to fasting values of plasma triglycerides and cholesterol. Stockholm prospective study, *Lancet*, 1, 865, 1972.
12. **Kannel, W. B., Castelli, W. P., and Gordon, T.**, Serum cholesterol lipoproteins and risk of coronary heart disease. The Framingham study, *Ann. Int. Med.*, 74, 1, 1971.
13. **Pelkonen, R., Nikkila, E. A., Koskinen, S., Penttinen, K., and Sarna, S.**, Association of serum lipids and obesity with cardiovascular mortality, *Br. Med. J.*, 11, 1185, 1977.
14. **Kannel, W. B. and Dawber, T. R.**, High density lipoprotein as a protective factor against coronary heart disease. The Framingham study, *Am. J. Med.*, 62, 707, 1977.
15. **Miller, N. E., Forder, C. H., Thelle, D. S., and Mjos, O. D.**, The Tromso heart study: high density lipoprotein and coronary heart disease: a prospective control study, *Lancet*, 1, 965, 1977.
16. **Rhodes, G. G., Gulbrandsen, C. L., and Kagan, A.**, Serum lipoprotein and coronary heart disease in a population study of Hawaii-Japanese men, *N. Engl. J. Med.*, 204, 293, 1976.
17. **Castelli, W. P., Doyle, J. T., Gordon, T., Haymes, C. G., Jortland, M. C., Hulley, S. B., Kagan, A., and Zukel, W. J.**, HDL cholesterol and other lipids in coronary heart disease: the co-operative lipoprotein. Phenotyping study, *Circulation*, 55(5), 767, 1977.
18. **Dayton, S., Pearce, M. L., Goldman, H., Harnish, A., Plotkin, D., Schickman, N., Windfield, M., Zagar, A., and Dickson, W.**, Controlled trial of a diet high in unsaturated fat for prevention of atherosclerotic complications, *Lancet*, 2, 1060, 1968.
19. **Leren, P.**, The effect of plasma cholesterol lowering diet in male survivors of myocardial infarction, *Acta Med. Scand.*, Suppl. 466, 1966.
20. **Miettinen, N., Turpeinen, O., Karvonen, M. J., Elosuo, R., and Paavilainen, E.**, Effect cholesterol-lowering diet on mortality from cornonary heart disease and other causes. A twelve-year clinical trial in men and women, *Lancet*, 2, 835, 1972.
21. Coronary Drug Research Project, Clofibrate and niacin in coronary heart disease, *JAMA*, 231, 360, 1975.
22. **Stamler, J.**, The coronary drug project. Findings with regard to estrogen, dextrothyroxine, clofibrate and niacin, *Adv. Exp. Med. Biol.*, 82, 52, 1977.
23. **Oliver, M. F., Heady, J. Y., Morris, J. N., and Cooper, J.**, A co-operative trial in the primary prevention of ischaemic heart disease using clofibrate. Report from the committee of principal investigators, *Br. Heart J.*, 40, 1069, 1978.
24. **Carlson, N. A.**, Nicotinic acid: its metabolism and its effects on free fatty acids, in *Metabolic Effects of Nicotinic Acid and its Derivatives*, Gey, K. F. and Carlson, L. A., Eds., Hans Huber, Bern, 1971, 157.
25. **Carlson, L. A. and Orö, L.**, The effect of nicotinic acid on the plasma free fatty acids. Demonstration of a metabolic type of sympathicolysis, *Acta Med. Scand.*, 172, 641, 1962.
26. **Carlström, S. and Laurell, S.**, The effect of nicotinic acid on the diurnal variation of the free fatty acids of plasma, *Acta Med. Scand.*, 184, 121, 1968.
27. **Carlson, L. A.**, The effect of nicotinic acid treatment on the chemical composition of plasma lipoprotein classes in man, *Adv. Exp. Biol. Med.*, 4, 327, 1969.
28. **Carlson, L. A., Orö, L., and Östman, J.**, Effect of nicotinic acid on plasma lipids in patients with hyperlipoproteinaemia during the first week of treatment, *J. Atheroscler. Res.*, 8, 67, 1968.
29. **Carlson, L. A., Orö, L. A., and Östman, J.**, Effect of a single dose of nicotinic acid on lipids in patients with hyperlipoproteinaemia, *Acta Med. Scand.*, 183, 457, 1968.
30. **Jackson, R. L., Morrison, J. D., and Gotto, A. M., Jr.**, Lipoprotein structure and metabolism, *Physiol. Rev.*, 56, 260, 1976.
31. **Levy, R. I. and Langer, T.**, Hypolipidemic drugs and lipoprotein metabolism, *Adv. Exp. Biol. Med.*, 26, 155, 1972.
32. **Kritchevsky, D.**, Effect of nicotinic acid and its derivatives on cholesterol metabolism: a review, in *Metabolic Effects of Nicotinic Acid and its Derivatives*, Gey, K. F. and Carlson, L. A., Eds., Hans Huber, Bern, 1971, 541.
33. **Shade, H. and Saltman, P.**, Influence of nicotinic acid on hepatic cholesterol synthesis, *Proc. Soc. Exp. Biol. Med.*, 102, 265, 1959.
34. **Gamble, W. and Wright, L. D.**, Effect of nicotinic acid and related compounds on incorporation of mevalonic acid into cholesterol, *Proc. Soc. Exp. Biol. Med.*, 107, 160, 1961.
35. **Nunn, S. E., Touxe, W. N., and Jurgens, J. L.**, Effect of nicotinic acid on human cholesterol biosynthesis, *Circulation*, 24, 1099, 1961.

36. **Parsons, W. B., Jr.**, Reduction in hepatic synthesis of cholesterol from C^{14} — acetate in hypercholesterolemic patients by nicotinic acid, *Circulation*, 24, 1099, 1961.
37. **Mahl, N. and Lance, K.**, Long term study of the effect of nicotinic acid medication on hypercholesterolemia, *Am. J. Med. Sci.*, 246, 673, 1963.
38. **Miller, O. N. and Hamilton, J. C.**, *Nicotinic Acid and its Derivatives in Lipid Pharmacology*, Vol. 2, Paoletti, R., Ed., Academic Press, New York, 1964, 275.
39. **Channan, R. C., Matthews, L. B., and Braxuler, C.**, Nicotinic acid in the treatment of hypercholesterolemia, *Angiology*, 23, 29, 1972.
40. **Miettinen, T. A.**, Influence of nicotinic acid on cholesterol synthesis in man, in *Metabolic Effects of Nicotinic Acid and its Derivatives*, Gey, K. F. and Carlson, L. A., Eds., Hans Huber, Bern, 1971, 649.
41. **Kritchevsky, D. and Tepper, S. A.**, Influence of nicotinic acid homology on oxidation of cholesterol — 26 — C^{14} by rat liver mitochondria, *Arch. Int. Pharmacodyn.*, 138, 149, 1962.
42. **Meittinen, T. A.**, Effect of nicotinic acid on the fecal excretion of neutral sterols and bile acids, in *Metabolic Effects of Nicotinic Acid and its Derivatives*, Gey, K. F. and Carlson, L. A., Eds., Hans Huber, Bern, 1971, 677.
43. **Moutafis, C. D., Myant, N. B., Mancini, M., and Oriente, P.**, Cholestryramine and nicotinic acid in the treatment of familial hyperbetalipoproteinemia in the homozygous form, *Atherosclerosis*, 14, 247, 1971.
44. **Einarsson, K., Hellström, N. K., and Leijd, B.**, Bile acid kinetics and steroid balance during nicotinci acid therapy in patients with hyperlipoproteinemia types II & IV, *J. Lab. Clin. Med.*, 90, 613, 1977.
45. **Sabine, J.**, *Cholesterol*, Marcel Dekker, New York, 1977.
46. **Fröberg, S. O., Boberg, J., Carlson, L. A., and Eriksson, M.**, Effect of nicotinic acid on the diurnal variation of plasma levels of glucose, free fatty acids, triglycerides and cholesterol and of urinary excretion of catecholamines, in *Metabolic Effects of Nicotinic Acid and its Derivatives*, Gey, K. F. and Carlson, L. A., Eds., Hans Huber, Bern, 1971, 167.
47. **Boberg, J., Carlson, L. A., Fröberg, S. O., Olsson, A., Orö, L., and Rössner, S.**, Effects of chronic treatment with nicotinic acid on intravenous fat tolerance and post-herapin lipoprotein lipase activity in man, in *Metabolic Effects of Nicotinic Acid and its Derivatives*, Gey, K. F. and Carlson, L. A., Eds., Hans Huber, Bern, 1971, 465.
48. **Nikkilä, E. A. and Pykälistö, O.**, Induction of adipose tissue lipoprotein lipase by nicotinic acid, *Biochem. Biophys. Acta*, 152, 421, 1968.
49. **Vik-Mo, H. and Mjos, O. D.**, Mechanism for inhibition of free fatty acid mobilization by nicotinic acid and sodium salicyate in canine subcutaenous adipose tissue in situ, *Scand. J. Clin. Lab. Invest.*, 38, 209, 1978.
50. **Skidmore, I. F., Kritchevsky, D., and Schönhüfer, P.**, Influence of nicotinic acid and nicotinic acid homologs on lipolysis, phosphodiesterase activity, adenyl cyclase activity and cyclic AMP synthesis in fat cell, in *Metabolic Effects of Nicotinic Acid and its Derivatives*, Gey, K. F. and Carlson, L. A., Eds., Hans Huber, Bern, 1971, 37.
51. **Anderson, R., Harthon, L., Hedstrom, M., and Lundholm, L.**, Inhibition of cyclic AMP formation and lipolysis in rat adipose tissue by nicotinic acid, *Atherosclerosis*, 18, 399, 1973.
52. **Kaijser, L.**, Nicotinic acid stimulates prostaglandin synthesis in the rabbit heart without releasing noradrenaline, *Acta Physiol. Scand.*, 102, 246, 1978.
53. **Vincent, J. E. and Zijlstra, F. J.**, Nicotinic acid inhibits thromboxane synthesis in platelets, *Prostaglandins*, 15, 629, 1979.
54. **Sutherland, W. H. F., Larkin, T. W., and Nye, E. R.**, Modification of nicotinic acid and prostaglandin E_1 antilipolytic acition in vitro, *Atherosclerosis*, 25, 45, 1976.
55. **Carlson, L. A.**, Determination of free nicotinic acid in blood plasma, *Clin. Chim. Acta*, 13, 349, 1966.
56. **Diab, A. M.**, Spectrophotometric assay of nicotinic acid in blood. Assessment of its daily profile in humans, *Arzneim. Forsch. Res.*, 27(2), 2134, 1977.
57. **Gravesen, J.**,pH metric method for the determination of nicotinic acid plasma, *J. Clin. Microbiol.*, 5, 390, 1977.
58. **Robinson, W. T., Cosyns, L., and Kram, L. M.**, An automated method for the analysis of nicotinic acid in serum, *Clin. Biochem.*, 11, 46, 1978.
59. **Svedmyr, N. and Harthon, L.**, Comparison between the absorption of nicotinic acid in pentaerythritol tetranicotinate (Pericyt®) from ordinary and enterocoated tablets, *Acta Pharmacol. Toxicol.*, 28, 66, 1970.
60. **Blum, C. B., Levy, R. I., Hall, M., Goebel, R., and Berman, M.**, Reciprocal changes in high density lipoprotein metabolism with nicotinic acid treatment and carbohydrate feeding, *Circulation*, 54, (Suppl. 2) 26, 1976.

61. **Blum, C. B., Levy, R. I., Eisenberg, S., Hall, M., Goebel, R. H., and Berman, M.,** High density lipoprotein metabolism in man, *J. Clin. Invest.*, 60, 795, 1977.
62. **Shepherd, J.,** The influence of polyunsaturated fat diets and nicotinic acid therapy on the metabolism and sub-fraction of human high density lipoproteins, in *High Density Lipoproteins and Atherosclorosis*, Gotto, A. M., Jr., Miller, N. E., and Oliver, M. F., Eds., Elsevier/North Holland Biomedical Press, Amsterdam, 1978, 193.
63. **Patsch, J. R., Yeshurin, B., Jackson, R. L., and Gotto, A. M.,** Effects of clofibrate, nicotinic acid and diet on the properties of plasma lipoproteins in a subject with type III hyperlipoproteinemia, *Am. J. Med.*, 63, 1001, 1977.
64. National Diet Heart Study Research Group: The National Diet-Heart Study Final Report, *Circulation*, (Suppl.), 37, 1968.
65. **Vessby, B., Lithell, H., and Gustafsson, I.,** Affects of dietary treatment on lipoprotein levels in hyperlipoproteinemia, *Postgrad. Med. J.*, 51(8), 52, 1975.
66. **Hansen, P. F.,** Kombinations-Behandling ved hyperlipidaemia, *Läkartidningen*, 68, 1769, 1971.
67. **Jones, R. J.,** The drug treatment of hyperlipidemia, *Adv. Exp. Biol. Med.*, 82, 656, 1977.
68. **Mann, J. I., Harding, P. A., Turner, R. C., and Wilkinson, R. H.,** A comparison of cholestyramine and nicotinic acid in the treatment of familial II hyperlipoproteinaemia, *Br. J. Clin. Pharmacol.*, 4, 305, 1977.
69. **Olsson, A. G., Orö, L., and Rössner, S.,** Clinical and metabolic effects of pentaerythritol tetranicotinate (Pericyt®) and a comparison with plain nicotinic acid, *Atherosclerosis*, 19, 61, 1974.
70. **Parsons, W. B., Jr.,** Treatment of hypercholesterolemia by nicotinic acid, *Arch. Int. Med.*, 107, 639, 1961.
71. **Parsons, W. B., Jr.,** Studies of nicotinic acid use in hypercholesterolemia, *Arch. Int. Med.*, 107, 653, 1961.
72. **Schleirf, G. and Hess, G.,** Inhibition of carbohydrate-induced hypertriglyceridemia by nicotinic acid, *Artery*, 3, 174, 1977.
73. **Howard, A. N. and Courtenay Evans, R. J.,** Secholex®, clofibrate and taurine in hyperlipidaemia, *Atherosclerosis*, 20, 105, 1974.
74. **Beaumont, J. L., Carlson Cooper, G. R., Fejfer, Z., Frederickson, D. S., and Strasser, T.,** W.H.O. Memorandum. Classification of hyperlipidemias and hyperlipoproteinemias, *Circulation*, 45, 501, 1972.
75. **Yeshurun, D. and Gotto, A. M., Jr.,** Drug treatment of hyperlipidemia, *Am. J. Med.*, 60, 379, 1976.
76. **Mahley, R. W., Weisgraber, K. H., Bersot, T. P., and Innerarity, T. L.,** Effects of cholesterol feeding on human and animal high density lipoproteins, in *High Density Lipoproteins and Atherosclerosis*, Gotto, A. M., Jr., Miller, N. E., and Oliver, M. F., Eds., Elsevier/North Holland Biomedical Press, Amsterdam, 1978, 149.
77. **Miller, G. J. and Miller, N. E.,** Do high density lipoproteins protect against coronary atherosclerosis?, in *High Density Lipoproteins and Atherosclerosis*, Gotto, A. M., Jr., Miller, N. E., and Oliver, M. F., Eds., Elsevier/North Holland Biomedical Press, Amsterdam, 1978, 95.
78. **Öst, C. R. and Stenson, S.,** Regression of atherosclerosis during nicotinic acid therapy. A study in man by means of repeated arteriographies, in *Niacin in Vascular Disorders and Hyperlipemia*, Altschul, R., Ed., Charles C Thomas, Springfield, Illinois, 1964, 245.
79. **Barndt, R., Jr., Blankenhorn, D. H., Crawford, D. W., and Brooks, S. H.,** Regression and progression of early femoral atherosclerosis in treated hyperlipoproteinemia patients, *Ann. Int. Med.*, 86, 139, 1977.
80. **Brattsand, R. and Lindström, E.,** Aterioskleros-Nicotinsyraterapi. Några Effekter hos Försöksdjur, *Läkartidningen*, 68, 1776, 1971.
81. **Brattsand, R. and Lundholm, L.,** The effect of nicotinic acid and pentaerythritoltetranicotinate upon experimental atherosclerosis in the rabbit, *Atherosclerosis*, 14, 91, 1971.
82. **Lundholm, L., Jacobsson, L., Brattsand, R., and Magnusson, O.,** Influence of nicotinic acid, niceritrol and β-pyrdiyl carbinol upon experimental hyperlipidemia and atherosclerosis in mini-pigs, *Atherosclerosis*, 29, 217, 1978.
83. **Portman, O. W.,** Arterial composition and metabolism: esterified fatty acids and cholesterol, *Adv. Lipid Res.*, 8, 41, 1978.
84. **Wahlqvist, M. L., Day, A. J., and Tume, R. K.,** Incorporation of oleic acid into lipid by foam cells in human atherosclerotic lesions, *Circ. Res.*, 24, 123, 1969.
85. **Wahlqvist, M. L. and Day, A. J.,** Phospholipid synthesis by foam cells in human atheroma, *Exp. Mol. Pathol.*, 11, 275, 1969.
86. **Lassers, B. W., Wahlqvist, M. L., Kaijser, L., and Carlson, L. A.,** Relationship in man between plasma free fatty acids and myocardial metabolism of carbohydrate substrates, *Lancet*, 2, 448, 1971.
87. **Wahlqvist, M. L., Kaijser, L., Lassers, B. W., and Carlson, L. A.,** Fatty acids as a determinant of myocardial substrate and oxygen metabolism in man at rest and during prolonged exercise, *Acta Med. Scand.*, 193, 89, 1973.

88. **Lassers, B. W., Wahlqvist, M. L., Kaijser, L., and Carlson, L. A.,** Effect of nicotinic acid on myocardial metabolism in man at rest and during exercise, *J. Appl. Physiol.*, 33, 72, 1972.
89. **Wahlqvist, M. L., Kaijser, L., Lassers, B. W., Löw, H., and Carlson, L. A.,** The role of fatty acid and hormones in the determination of myocardial carbohydrate metabolism in healthy fasting men, *Eur. J. Clin. Invest.*, 3, 57, 1973.
90. **Balasse, E. O. and Neef, M. A.,** Influence of nicotinic acid on the role of turnover and oxidation of plasma glucose in man, *Metabolism*, 22, 1193, 1973.
91. **Kaijser, L. A., Carlson, L. A., Eklund, B., Nye, E. R., Rössner, S., and Wahlqvist, M. L.,** Substrate uptake by the ischaemic human heart during angina induced by atrial pacing, in *Effect of Acute Ischaemia on Mycardial Function*, Proc. 7th Pfizer Int. Symp., Oliver, M. F., Julian, D. G., and Donald, K. W., Eds., Churchill Livingston, 1972, 223.
92. **Vik-Mo, H.,** Distribution of coronary blood flow during acute coronary oclusion in dogs. Effect of nicotinic acid and sodium salicylate, *Scand. J. Clin. Invest.*, 37, 697, 1977.
93. **Russell, D. C. and Oliver, M. F.,** Effect of anti-lipolytic therapy on ST segment elevation during myocardial ischaemia in man, *Br. Heart J.*, 40, 117, 1978.
94. **Carlson, L. A., Danielson, M., Eckberg, I., Klintemar, B., and Rosenhamer, G.,** Reduction of myocardial reinfarction by the combined treatment with clofibrate and nicotinic acid, *Atherosclerosis*, 28, 81, 1977.
95. **West, R. J., Fusbrooke, A. S., and Lloyd, J. K.,** Treatment of children with familial hypercholesterolaemia, *Postgrad. Med. J.*, 51, (Suppl. 8), 82, 1975.
96. **Wahlqvist, M. L. and Hurley, B. P.,** Hyperlipoproteinaemia and dietary fat modification in haemodialysis and renal transplant patients, *Med. J. Aust.*, 2, 207, 1977.
97. **Gokal, R., Mann, J. I., Oliver, D. O., Ledingham, J. G., and Carter, R. D.,** Treatment of hyperlipidemia in patients on chronic haemodialysis, *Br. Med. J.*, 1, 82, 1978.
98. **Barboriak, J. J. and Mead, R. C.,** Nicotinic acid and alcohol-induced lipemia, *Atherosclerosis*, 13, 199, 1971.
99. **Sorrell, M. F., Baker, H., Tuma, D. J., Frank, O., and Barak, A. J.,** Potentiation of ethanol fatty liver in rats by chronic administration of nicotinic acid, *Biochem. Biophys. Acta*, 450, 231, 1976.
100. **Åberg, G. and Svedmyr, N.,** Thermographic registration of flush, *Arzn. Forsch.*, 21, 795, 1971.
101. **Nordqvist, P. and Wahlander, L.,** Nicangin®-Pericyt®: a comparison with regard to the frequency and intensity of flush, *Nobel-Pharma.*, 5, 1970.
102. **Svedmyr, N., Heggelund, A., and Aberg, G.,** Influence of indomethacin on flush induced by nicotinic acid in man, *Acta Pharmacol. Toxicol.*, 41, 397, 1977.
103. **Berge, K. G.,** Side effects of nicotinic acid in treatment of hyperlipidemia, *Geriatrics*, 16, 416, 1961.
104. **Parsons, W. B., Jr.,** Activation of peptic ulcer by nicotinic acid: report of five cases, *JAMA*, 173, 1466, 1960.
105. **Tromovitch, T. A., Jacobs, P. H., and Kerr, S.,** Acanthosis nigricans-like lesions from nicotinic acid, *Arch. Dermatol.*, 89, 222, 1964.
106. **Gurian, H. and Adlersberg, D.,** The effect of large doses of nicotinic acid on circulating lipid and carbohydrate tolerance, *Am. J. Med. Sci.*, 237, 12, 1959.
107. **Berge, K. G., Achor, R. W. P., Christensen, N. A., Mason, H. L., and Barker, N. W.,** Hypercholesterolemia and nicotinic acid. A long term study, *Am. J. Med.*, 31(1), 24, 1961.
108. **Molnar, G. D., Berge, K. G., and Rosevear, J. W.,** The effect of nicotinic acid in diabetes mellitus, *Metab. Clin. Exp.*, 13, 181, 1965.
109. **Miettinen, T. A., Taskinen, M. R., Pelkonen, R., and Nikkila, E. A.,** Glucose tolerance & plasma insulin in man during acute & chronic administration of nicotinic acid, *Acta Med. Scand.*, 186, 247, 1969.
110. **Stowers, J. M., Bewsher, P. D., Stein, J. M., and Mowat, J.,** Studies on the effects of nicotinic acid given orally or intravenously on oral and intravenous glucose tolerance in man, in *Metabolic Effects of Nicotinic Acid and its Derivatives*, Gey, K. F. and Carlson, L. A., Eds., Hans Huber, Bern, 1971, 723.
111. **Carlson, L. A.,** Antilipolysis as a tool in the study of clinical and experimental diabetes. Lecture for the Minkowski Award, *Diabetologia*, 5, 361, 1969.
112. **Mechaelis, D., Hahn, H. J., Michael, R., Knospe, S., Schäfer, S., Jutzi, F., and Wulfert, P.,** Effekte der Intravenösen Nikotinsäureinfusion Auf Substrat, Metabolit- Und Hormonkonzentrationen in Blut bein Juvenil-manifestierten Diabetes Mellitus, *Diabetologia*, 6, 550, 1970.
113. **Zöllner, N.,** Effect of nicotinic acid derivatives on the glucose metabolism, in *Metabolic Effect of Nicotinic Acid and its Derivatives*, Gey, K. F. and Carlson, L. A., Eds., Hans Huber, Bern, 1971, 719.
114. **Creutzfeldt, W., Frerichs, H., and Sickinger, K.,** Liver disease and diabetes mellitus, in *Progress in Liver Disease*, Vol. 3, Popper, H. and Schaffner, F., Eds., Grune & Stratton, New York, 1970, 371.

115. **Carlson, L. A., Freyschuss, U., Kjellberg, J., and Östman, J.,** Suppression of splanchnic ketone body production in man by nicotinic acid, *Diabetologia,* 3, 494, 1967.
116. **Mertz, W.,** Effects and metabolism of glucose tolerance factor, in *Nutrition Reviews. Present Knowledge in Nutrition,* Hegstead, D. M., Chichester, C. O., Darby, W. J., McNutt, K. W., Stalvey, R. M., and Stutz, E. H., Eds., The Nutrition Foundation, New York, 1976, 365.
117. **Doisy, R. J., Streeten, D. H. P., Freiberg, J. M., and Schneider, A. J.,** Chromium metabolism in man and biochemical effects, in *Trace Elements in Human Health and Disease,* Vol. 2, Prasad, A. S. and Oberleas, D., Eds., Academic Press, New York, 1976, 79.
118. **Pardue, W. O.,** Severe liver dysfunction during nicotinic acid therapy, *JAMA,* 175, 137, 1961.
119. **Christensen, N. A., Achor, R. W. P., Burge, K. G., and Mason, H. L.,** Nicotinic acid treatment of hypercholesterolemia, *JAMA,* 77, 546, 1961.
120. **Rivin, A. U.,** Jaundice occurring during nicotinic acid therapy for hypercholesterolemia, *JAMA,* 170, 2088, 1959.
121. **Winter, S. L. and Boyer, J. L.,** Hepatic toxicity from large doses from vitamin B_3 (nicotinamide), *N. Engl. J. Med.,* 289, 1180, 1973.
122. **Baggenstoss, A. H., Christensen, N. A., Burge, K. G., Baldus, W. P., Spiekerman, R. E., and Eccefeson, R. D.,** Fine structural changes in the liver in hypercholesterolemic patients receiving long-term nicotinic acid therapy, *Mayo Clin. Proc.,* 42, 385 1967.
123. **Kohn, R. M. and Montes, M.,** Hepatic fibrosis following long-acting nicotinic acid therapy. A case report, *Am. J. Med. Sci.,* 258, 94, 1969.
124. **Rowe, N. J., Dolder, M. A., Kirby, B. J., and Oliver, M. F.,** Effect of a nicotinic acid analogue on raised plasma free fatty acids after acute myocardial infarction, *Lancet,* 2, 814, 1973.
125. **Gass, J. D. M.,** Nicotinic acid maculopathy, *Am. J. Ophthalmol.,* 75, 500, 1973.
126. **Parsons, W. B., Jr.,** Use of aluminum nocitinate in hypercholesterolemia, *Cur. Ther. Res.,* 2, 137, 1960.
127. **McCabe, E. S.,** Use of aluminum nicotinate in the long term reduction of serum cholesterol, *Del. Med. J.,* 38, 49, 1966.
128. **Witte, E. C.,** Anti hyperlipdaemic agents, in *Progress in Medicinal Chemistry,* Vol. 2, Ellis, G. P. and West, G. B., Eds., North Holland, Amsterdam, 1975, 119.
129. **Avogara, E., Bittolo-Bon, G., Paris, M., and Taroni, G. C.,** Effect of a new niacin derivative (nicotinic hexaester of D-glucitol) on type II A, II B, and IV hyperlipoproteinemia in man, *Pharmacol. Res. Comm.,* 9, 599, 1977.
130. **Brox, D. and Selvaag, O.,** The effect of erythritol tetranicotinate on serum cholesterol levels in man, *Acta Med. Scand.,* 182, 437, 1967.
131. **Olsson, A. G., Orö, L., and Rössner, S.,** Clinical & metabolic effects with pentaerythritoltetranicotinate in combination with cholesolvin or clofibrate, *Atherosclerosis,* 19, 407, 1974.
132. **Olsson, A. G., Orö, L., and Rössner, S.,** Dose-response effect of single and combined clofibrate (atromidin®) and niceritrol (Pericyt®) treatment on serum lipids and lipoproteins in Type II hyperlipoproteinaemia, *Atherosclerosis,* 22, 91, 1975.
133. **Orö, L., Olsson, A. G., Rössner, S., and Carlson, L. A.,** Cholestyramine clofibrate and nicotinic acid as single or combined treatment of Type IIa and IIb hyperlipoproteinaemia, *Postgrad. Med. J.,* 51, (Suppl. 8), 76, 1975.
134. **Rössner, S., Olsson, A. G., and Orö, L.,** The effect of different dose regimens of niceritrol on serum lipid concentrations in man, *Acta Med. Scand.,* 200, 269, 1976.
135. **Marks, V.,** Effect of β-pyridyal carbinol on fasting plasma cholesterol levels in hyperlipoproteinaemic subjects, in *Metabolic Effects of Nicotinic Acid and its Derivatives,* Gey, K. F. and Carlson, L. A., Eds., Han Huber, Bern, 1971, 563.
136. **Nye, E. R. and MacBeth, W. A. A. G.,** The treatment of intermittent claudication with beta pyridyl carbinol over two years, *Atherosclerosis,* 17, 95, 1973.
137. **McInnis, D. L., Wahlqvist, M. L., and Balazs, N. D.,** Use of nicotinic acid and clofibrate in the management of hyperlipoproteinaemic patients, *Proc. Nutr. Soc. Aust.,* 5, 217, 1980.
138. **Wahlqvist, M. L. and McInnis,** unpublished data.

Chapter 5

PLASMA LIPIDS AND VITAMIN C

Emil Ginter

TABLE OF CONTENTS

I. Introduction 96

II. Biochemical Background 96
A. Vitamin C and Cholesterol Metabolism 96
B. Vitamin C and Triglyceride Metabolism 96

III. Vitamin C and Plasma Lipids in Man 97
A. Problem of Optimum Dosage of Vitamin C 97
B. Hypolipemic Action of Ascorbate in Man 99

IV. Conclusions 102

References 103

I. INTRODUCTION

Clinical studies describing the effect of vitamin C (ascorbic acid, ascorbate) on plasma lipids were carried out under very different conditions (doses and time of ascorbate administration) on humans with very different starting values of plasma lipids, with a different and often unknown vitamin C-status. It is thus only natural that their results should be so divergent, even contradictory. Earlier works[1-3] reporting an evident hypolipemic effect of vitamin C are rightly reproached that they were carried out on hospitalized patients in whom plasma lipids were affected alongside vitamin C by numerous other factors, e.g., altered diet, a change in daily regimen, exclusion of alcohol, etc. But studies describing ineffectiveness of vitamin C on plasma lipids are also open to criticism: it is, in fact, improbable that ascorbate be effective in subjects with low starting values of cholesterol (below 200 mg%)[4-6] or in hyperlipemic subjects with a high vitamin C intake prior to the beginning of the test.[7,8] Nor can a manifest hypolipemic effect be expected following a few days of ascorbate intake.[9] A correct understanding of the possibilities of plasma lipid levels being affected requires a knowledge of the mechanism by which vitamin C intervenes into the cholesterol and triglyceride metabolism.

II. BIOCHEMICAL BACKGROUND

A. Vitamin C and Cholesterol Metabolism

The use of the guinea pig model of a chronic latent vitamin C deficiency,[10] which simulates a marginal vitamin C deficiency in humans, has shown cholesterol accumulation in the liver and hypercholesterolemia to occur in deficient animals.[11] The degree of hypercholesterolemia depends on the duration of vitamin C deficiency, on the triglyceride and cholesterol content in the diet, and on the composition of dietary fatty acids.[12-16]

The investigation of this phenomenon,[17,18] designed to follow cholesterol distribution between blood and tissues, absorption of exogenous and synthesis of endogenous cholesterol, excretion of neutral sterols and bile acids, yielded an unequivocal conclusion: ascorbate is necessary for cholesterol transformation into its principal catabolic product — bile acids.[19] In a chronic latent vitamin C deficiency, this process becomes slowed down, with the subsequent decline of bile acid body pool,[20] bile acid concentration in the bile,[21] fecal excretion of bile acids,[22] cholesterol accumulation in the liver, hypercholesterolemia, and prolongation of the half-life of plasma cholesterol.[23] Cholesterol accumulation, detectable chemically and radiochemically, occurs also in the aorta of ascorbate-deficient animals during chronic experiments.[13,16,24]

The impact of vitamin C deficiency on the biogenesis of bile acids was localized in vivo[11] and in tests at the subcellular level:[25] ascorbate is required for the rate-limiting reaction of cholesterol transformation to bile acids, microsomal 7α-hydroxylation of cholesterol (Figure 1). Participation of ascorbate in this reaction is mediated by the vitamin C intervention, as yet incompletely understood, into the turnover of microsomal cytochrome P-450.[18,25-27] By a similar mechanism, ascorbate also affects the catabolism of xenobiotics. The rate of cholesterol transformation to bile acids is a function of ascorbate concentration in the liver.[20,23] The highest rate of bile acid formation has been found in guinea pigs with a maximal steady-state levels of ascorbate in hepatal tissue.[28]

B. Vitamin C and Triglyceride Metabolism

Even a short-term vitamin C deficiency is capable of inducing hypertriglyceridemia

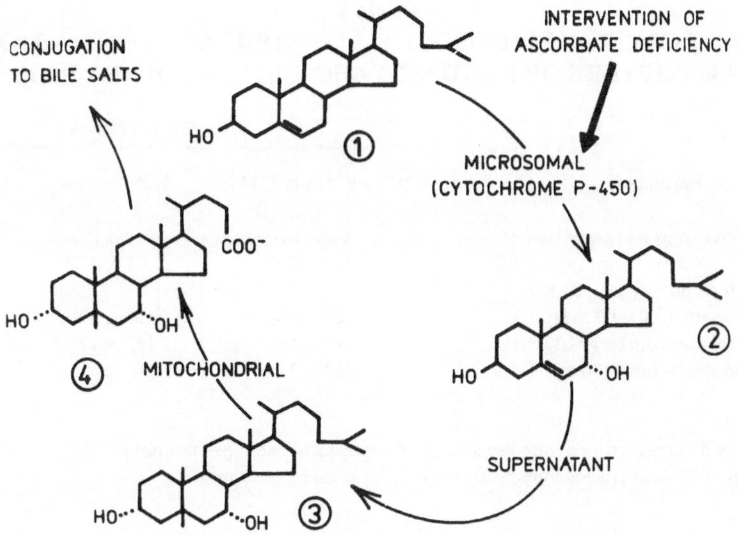

FIGURE 1. Localization of the interference of chronic marginal vitamin C deficiency with reactions transforming cholesterol to chenodeoxycholic acid in the liver cell. (1) Cholesterol; (2) 5-cholestene-3β,7α-diol (7α-hydroxycholesterol); (3) 5β-cholestane-3α,7α-diol; (4) chenodeoxycholic acid. The process of cholesterol catabolism to bile acids is intentionally simplified in the diagram. (From Ginter, E., *Wld. Rev. Nutr. Diet.*, 33, 104, 1979. With permission.)

in guinea pigs,[14] which augments in chronic experiments and leads to an accumulation of triglycerides in the liver and aorta.[15,24,29] An enhanced ascorbate intake depresses the level of plasma triglycerides in various animal species,[3,15,16,30-32] which is conditioned, among other factors, by activation of the lipolytic systems.[3,14,33,34] However, the intervention of ascorbate into these processes represents a complex phenomenon, for while activating the postheparin lipoprotein lipase in plasma, ascorbate inhibits the lipoprotein lipase in the heart and the hormone-sensitive lipase in the adipose tissue.[33,35] Some authors are of the opinion that ascorbate intervenes into these processes by influencing cAMP and cGMP metabolism.[36,37]

Although the mechanism of intervention of ascorbate into plasma triglyceride turnover has not been fully clarified so far, a hypotriglyceridemic action of ascorbate in various animal species, including primates, is irrefutable. Similarly, as in the case of cholesterol, the most striking hypotriglyceridemic effect of vitamin C is exercised by doses that ensure a maximal steady-state levels of ascorbate in tissues.[28]

III. VITAMIN C AND PLASMA LIPIDS IN MAN

From experiments on animals, it ensues that the plasma lipid levels are a function of vitamin C-status. A chronic latent ascorbate deficiency brings about hyperlipemia; on the other hand, vitamin C doses ensuring a maximal steady-state levels of ascorbate in the tissues ("saturation") are optimal for achieving a hypolipemic effect (Table 1). So far we do not know exactly what is the corresponding optimum dose of vitamin C for humans.

A. Problem of Optimum Dosage of Vitamin C

Recently we published a critical analysis of the contemporary concepts relating to optimum doses of vitamin C,[38] e.g., the officially recommended doses (tenths of mg

Table 1
EFFECT OF GRADED DOSES OF ASCORBIC ACID ON VARIOUS PARAMETERS OF LIPID METABOLISM IN GUINEA PIGS[28]

Parameter	Ascorbic acid intake 0.5 mg/animal/24 hr	0.05% in diet	0.5% in diet
Vitamin C in liver (mg/100 g wet tissue)	1.6 ± 0.2[a]	7.5 ± 0.3[b]	44.9 ± 4.9[b,c]
Rate of cholesterol transformation to bile acids (mg/500 g body weight/24 hr)	8.2 ± 0.6	13.5 ± 1.3[b]	19.0 ± 2.0[b,c]
Total lipids in serum (mg/100 mℓ)	527 ± 59	400 ± 29	307 ± 22[b,c]
Total cholesterol in serum (mg/100 mℓ)	170 ± 13	103 ± 12[b]	86 ± 11[b]
Triglycerides in serum (mg/100 mℓ)	229 ± 27	168 ± 8[b]	99 ± 8[b,c]

[a] Mean ± SEM.
[b] Significantly different from group receiving 0.5 mg ascorbic acid per animal per 24 hr.
[c] Significantly different from group receiving 0.05% ascorbic acid in diet.

per day), the saturation theory (approximately 100 mg/day) and the theory of megadoses (several thousand mg per day). The officially recommended levels are inadequate to achieve maximum levels, even in blood serum, and less so in tissues. Thus, for instance, with a daily intake of 30 mg (officially recommended dose by WHO/FAO) the serum ascorbate level amounts to about 0.4 mg %[39] which is less than one third of the maximum levels limited by the renal threshold. A daily intake of about 100 mg may achieve maximum levels in the serum, but not in the leukocytes,[38] which represent a good indicator of ascorbate levels in tissues.[40] According to our approximate estimates, the maximum value of the ascorbate pool in the body of an adult man is around 5 g.[38] The fractional turnover rate of ascorbate determined by the isotope technique is in the region of 3%/day,[41,42] which, with a 5 g body pool, corresponds to 150 mg/24 hr. If account is made of the partial degradation of ascorbate in the gastrointestinal tract, its incomplete absorption[43,44] and the possibility of part of it being excreted in urine before reaching the tissue pool,[41,42] then, to achieve and maintain a maximum size of the pool, a daily intake of 200 to 300 mg of vitamin C would probably be required. This dose will plausibly be even higher in elderly people with a diminished absorbing capacity, in various pathologic states (e.g., hypercholesterolemia, atherosclerosis, diabetes mellitus),[45,46] in strong smokers,[47] and in women using oral contraceptives.[48] Our suggestion of optimum doses should be regarded as an approximative estimate, for the human body has at least three ascorbate compartments, each with a different turnover,[49] and the maximum body pool has not as yet been experimentally determined in man.

If our conception is correct, then the suggested daily intake of a few hundred mg of ascorbic acid is a physiological dose, for it represents a correction of a genetic disorder, e.g., the loss of gulonolactone oxidase, a key enzyme in ascorbate biogenesis. The supply of such doses of vitamin C ensures in man maximal tissue steady-state levels of ascorbate which the other mammalian species (with the exception of monkeys, guinea pigs, and certain bats) achieve by a permanent provision of endogenous ascorbate synthesized in the liver. The application of megadoses of ascorbate, on the other hand, is to be understood as a pharmacological intervention which may have a positive,[50,51] but under certain conditions also a negative impact.[52-54]

A verification of the hypothesis on optimal steady-state levels of ascorbate in tissues would require studies designed to correlate vitamin C-status in humans with biochemical processes dependent on vitamin C, for example, with the metabolism of collagen,[55,56] cholesterol,[57] and detoxication of xenobiotics. A survey of 600 healthy

Table 2
NEGATIVE CORRELATION BETWEEN VITAMIN C CONCENTRATION IN LEUKOCYTES, AND CHOLESTEROL AND TRIGLYCERIDE LEVEL IN BLOOD SERUM IN 600 HEALTHY BLOOD DONORS[58]

Parameter		Vitamin C concentration in leukocytes Low (< 20 mg%)	Middle (20—30 mg%)	High (> 30 mg%)
Vitamin C in leukocytes (mg/100 g)		16.7 ± 0.2[a]	24.9 ± 0.2[b]	36.4 ± 0.6[b]
Cholesterol in serum (mg/100 mℓ)		217 ± 3.3	210 ± 2.4	196 ± 3.4[b]
Triglycerides in serum (mg/100 mℓ)		134 ± 5.8	121 ± 3.3[b]	103 ± 4.1[b]
Cholesterolemia	Normal	63.7%	64.1%	80.4%[b]
(percentage occurence	Moderately enhanced	24.0%	30.4%	16.2%
in the whole group)	High	12.3%	5.5%[b]	3.4%[b]
Triglyceridemia	Normal	75.4%	82.7%	90.6%[b]
(percentage occurence	Moderately enhanced	18.1%	13.8%	8.5%[b]
in the whole group)	High	6.5%	3.5%	0.9%[b]

[a] Mean ± SEM.
[b] Significantly different from group with low vitamin C concentration ($P < 0.05$—0.001).

subjects yielded a negative correlation between ascorbate concentration in leukocytes and cholesterol and triglyceride levels in blood serum.[58] When this group was subdivided into those with a low, middle, or high vitamin C levels in the leukocytes, the triglyceride and cholesterol level was noted to decline with an increasing ascorbate concentration (Table 2). The probability of enhanced cholesterol or triglyceride level being found in subjects with high ascorbate levels was two to threefold lower than in those with low ascorbate levels.

B. Hypolipemic Action of Ascorbate in Man

The total exchangeable body pool of cholesterol in normal subjects ranges around 200 mmol, but in hypercholesterolemic patients it may exceed 300 mmol. On the other hand, human liver catabolizes only about 1 mmol of cholesterol to bile acids per day, so that an acceleration of this process under the action of vitamin C may become manifest on the plasma or the total body pool of cholesterol only after a prolonged lapse of time. In addition, the possibility must be envisaged that a stimulation of 7α-hydroxylation of cholesterol will induce an enhanced synthesis of endogenous cholesterol through negative feedback. Such an enhancement of cholesterol synthesis was experimentally proved in the liver of monkeys fed high doses of vitamin C.[59] The hypocholesterolemic effect of vitamin C may become manifest only if the ascorbate-induced accelerated rate of cholesterol transformation into bile acids exceeds the liver's capacity to increase cholesterol synthesis. In subjects with a low cholesterolemia (below 200 mg %), regulation of cholesterol turnover is evidently very stable and is not affected even by high ascorbate doses, as was shown by numerous authors[3-6,60,61] and confirmed also by us (Table 3). Ascorbate concentration in the liver of persons with a permanently high vitamin C intake is close to maximum steady-state levels, so that insofar as one may judge from data obtained on guinea pigs, the rate of cholesterol catabolism is not affected in them by a further increase of ascorbate supply. If hypercholesterolemia does develop in persons with a permanently high vitamin C supply, its cause resides in factors other than a deficiency of this vitamin and an ascorbate-based therapy will prove ineffective, as was shown by two carefully performed stud-

Table 3
EFFECT OF ASCORBIC ACID ON CHOLESTEROLEMIA IN SUBJECTS WITH LOW AND ENHANCED STARTING CHOLESTEROL LEVELS[46,64,78]

Group characteristic	Number of subjects in group	Dose of ascorbic acid (mg/24 hr)	Duration of experiment (weeks)	Cholesterolemia (mg/100 mℓ)		P
				Beginning of experiment	End of experiment	
Healthy students, 18—19 years, male	20	100	8	192 ± 7[a]	183 ± 8	NS
Healthy students, 18—19 years, male	20	200	8	184 ± 6	189 ± 7	NS
Healthy students, 18—19 years, male	20	500	8	194 ± 7	203 ± 8	NS
Healthy students, 18—19 years, male	20	2000	8	189 ± 8	182 ± 7	NS
Hospitalized alcoholics, 27—48 years, male	14	3 × 300	3	205 ± 9	216 ± 14	NS
Long-term hospitalized patients, 42—92 years, men + women	39	3 × 300	3	217 ± 8	190 ± 7	<0.02
Healthy country population with moderate hypercholesterolemia, 42—65 years, men + women	24	300	7	255 ± 2	238 ± 8	<0.05
Inmates of a pensioners home with moderate hypercholesterolemia, 50—75 years, predominantly women	19	2 × 500	52	263 ± 6	229 ± 8	<0.002
Maturity-onset diabetes mellitus, hypercholesterolemic outpatients, 50—60 years, men + women	35	500	52	336 ± 7	273 ± 8	<0.001

[a] Mean ± SEM.
NS = not significant.

ies.[7,8] Genetically conditioned hypercholesterolemia is due to an absence of receptors for plasma low density lipoproteins on the cell surfaces,[62] and it seems unlikely that vitamin C would be capable of normalizing this situation.

An evident hypocholesterolemic effect of vitamin C may be expected to occur in persons with low tissue ascorbate levels in which hypercholesterolemia ensued as a result of a disbalance between cholesterol input into the system (high intake of exogenous or enhanced synthesis of endogenous cholesterol) and its irreversible elimination in the form of bile acids. Such an imbalance may be brought about by a number of factors, e.g., a nutritionally unbalanced diet or various metabolic disorders, such as diabetes mellitus.[63] The lower part of Table 3 summarizes the results in persons with a moderate or enhanced hypercholesterolemia in whom the initial vitamin C-status had been generally low. The most striking hypocholesterolemic effect was achieved in institutionalized elderly persons and in hypercholesterolemic maturity-onset diabetics after a 1-year administration of ascorbate in doses of 500 to 1000 mg/day.[46,64] In about 60% of the subjects, a clinically noteworthy decrease of cholesterolemia was achieved (by 40 mg % and more); in certain cases, this decline exceeded 100 mg %. Similar results were obtained in atherosclerotic patients[65,66] and in healthy subjects with a moderate hypercholesterolemia.[67]

All the studies referred to previously were designed to follow the concentration of total cholesterol in plasma. New observations have made it necessary to differentiate between cholesterol present in low density lipoproteins (LDL) and high density lipoproteins (HDL): a high LDL-cholesterol concentration is associated with a high risk of atherosclerosis,[68] while a high HDL-cholesterol level entails a low risk.[69] Bates et al.[70] found a positive correlation in elderly men between vitamin C-status and HDL-cholesterol concentration: the HDL-cholesterol level was noted to rise with a rising ascorbate level in plasma. A preliminary report described a slight reduction in total cholesterol and an increase in HDL-levels following administration of ascorbate.[71] The mechanism of relation between vitamin C and HDL-cholesterol requires further study. Unfortunately, guinea pigs are unsuitable for this study as, under normal conditions, their plasma contains no HDL at all.[72] Induction of liver microsomes may stimulate HDL synthesis;[73] abundant evidence is available on the activating effect of vitamin C on microsomal cytochrome P-450 in the liver.[18,25-27]

Vitamin C fails to affect plasma triglyceride concentration in persons with low starting values (around 150 mg % and lower).[3,7,29] However, a striking hypotriglyceridemic effect was achieved in hyperlipemic and/or cardiac patients receiving 2 to 3 g of ascorbic acid daily for 12 to 30 months.[3] Table 4 presents the results of two long-term studies,[29,46] which likewise brought support to the hypotriglyceridemic effect of vitamin C in man. This effect proved very evident in elderly people with pronounced hypertriglyceridemia. In maturity-onset diabetics with moderate hypertriglyceridemia, the clinical significance of the reduction achieved is questionable. An obstacle to a rational utilization of vitamin C in the therapy of hypertriglyceridemia is the unknown mechanism by which ascorbate intervenes into the turnover of plasma triglycerides.

An issue of interest, though as yet not investigated in humans, is the possible synergic action of ascorbate and hypolipemic drugs. Our preliminary results imply that a simultaneous administration of cholestyramine and ascorbate to guinea pigs has a very striking hypocholesterolemic effect and prevents the onset of hypertriglyceridemia which often ensues in patients treated with cholestyramine. Administration of clofibrate, but also of insulin has an adverse effect on the composition of bile in the sense of an enhanced risk of cholesterol cholelithiasis (increase in the cholesterol to bile acid ratio in the bile).[63,74] A simultaneous addition of ascorbate in these circumstances might have a favorable effect, for ascorbate given to experimental animals increases the quantity of chenodeoxycholic acid excreted into the bile, depresses the cholesterol

Table 4
HYPOTRIGLYCERIDEMIC ACTION OF ASCORBIC ACID IN LONG-TERM EXPERIMENTS[29,46]

Group characteristic	Parameter (mg/100 mℓ)	Number of subjects in group	Initial status	Administration of ascorbic acid: 6 months	Administration of ascorbic acid: 12 months
Inmates of a pensioners home with manifest hypertriglyceridemia, predominantly women aged 50—75 yrs	Triglycerides in serum	24	331 ± 22[a]	173 ± 19[b]	188 ± 10[b]
Ascorbic acid dose: 2 × 500 mg/day	Vitamin C in whole blood	19	0.61 ± 0.08	1.39 ± 0.10[b]	1.42 ± 0.11[b]
Maturity-onset diabetes mellitus, outpatients on a diabetic diet without insulin or peroral antidiabetic drugs, men + women, aged 50—60 yrs	Triglycerides in serum	35	227 ± 9	194 ± 8[b]	187 ± 12[b]
Ascorbic acid dose: 500 mg/day	Vitamin C in whole blood	35	0.37 ± 0.04	1.11 ± 0.07[b]	0.91 ± 0.05[b]

[a] Mean ± SEM.
[b] Significantly different from initial states ($P < 0.01$—0.001).

to bile acid ratio in the gall bladder bile, and has a preventive effect on experimental cholelithiasis.[21,75-77]

IV. CONCLUSIONS

Vitamin C is necessary for the transformation of cholesterol to bile acids, for it affects the rate-limiting reaction of cholesterol catabolism, the microsomal 7α-hydroxylation of cholesterol in the liver. In a chronic marginal vitamin C deficiency, this reaction becomes slowed down, thus bringing about an accumulation of cholesterol in the liver, hypercholesterolemia, prolongation of the half-life of plasma cholesterol, an enhanced cholesterol to bile acid ratio in gallbladder bile, formation of cholesterol gallstones, and storage of cholesterol in blood vessels of experimental animals. This is accompanied by hypertriglyceridemia, but the mechanism of this phenomenon is as yet obscure. The most effective means for preventing these changes are vitamin C doses, ensuring a maximal steady-state levels of ascorbate in tissues.

Prevention of subclinical hypovitaminosis C opens prospects of hyperlipidemia being possibly controlled in a very large number of people. A hypocholesterolemic action of ascorbate may become manifest in persons with a low vitamin C intake in whom hypercholesterolemia had ensued as a result of a disbalance between cholesterol input into the organism (exogenous and/or endogenous cholesterol), and its elimination in the form of bile acids. In every form of therapy of hyperlipidemia (dietary and pharmacological), an adequate vitamin C supply should be ensured in doses capable of creating maximal tissue steady-state levels of ascorbate. It appears very probable that these doses are several times higher than the officially recommended doses of vitamin C.

REFERENCES

1. **Sedov, K. R.,** Prevention and therapy of atherosclerosis with ascorbic acid (in Russian), *Ter. Arkh.*, 28(2), 58, 1956.
2. **Bukovskaya, A. V.,** The use of ascorbic acid in atherosclerotic patients (in Russian), *Sov. Med.*, 21(1), 77, 1957.
3. **Sokoloff, B., Hori, M., Saelhof, C. C., Wrzolek, T., and Imai, T.,** Aging, atherosclerosis and ascorbic acid metabolism, *J. Am. Geriatr. Soc.*, 14, 1239, 1966.
4. **Anderson, J., Grande, F., and Keys, A.,** Dietary ascorbic acid and serum cholesterol, *Fed. Proc. Fed. Am. Soc. Exp. Biol.*, 17, 468, 1958.
5. **Anderson, T. W., Reid, D. B. W., and Beaton, G. H.,** Vitamin C and serum-cholesterol, *Lancet*, 2, 876, 1972.
6. **Crawford, G. P. M., Warlow, C. P., Bennet, B., Dawson, A. A., Douglas, A. S., Kerridge, D. F., and Ogston, D.,** The effect of vitamin C supplements on serum cholesterol, coagulation, fibrinolysis and platelet adhesiveness, *Atherosclerosis*, 21, 451, 1975.
7. **Peterson, V. E., Crapo, P. A., Weininger, J., Ginsberg, H., and Olefsky, J.,** Quantification of plasma cholesterol and triglyceride levels in hypercholesterolemic subjects receiving ascorbic acid supplements, *Am. J. Clin. Nutr.*, 28, 584, 1975.
8. **Naito, H. K., De Wolfe, V. G., Brown, H. B., Wilcoxen, K., and Raulinaitis, I.,** Effect of ascorbate on serum lipid concentration in patients with hyper-β-lipoproteinemia, *Fed. Proc. Fed. Am. Soc. Exp. Biol.*, 36, 1158, 1977.
9. **Ratzmann, K. P., Wulfert, P., Ratzmann, M. L., and Hildmann, W.,** Wirkung einer hochdosierten Vitamin-C-Behandlung auf den Kohlenhydrat- und Fettstoffwechsel bei insulinpflichtigen Diabetikern, *Z. Inn. Med.*, 32, 571, 1977.
10. **Ginter, E., Bobek, P., and Ovečka, M.,** Model of chronic hypovitaminosis C in guinea pigs, *Int. J. Vitam. Res.*, 38, 104, 1968.
11. **Ginter, E.,** Ascorbic acid in cholesterol and bile acid metabolism, *Ann. N.Y. Acad. Sci.*, 258, 410, 1975.
12. **Ginter, E., Ondreička, R., Bobek, P., and Šimko, V.,** The influence of chronic vitamin C deficiency on fatty acid composition of blood serum, liver triglycerides and cholesterol esters in guinea pigs, *J. Nutr.*, 99, 261, 1969.
13. **Ginter, E., Červeň, J., Nemec, R., and Mikuš, L.,** Lowered cholesterol catabolism in guinea pigs with chronic ascorbic acid deficiency, *Am. J. Clin. Nutr.*, 24, 1238, 1971.
14. **Fujinami, T., Okado, K., Senda, K., Sugimura, M., and Kishikawa, M.,** Experimental atherosclerosis with ascorbic acid deficiency, *Jap. Circ. J.*, 35, 1559, 1971.
15. **Nambisan, B. and Kurup, P. A.,** Ascorbic acid and glycosaminoglycan and lipid metabolism in guinea pigs fed normal and atherogenic diets, *Atherosclerosis*, 22, 447, 1975.
16. **Hanck, A. and Weiser, H.,** Vitamin C and lipid metabolism, *Int. J. Vitam. Nutr. Res.*, Suppl. 16, 67, 1977.
17. **Turley, S. D., West, C. E., and Horton, B. J.,** The role of ascorbic acid in the regulation of cholesterol metabolism and in the pathogenesis of atherosclerosis, *Atherosclerosis*, 24, 1, 1976.
18. **Ginter, E.,** Marginal vitamin C deficiency, lipid metabolism, and atherogenesis, *Adv. Lipid Res.*, 16, 167, 1978.
19. **Ginter, E.,** Cholesterol: vitamin C controls its transformation to bile acids, *Science*, 179, 702, 1973.
20. **Hornig, D. and Weiser, H.,** Ascorbic acid and cholesterol: effect of graded oral intakes on cholesterol conversion to bile acids in guinea-pigs, *Experientia*, 32, 687, 1976.
21. **Jenkins, S. A.,** Vitamin C and gallstone formation: a preliminary report, *Experientia*, 33, 1616, 1977.
22. **Harris, W. S. and Kottke, B. A.,** The effect of ascorbic acid deficiency on cholesterol excretion in the guinea pig, *Fed. Proc. Fed. Am. Soc. Exp. Biol.*, 36, 1177, 1977.
23. **Ginter, E.,** Vitamin C in lipid metabolism and atherosclerosis, in Vitamin C, Birch, G. G. and Parker, K., Eds., *Appl. Sci. Publ. London*, 179, 1974.
24. **Fujinami, T., Okado, K., Senda, K., Nakano, S., Higuchi, R., Nakayama, K., Hayashi, K., and Sakuma, N.,** Experimental atherosclerosis with chronic covert ascorbic acid deficiency, *Jap. J. Atheroscler.*, 3, 117, 1975.
25. **Björkhem, I. and Kallner, A.,** Hepatic 7α-hydroxylation of cholesterol in ascorbate-deficient and ascorbate-supplemented guinea pigs, *J. Lipid Res.*, 17, 360, 1976.
26. **Degkwitz, E., Walsch, S., Dubberstein, M., and Winter, J.,** Ascorbic acid and cytochromes, *Ann. N.Y. Acad. Sci.*, 258, 201, 1975.
27. **Zannoni, V. G., Smith, C. R., and Rikans, L. E.,** Drug metabolism and ascorbic acid, *Int. J. Vitam. Nutr. Res.*, Suppl. 16, 99, 1977.

28. **Ginter, E., Bobek, P., and Vargová, D.**, Tissue levels and optimum dosage of vitamin C in guinea pigs, *Nutr. Metabol.*, 23, 217, 1979.
29. **Ginter, E., Černá, O., Ondreička, R., Roch, V., and Baláž, V.**, Vitamin C and plasma triglycerides in experimental animals and in humans, *Food Chem.*, 1, 23, 1976.
30. **Higuchi, R., Fujinami, T., Nakano, S., Nakayama, K., Hayashi, K., Sakuma, N., and Takada, K.**, Aortic glycosaminoglycan of the chronic hypovitamin C with or without coconut oil administration, *Jap. J. Atheroscler.*, 3, 303, 1975.
31. **Kotzé, J. P., Menne, I. V., Spies, J. H., and DeKlerk, W. A.**, Effect of ascorbic acid on serum lipid levels and depot cholesterol in the baboon (Papio ursinus), *S. Afr. Med. J.*, 49, 906, 1975.
32. **Machlin, L. J., Garcia, F., Kuenzig, W., Richter, C. B., Spiegel, H. E., and Brin, M.**, Lack of antiscorbutic activity of ascorbate 2-sulfate in the rhesus monkey, *Am. J. Clin. Nutr.*, 29, 825, 1976.
33. **Kotzé, J. P. and Spies, J. H.**, The effect of ascorbic acid on the activity of lipoprotein lipase in the baboon (Papio ursinus), *S. Afr. Med. J.*, 50, 1760, 1976.
34. **Bobek, P. and Ginter, E.**, Serum triglycerides and post-heparin lipolytic activity in guinea-pigs with latent vitamin C deficiency, *Experientia*, 34, 1554, 1978.
35. **Tsai, S., Fales, H. M., and Vaughan, M.**, Inactivation of hormone-sensitive lipase from adipose tissue with adenosine triphosphate, magnesium, and ascorbic acid, *J. Biol. Chem.*, 248, 5278, 1973.
36. **Hynie, S., Černohorský, M., and Čepelík, J.**, Actions of ascorbic acid on hormone induced lipolysis in vitro, *Eur. J. Pharmacol.*, 10, 111, 1970.
37. **Lewin, S.**, *Vitamin C: Its Molecular Biology and Medical Potential*, Academic Press, New York, 1976.
38. **Ginter, E.**, Chronic marginal vitamin C deficiency: biochemistry and pathophysiology, *World Rev. Nutr. Diet.*, 33, 104, 1979.
39. **Sauberlich, H. E.**, Vitamin C status: methods and findings, *Ann. N.Y. Acad. Sci.*, 258, 438, 1975.
40. **Beattie, A. D. and Sherlock, S.**, Ascorbic acid deficiency in liver disease, *Gut*, 17, 571, 1976.
41. **Baker, E. M., Hodges, R. E., Hood, J., Sauberlich, H. E., March, S. C., and Canham, J. E.**, Metabolism of ^{14}C- and ^{3}H-labeled L-ascorbic acid in human scurvy, *Am. J. Clin. Nutr.*, 24, 444, 1971.
42. **Hodges, R. E., Hood, J., Canham, J. E., Sauberlich, H. E., and Baker, E. M.**, Clinical manifestations of ascorbic acid deficiency in man, *Am. J. Clin. Nutr.*, 24, 432, 1971.
43. **Kallner, A., Hartmann, D., and Hornig, D.**, On the absorption of ascorbic acid in man, *Int. J. Vitam. Nutr. Res.*, 47, 383, 1977.
44. **Zetler, G., Seidel, G., Siegers, C. P., and Iven, H.**, Pharmacokinetics of ascorbic acid in man, *Eur. J. Clin. Pharmacol.*, 10, 273, 1976.
45. **Ginter, E. and Zloch, Z.**, Raised ascorbic acid consumption in cholesterol-fed guinea pigs, *Int. J. Vitam. Nutr. Res.*, 42, 72, 1972.
46. **Ginter, E., Ždichynec, B., Holžerová, O., Tichá, E., Kobza, R., Kožiaková, M., Černá, O., Ozdín, Ľ., Hrubá, F., Nováková, V., Šaško, E., and Gaher, M.**, Hypocholesterolemic effect of ascorbic acid in maturity-onset diabetes mellitus, *Int. J. Vitam. Nutr. Res.*, 48, 368, 1978.
47. **Pelletier, O.**, Vitamin C and cigarette smokers, *Ann. N.Y. Acad. Sci.*, 258, 156, 1975.
48. **Briggs, M. and Briggs, M.**, Vitamin C requirements and oral contraceptives, *Nature (London)*, 238, 277, 1972.
49. **Kallner, A., Hartmann, D., and Hornig, D.**, Determination of bodypool size and turnover rate of ascorbic acid in man, *Nutr. Metabol.*, 21 (Suppl. 1), 31, 1977.
50. **Pauling, L.**, *Vitamin C, the Common Cold and the Flu*, Freeman and Co., San Francisco, 1976.
51. **Stone, I.**, *The Healing Factor. Vitamin C Against Disease*, Grosset and Dunlap, New York, 1972.
52. **Hughes, R. E.**, Use and abuse of ascorbic acid — a review, *Food Chem.*, 2, 119, 1977.
53. **Mašek, J.**, Recommended nutrient allowances, *World Rev. Nutr. Diet.*, 25, 1, 1976.
54. **Basu, T. K.**, Possible toxicological aspects of megadoses of ascorbic acid, *Chem. Biol. Interact.*, 16, 247, 1977.
55. **Windsor, A. C. W. and Williams, C. B.**, Urinary hydroxyproline in the elderly with low leucocyte ascorbic acid levels, *Br. Med. J.*, 1, 732, 1970.
56. **Bates, C. J.**, Proline and hydroxyproline excretion and vitamin C status in elderly human subjects, *Clin. Sci. Mole. Med.*, 52, 535, 1977.
57. **Cheraskin, E. and Ringsdorf, W. M.**, Relationship of nonfasting serum cholesterol and vitamin C state, *Int. J. Vitam. Res.*, 38, 415, 1968.
58. **Černá, O. and Ginter, E.**, Blood lipids and vitamin-C status, *Lancet*, 1, 1055, 1978.
59. **Kotzé, J. P.**, The effects of vitamin C on lipid metabolism, *S. Afr. Med. J.*, 49, 1651, 1975.
60. **Hrubá, F. and Mašek, J.**, Einige Aspekte der Wirkung hoher Dosen von L-Ascorbinsäure auf den gesunden Menschen, *Nahrung*, 6, 507, 1962.
61. **Bronte-Stewart, B., Roberts, B., and Wells, V. M.**, Serum cholesterol in vitamin C deficiency in man, *Br. J. Nutr.*, 17, 61, 1963.

62. **Brown, M. S. and Goldstein, J. L.,** Receptor-mediated control of cholesterol metabolism, *Science,* 191, 150, 1976.
63. **Bennion, L. J. and Grundy, S. M.,** Effects of diabetes mellitus on cholesterol metabolism in man, *N. Engl. J. Med.,* 296, 1365, 1977.
64. **Ginter, E., Černá, O., Budlovsky, J., Baláž, V., Hrubá, F., Roch, V., and Šaško, E.,** Effect of ascorbic acid on plasma cholesterol in humans in a long-term experiment, *Int. J. Vitam. Nutr. Res.,* 47, 123, 1977.
65. **Novitski, A. A.,** Relationship between the metabolism of ascorbic acid and lipids, and the function of the system hypophysis-adrenals in atherosclerosis (in Russian), *Diss. Abstr.,* 1971.
66. **Norden, C., Heine, H., Schmidt, H. H., and Prockat, U.,** Die Hypercholesterinämie bei Patienten mit organischen arteriellen Durchblutungsstörungen und ihre Behandlung mit Vitamin C, *Dtsch. Gesundheltswes.,* 32, 160, 1977.
67. **Kothari, L. K. and Jain, K.,** Effect of vitamin C administration on blood cholesterol level in man, *Acta Biol. Acad. Sci. Hung.,* 28, 111, 1977.
68. **Castelli, W. P., Doyle, J. T., and Gordon, T.,** HDL cholesterol and other lipids in coronary heart disease: the cooperative lipoprotein phenotyping study, *Circulation,* 55, 767, 1977.
69. **Gordon, T., Castelli, W. P., and Jhortland, M. C.,** High density lipoprotein as a protective factor against coronary heart disease: the Framingham study, *Am. J. Med.,* 62, 707, 1977.
70. **Bates, C. J., Mandal, A. R., and Cole, T. J.,** H.D.L. cholesterol and vitamin-C status, *Lancet,* 2, 611, 1977.
71. **Lopez-S, A., Yates, B., Hardon, C., and Mellert, H.,** Effects of ascorbic acid on platelet aggregation and serum lipids and lipoproteins, *Am. J. Clin. Nutr.,* 31, 712, 1978.
72. **Puppione, D. L., Sardet, C., Yamanaka, W., Ostwald, R., and Nichols, A. V.,** Plasma lipoproteins of cholesterol-fed guinea pigs, *Biochim. Biophys. Acta,* 231, 295, 1971.
73. **Carlson, L. A. and Kolmodin-Hedman, B.,** Hyper-alpha-lipoproteinemia in men exposed to chlorinated hydrocarbon pesticides, *Acta Med. Scand.,* 192, 29, 1972.
74. **Grundy, S. M., Ahrens, E. H., Salen, G., Schreibman, P. H., and Nestel, P. J.,** Mechanisms of action of clofibrate on cholesterol metabolism in patients with hyperlipidemia, *J. Lipid Res.,* 13, 531, 1972.
75. **Ginter, E.,** Vitamin C and cholesterol, *Int. J. Vitam. Nutr. Res.,* Suppl. 16, 53, 1977.
76. **Iwamoto, K., Ozawa, N., Ito, F., Okamoto, N., and Watanabe, J.,** Effect of ascorbic acid on the intestinal absorption of bile salts and metabolism of cholesterol in guinea pigs, *Chem. Pharm. Bull.,* 24, 2014, 1976.
77. **Ginter, E. and Mikuš, L.,** Reduction of gallstone formation by ascorbic acid in hamsters, *Experientia,* 33, 716, 1977.
78. **Ginter, E., Kajaba, I., and Nizner, O.,** The effect of ascorbic acid on cholesterolemia in healthy subjects with seasonal deficit on vitamin C, *Nutr. Metabol.,* 12, 76, 1970.

Chapter 6

THIAMIN AND THE WERNICKE-KORSAKOFF SYNDROME

John P. Blass

TABLE OF CONTENTS

I. Introduction 108

II. Wernicke-Korsakoff Syndrome 108
A. Wernicke Syndrome 108
1. Discovery 108
2. Clinical Manifestations 109
3. Pathological Manifestations 109
B. Korsakoff Psychosis 110
1. Discovery 110
2. Clinical Manifestations 110
3. Pathological Manifestations 111
C. Relation of Wernicke and Korsakoff Syndromes 111
D. Thiamin Deficiency in Wernicke-Korsakoff Syndrome 111
1. General Observations 111
2. Thiamin Therapy 111

III. Thiamin and Its Functions 112
A. Discovery of Thiamin 112
B. Structure 112
C. Enzymatic Actions of TPP 113
1. Pyruvate Dehydrogenase (EC 1.2.4.1) 114
2. Ketoglutarate Dehydrogenase (EC 1.2.4.2) 115
3. Transketolase (EC 2.2.1.1) 115
4. Branched Chain Dehydrogenase (BCDHC) 115
5. Thiamin Pyrophosphokinase (EC 2.7.4.15) 116
D. Metabolic Roles of Thiamin 116
1. Carbohydrate Metabolism 117
a. Oxidation of Carohydrates 117
b. Pentose Phosphate Pathway 118
2. Lipid Metabolism 119
3. Amino Acids and Proteins 119
a. Amino Acid Metabolism 119
b. Protein Synthesis 120
4. Nucleic Acids 120
5. Neurotransmitters 120
a. Acetylcholine 120
b. Catecholamines 121
c. Serotonin 121
d. Other Neurotransmitters 121
6. Membrane Function 121
E. Experimental Thiamin Deficiency 122
1. Species 122

2. Induction of Thiamin Deficiency 122
a. Diet 122
b. Antagonists 123
3. Pathological Lesions 123

IV. Problems 123
A. Wernicke-Korsakoff-Like Syndromes 123
1. Other Acquired Disorders 124
2. Leigh's Disease 124
B. Varying Presentations of Thiamin Deficiency 124
C. Genetic Aspects 125
1. Epidemiologic Observations 125
2. Biochemical Genetic Studies 125
a. Transketolase 125
b. Pyruvate Dehydrogenase 126
c. Ketoglutarate Dehydrogenase 127
d. Branched Chain Dehydrogenase 127
e. General Genetic Considerations 127
D. Metabolic Inhibitors 128

V. Formulation of Pathophysiology 128

VI. Future Directions 129

References 129

I. INTRODUCTION

Elucidation of the role of thiamin deficiency in the Wernicke-Korsakoff syndrome is a striking example of fruitful interaction between clinical medicine and laboratory investigation.[1] Careful, creative clinical observations led not only to the discovery of thiamin, but to that of vitamins in general. Chemical and biochemical laboratory studies elucidated the structure, mechanism of action, and many aspects of the physiological role of thiamin. This information not only allowed more effective prevention and treatment of the clinical disorder, but also led to more precise analysis and classification of the clinical problems. However, although it is possible to describe the relation between Wernicke-Korsakoff syndrome and thiamin deficiency with great clarity, there are a number of relevant observations which remain unexplained.[2]

The discussion below reviews the recognition and description of the classical syndrome and of the role of thiamin in it, of the discovery of thiamin and of its molecular and metabolic roles, and then reviews some of the unsolved problems.

II. WERNICKE-KORSAKOFF SYNDROME

Neurologists in the U.S. usually use the term Wernicke-Korsakoff Syndrome to refer to the central nervous system manifestations of thiamin deficiency.[1] Europeans often distinguish between Wernicke's Syndrome and Korsakoff's Psychosis.[3]

A. Wernicke Syndrome

1. Discovery

Although earlier reports of patients with what is now recognized as Wernicke's syndrome may have appeared,[1] the first clear description appeared in Carl Wernicke's

Textbook of Brain Diseases for Physicians and Students, published in 1881. He described three such patients. Two were alcoholic men who presented in delirium. One was a seamstress who had 4 weeks of pernicious vomiting secondary to poisoning with sulfuric acid and then developed a neuropsychiatric syndrome. All three patients had the eye signs, staggering gait, and impaired consciousness now recognized as characteristic of the disorder. All died, and at autopsy had punctate periventricular hemorrhages. Wernicke characterized this state as a polioencephalitis hemorrhagica.[4]

The existence of this clinical-pathological entity was promptly and repeatedly confirmed,[1,5] and is now universally accepted.

2. Clinical Manifestation[1]

The characteristic clinical features are the Wernicke triad: opthalmoplegia, ataxia, and altered state of consciousness. They occur simultaneously, or one may precede the others by days or weeks.

Opthalmoplegia — The eye signs generally consist of nystagmus (vertical or horizontal), weakness or complete paralysis of lateral gaze (i.e., of VI nerve function), and weakness or paralysis of conjugate gaze (i.e., movement of both eyes together). Usually, but not always, all three abnormalities occur together. A number of other abnormalities can occur but are less characteristic.

Ataxia — The disorder of coordination generally manifests itself as a staggering, drunken gait. Other signs of dysfunction of the cerebellum can often be elicited but are typically less obvious. The gait abnormality is identical to that induced by alcohol excess, and recognition that it is due to thiamin deficiency can be difficult in alcoholics who are the major population at risk for this disorder in the developed world.

Altered state of consciousness — The most common form of mental abnormality in Wernicke's Syndrome is the global confusional state. Patients are sleepy and pay little attention to their surroundings. Frank stupor or coma are rare. Some patients present with more or fewer signs of delerium tremens and others with true Korsakoff Psychosis as described below.

Other signs and symptoms — Peripheral neuropathy typically occurs in more than 80% of the patients (189/230 in the series of Victor et al.).[1] Both motor and sensory nerves can be involved. Cardiovascular abnormalities are frequent but usually minor. From published reports, the coincident occurrence of frank neurological and cardiac disorders (i.e., of dry and wet beriberi) together in thiamin-deficient patients appears to be rare (see below, Section IVB). Other clinical characteristics, including age and sex incidence, are discussed in detail by Victor et al.[1]

Clinical laboratory studies[1] — Among tests which are typically abnormal are

1. Elevations of blood pyruvate
2. Decreases of red blood cell transketolase activity with an elevation of the stimulation by thiamin pyrophosphate (TPP effect — see below, Section IVC2)
3. Decreased response on ice water caloric testing (vestibular paralysis)
4. Electroencephalographic (EEG) decreases in frequency in about half the patients
5. Decreases in cereberal blood flow which may be persistent despite apparently successful treatment[6]

Clinical course and treatment — Thiamin is effective therapy, as discussed below (Section IID2).

3. Pathological Manifestations

The characteristic neuropathological changes of Wernicke-Korskoff syndrome are symmetrical lesions of structures surrounding the third and fourth ventricles. In milder lesions, loss of myelin often appears to be more marked than that of neurones. In more severe lesions, nerve cells and axis cylinders are lost as well as myelin sheaths. Relative persistence of glia and of reactive cells prevents cavitation. Relative persistence of the blood vessels may render them prominent. However, in contrast to frequent statements in the earlier literature, Victor et al. found hemorrhages on gross examination in only 6 of 62 cases.[1]

Lesions of the mammillary bodies are common (e.g., in 46 of 62 patients of Victor et al.).[1] The mammillary body receives input through the fornix from the hippocampus, a structure which has been related to "memory function." These lesions have been invoked to explain some of the memory disturbance (see below, Section IIB2).[1]

The pathology including detailed clinicopathological correlations has been discussed extensively by Victor et al.[1]

B. Korsakoff Psychosis

1. Discovery

Between 1887 and 1891, S. S. Korsakoff, a Russian physician, published a series of articles pointing out the existence of a group of patients with profound disturbances of memory and more or less peripheral neuropathy.[7-9] Others had noted this association previously, particularly in alcoholics, and there was subsequent controversy about whether or not there was any significant relationship between the neuropathy and the mental changes.[10] Eventually, recognition of the role of thiamin deficiency in producing both the neuropathy and the mental changes ended the dispute.[1]

However, it should be noted that the coexistence of neuropathy and dementia is not limited to Korsakoff's syndrome and to thiamin deficiency. For instance, both characteristically occur in the neurological form of vitamin B_{12} deficiency, combined system disease.[11] Dementia commonly occurs in neuromuscular disorders associated with abnormal mitochondria in Kearn-Sayre syndrome.[12] Neuropathic abnormalities can occur in patients with the diagnosis of chronic, process schizophrenia.[13] Indeed, any systemic disease which affects both peripheral nerves and portions of the brain associated with higher cortical function can lead to a combination of neuropathic and mental abnormalities with impaired performance on tests of memory.

Even in retrospect, Korsakoff's insight in recognizing the syndrome named for him was inspired.

2. Clinical Manifestations

The cardinal manifestation of Korsakoff Psychosis is a profound impairment in learning and retaining new information. There is also typically some anterograde amnesia, but immediate registeration of new information is relatively spared. On the basis of detailed psychological examinations, Talland proposed that the prime difficulty is in orienting to new situations.[14]

The nature of the deficit is illustrated by two common ploys used to demonstrate the problem to medical students. The first is for the examiner to put a ten dollar bill under a book and promise to give it to the patient if the patient remembers where the bill is 5 min later. If the diagnosis is correct, the examiner keeps his money. The second ploy is to ask a question mildly and then much more forcibly. The patient may be able to answer the same question when posed vigorously that he did not answer when posed mildly, illustrating the "attentional" component of the disorder.

Confabulation has been described as a characteristic aspect of Korsakoff psychosis. Indeed, it has been referred to as confabulatory psychosis. Whether patients with this disorder confabulate more than patients with other chronic organic brain syndromes without aphasia is a moot point. It certainly has not been well documented in the literature.

The course and treatment of Korsakoff Psychosis are discussed below (Section II D2.)

3. *Pathological Manifestations*

Pathological manifestations are essentially the same as in Wernicke's Syndrome. Adams and Victor suggest that the memory disturbance correlates better to lesions of the medial dorsal nucleus of the thalamus and perhaps the posterior nucleus as well than to lesions of the mammillary bodies.[15]

C. Relation of Wernicke and Korsakoff Syndromes

The close relation of Wernicke and Korsakoff syndromes is shown by clinical, pathological, and etiologic evidence.

Clinically, a majority of patients with Wernicke's Syndrome who do not die in the acute phase demonstrate residual Korsakoff psychosis (157 of 186 patients in the series of Victor et al.).[1,15] Although this association was not recognized by either Wernicke or Korsakoff, it was pointed out as early as 1897 and is now well established.[16] To quote Adams and Victor, "the symptoms of Wernicke's disease and Koraskoff's psychosis are not separate clinical events ... (but) ... successive states in ... a single disease process."[15]

Pathologically, the fundamental lesions in both disorders are identical, although different clinical manifestations may relate to geographic differences in the specific sites of damage in the central nervous system.

Etiologically, the role of thiamin deficiency in both disorders is well established and is discussed in the next section.

D. Thiamin Deficiency in Wernicke-Korsakoff Syndrome

The role of nutritional factors in the Wernicke-Korsakoff syndrome was first recognized on clinical grounds. Indeed, as discussed below (Section IIIA) it was one of the stimuli for precise biochemical characterization of the nutritional factor involved. When pure crystalline thiamin was isolated, its efficacy in treating the syndrome documented its role unequivocally.

1. *General Observations*

In the original studies, Wernicke's Syndrome was thought to be inflammatory and Korsakoff's Psychosis to be toxic in origin. However, it was recognized early that the changes were in some sense due to perverted nutrition.[17] This recognition arose, in large part, because the Wernicke-Korsakoff syndrome typically develops in patients with either gastrointestinal problems, poor diet, or both. In the developed world, it occurs most frequently in chronic alcoholics, perhaps most frequently in severe alcoholics who drink steadily and eat very poorly rather than in binge drinkers who eat a more or less reasonable diet between bouts.[1] However, well-documented cases have occurred associated with pernicious vomiting secondary to pregnancy,[18,19] carcinoma,[20,21] intestinal obstruction,[22] and more recently in patients on chronic renal dialysis.[23] It also occurred in prisoners of war on a diet consisting primarily of polished rice.[5] The importance of diet in treating this, as well as other complications of alcoholism, was also recognized early.

2. *Thiamin Therapy*

A specific role of thiamin (vitamin B_1) in the syndrome was suggested in the 1930s, on the basis of studies of experimental animals. The chemical isolation and characterization of thiamin allowed the controlled production of B_1-deficiency in experimental animals, some of whom developed neuropathological lesions similar to those of Wernicke-Korsakoff syndrome.[24-26] Jolliffe and co-workers clearly demonstrated in large groups of patients that prompt treatment with thiamin reversed the opthalmoplegia,[27,28] ataxia, and acute confusional state, although it did not consistently ameliorate the residual memory defect.[27,28] These observations were promptly and widely confirmed and accepted. Indeed, within a year of the publication of that original report, de Wardener and Lenox prepared vegetable extracts rich in thiamin to treat patients in a prisoner-of-war camp where more sophisticated medications were not available.[5]

Victor et al. documented the specific effects of thiamin in detailed clinical investigations.[1] They demonstrated that addition of thiamin to a diet of glucose, minerals, and water or to an artificial rice diet promptly benefited the signs of Wernicke's syndrome which then cleared completely within days or weeks, whether or not the patients continued to drink alcohol.[29] On long-term treatment with thiamin, there was an improvement in memory in 20% or more of the patients.[1,30,31] Treatment with newer, better absorbed thiamin deivatives such as tetrafurfuryl thiamin disulfide has not markedly altered the outcome.[32,33]

At the present time, Wernicke-Korsakoff syndrome is recognized as an acute medical emergency and is routinely teated successfully with 50 to 100 mg of thiamin intravenously followed by 50 to 100 mg by intramuscular injection daily.[1]

III. THIAMIN AND ITS FUNCTIONS

A. Discovery of Thiamin[34]

As noted above, it was recognized during the 19th century that diets containing meat and vegetables prevented and often cured Wernicke-Korsakoff syndrome and other manifestations of what is now recognized to be thiamin deficiency. In the first decade of this century, Eijkman, a Dutch physician working in Java, made the acute clinical observation that the chickens in his yard developed a neuropathy on a diet of polished rice, which was prevented by feeding them whole rice or rice husks.[34,35] This observation not only suggested an effective treatment for the human disorder, namely, the feeding of rice husks, but also provided an animal model. It also provided an assay system for the missing food factor, and thus allowed its isolation and the determination of its structure. This task — formidable with the chemical methods of the time — was accomplished by R. R. Williams in 1936.[34,36] Thiamin played a critical role in the elucidation of the vitamin theory by Hopkins and Funk in 1912.[37] Even the term vitamin is derived from the belief that thiamin was an amine vital for health. The discovery of thiamin has thus had immense implications for human health.

B. Structure

Thiamin consists of a substituted pyrimidine ring and a substituted thiazole ring joined by a methylene bridge (Figure 1). The carbon located between the sulfur and the nitrogen in the thiazole ring is relatively acidic, i.e., it readily loses its hydrogen to form a negatively charged carbanion. This compound catalyzes nonenzymatic reactions involving rupture or formation of carbon-carbon bonds with a variety of compounds containing aldehyde or ketone moieties (Figure 1).[38]

The derivative of thiamin which acts as a coenzyme in several enzymatic reactions is thiamin pyrophosphate (TPP or TDP). It is formed by a pyrophosphorylation reaction involving ATP (Figure 2).[39] Removal of one phosphate by a phosphatase gives

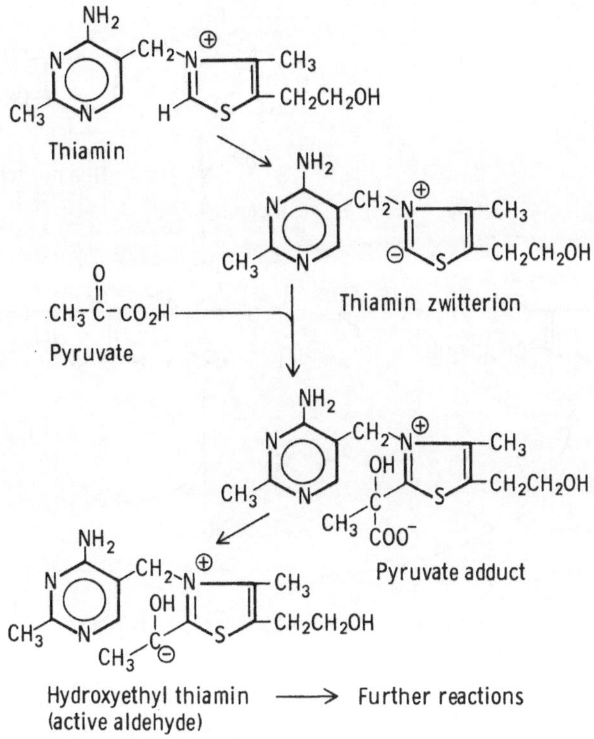

FIGURE 1. Nonenzymatic reaction. The nonenzymatic decarboxylation of pyruvate is typical of the reactions catalyzed by thiamin.[38] The nature of the final product depends on the acceptor molecule. If the acceptor is acetaldehyde ($CH_3CH{:}O$), the product is acetoin ($CH_3CHOHC{:}OCH_3$). Thiamin is regenerated. The nonenzymatic reactions catalyzed by thiamin are very slow compared to the analogous enzymatic reactions catalyzed by thiamin pyrophosphate (TPP).

rise to thiamin monophosphate (TMP).[40-42] Addition of a third phosphate to form thiamin triphosphate (TTP) is reportedly catalyzed by a specific kinase which transfers the terminal phosphate from ATP.[43,44]

Thiamin is not synthesized by human tissues. (Indeed, if it was it would not be a vitamin.) The pathways of its synthesis in yeast and other plants are well worked out. Thiamin is found in many plant and animal foodstuffs, notably the outer layers of many seeds. In man, it is excreted unchanged or after cleavage between the ring systems by a thiaminase present in intestinal microorganisms rather than in the mammalian tissues.[45]

C. Enzymatic Actions of TPP

The enzymatic reactions in which TDP participates as a cofactor are analogous to the nonenzymatic reactions catalyzed by thiamin itself, but proceed much more efficiently. In mammalian tissues, TDP has been demonstrated to be a cofactor for four enzymes: pyruvate dehydrogenase, ketoglutarate dehydrogenase, transketolase, and branched chain dehydrogenases. It has also been reported to be a substrate for a specific pyrophosphokinase.

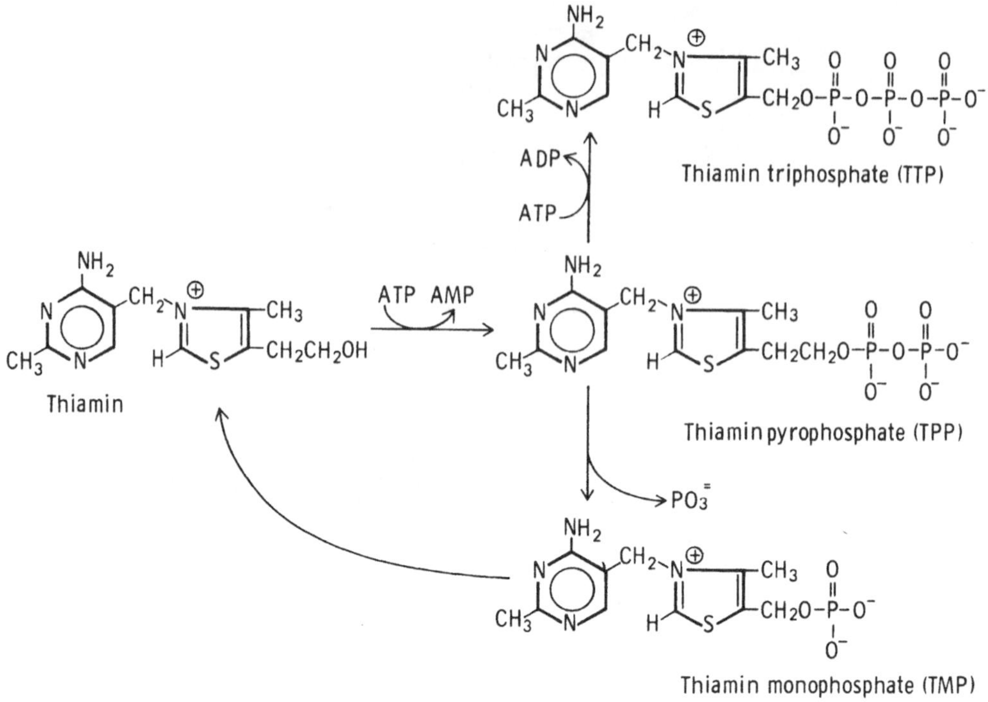

FIGURE 2. Metabolism of thiamin. The diagram shows the reactions by which the vitamin thiamin is converted to biologically active phosphorylated derivatives. ATP, ADP, and AMP are adenosine triphosphate, diphosphate, and monophosphate, respectively.

1. Pyruvate Dehydrogenase (EC 1.2.4.1)

This enzyme is part of a large and highly regulated multienzyme complex (PDHC) which carries out the oxidation of pyruvate to acetyl coenzyme A and CO_2:

$$\begin{array}{c} CO_2H \\ | \\ C{=}O \\ | \\ CH_3 \end{array} \longrightarrow \begin{array}{c} SCoA \\ | \\ C{=}O \\ | \\ CH_3 \end{array} + CO_2$$

PDHC has been isolated from bacteria, other plants, and several mammalian tissues including brain.[46-48] Its structure and many aspects of its regulation are known in detail. It undergoes product inhibition and is acted on by a number of effectors. A major mechanism of control of the activity of the enzyme is covalent modification by phosphorylation of the α-peptide of the thiamin-dependent enzyme. Recent studies have shown that there are three sites which are phosphorylated. Site one is a serine, and phosphorylation there effectively inactivates PDHC and dephosphorylation reactivates it. Excess amounts of thiamin or of thiamin pyrophosphate inhibit phosphorylation at this site and thus favor activation.[49,50] Recent studies have indicated that phosphorylation at sites two and three may affect activation, but the precise mechanisms remain to be worked out.[51] Regulation of PDHC is now a very active area of research and new layers of intricacy are being uncovered.[52-55]

The apparent K_m for binding of TPP and PDHC is about 0.1 to 0.5 μM for the animal enzyme,[47] but was about 2 μM for a preparation of PDHC from a human liver.[56]

2. Ketoglutarate Dehydrogenase (EC 1.2.4.2.)

This enzyme is also part of a large and complex multienzyme complex (KGDHC), which is a close structural analogue of PDHC and carries out an analogous reaction, the oxidative decarboxylation of 2-ketoglutarate to succinyl coenzyme A and CO_2:

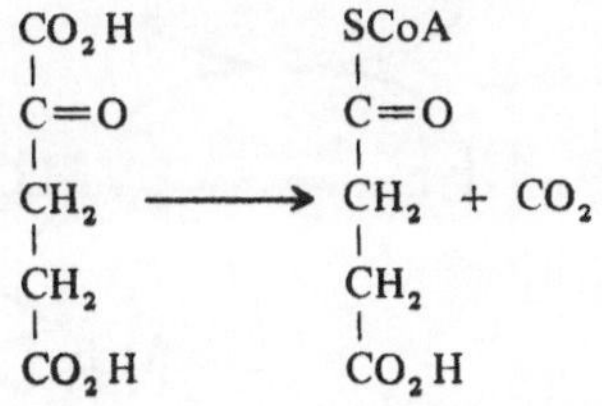

KGDH has also been isolated from plant and animal tissues and its structure and regulation studied.[46] There does not appear to be a phosphorylation-dephosphorylation mechanism for KGDHC.

The binding of TPP to KGDHC is very tight, and removal of TPP from the enzyme without inactivating it has not been reported. The K_m must be very low.

3. Transketolase (EC 2.2.1.1)

This enzyme catalyzes the reversible intraconversions of sugars, for instance that of ribose-5-phosphate and xylulose-5-phosphate to glycerol-3-phosphate and seduheptalose phosphate:

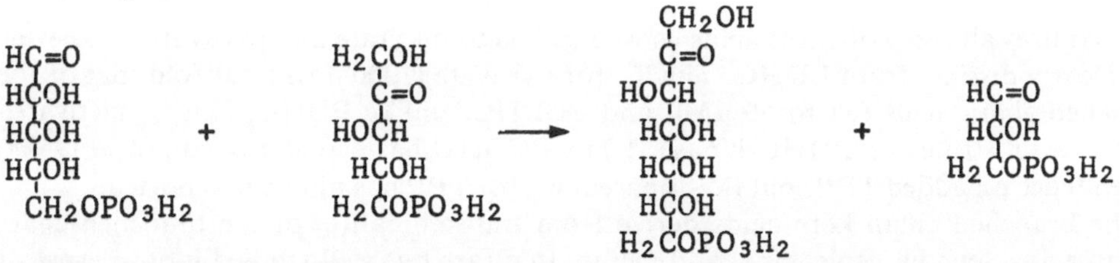

The enzyme was originally isolated from yeast and spinach, but has recently been isolated from mammalian liver.[57-59]

The apparent K_m for binding of TPP to transketolase from human fibroblasts and red cells is about 15 μM, varying from less than 1 to over 20 μM in different individuals. The separate kinetic constants for association (K_a) and dissociation (K_d) have not yet been measured. Doing so is important because for transketolase, as for PDHC, measurement of apparent K_m is somewhat artificial. For both enzymes, removal of TPP requires conditions of pH and ionic strength which partially unfold the proteins.[33,57,58] In vivo, the rate of association of TPP to the newly synthesized enzyme(s) may be more relevant in dietary thiamin deficiency than are the overall K_m values.

4. Branched Chain Dehydrogenases (BCDHC)

The enzyme complex which catalyzes the oxidation of the branched chain 2-keto acids has been successfully purified only very recently.[60] The reactions are

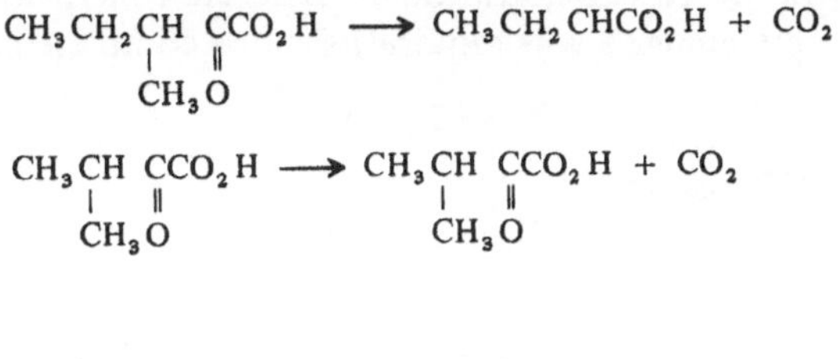

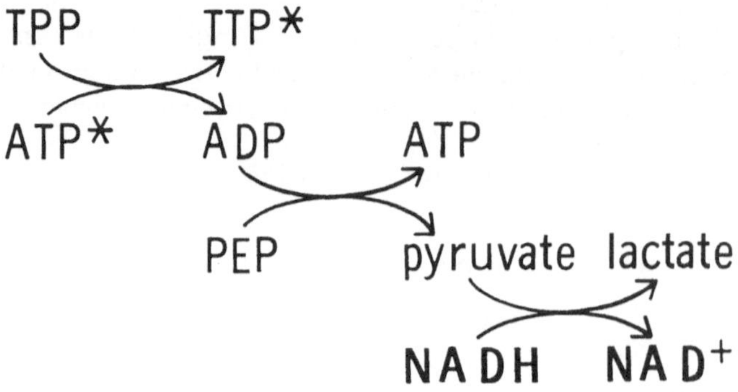

FIGURE 3. Assay system for TTP synthesis. The conventional assay system for the formation of TTP (thiamin triphosphate) is the coupled enzyme system shown.[150,151] The actual measurement made is of the TTP-dependent oxidation of NADH to NAD^+, which is followed by the decrease in absorption at 340 nm; inhibition of this process is used to detect the inhibitor-protein discussed by Pincus et al.[150,151] Incorporation of radioactivity from the γ-phosphate of ATP into TPP has been demonstrated for endogenous TPP but not for exogenous TPP added to the preparation.[43] TPP, thiamin pyrophosphate; TTP, thiamin triphosphate; ATP, adenosine triphosphate; ADP, adenosine diphosphate; PEP, phosphoenolpyruvate.

It oxidizes all three of these acids as well as 2-keto butyrate and pyruvate. However, it is clearly distinct from PDHC. The K_m for pyruvate (1000 μM) is 20-fold that of the branched chain acids (37 to 50 μM), and BCDHC, unlike PDHC, is not inactivated by nor a substrate for PDHC kinase.[60] The BCDHC has essentially complete (95%) dependence on added TPP, but the apparent K_m for TPP has not been reported.

The branched chain keto acids derive from transamination of the branched chain amino acids, leucine, isoleucine, and valine. In a rare but well-studied inborn error of metabolism, the three branched-chain amino acids and the three keto acids accumulate. The defect appears to be in the dehydrogenase(s) which carry out the TPP dependent oxidative decarboxylations. (see Section IVC2c).

5. Thiamin Pyrophosphokinase (EC 2.7.4.15)

This enzyme has been reported to carry out the phosphorylation of TPP to TTP (Figure 2).[43,44] It has not yet been highly purified nor its properties well-defined. The usual assay system for the enzyme actually measures the TPP-stimulated hydrolysis of ATP (Figure 3). It can be interpreted as measuring the effect of TPP on ATPases rather than the synthesis of TTP. Furthermore, under the usual assay conditions neither net synthesis of product nor incorporation of the γ-phosphate from appropriately labeled ATP into exogenous TPP can be demonstrated. Schrijver has attributed previous reports of the synthesis of TTP by preparations of brain and other tissues to contamination of the substrate, and has found that most commercial preparations of TPP contain about 1% of material which cochromatographs with TTP.[44] This material may be a complex of TPP and phosphate.[44] Recently Ruenwongsa and Cooper reported incorporation of the ^{32}P-labeled-γ-phosphate moiety of ATP into endogenous protein-bound TPP in rat liver preparations.[43] Discrete conclusions about the presence and significance of this enzyme will require its more complete chemical characterization.

D. Metabolic Roles of Thiamin

Thiamin plays important roles, directly or indirectly, in carbohydrate, lipid, amino acid, neurotransmitter, and membrane metabolism. Available evidence is consistent with the assumption that the functional abnormalities in Wernicke-Korsakoff syndrome, which respond promptly to treatment with thiamin, reflect abnormalities of neurotransmitter metabolism. The precise origins of the anatomic lesions are less clear.

1. Carbohydrate Metabolism

Thiamin pyrophosphate (TPP) is a cofactor for key enzymes in both of the major pathways of carbohydrate utilization — the oxidative pathway through the Krebs tricarboxylic acid cycle and the pentose phosphate pathway.

a. Oxidation of Carbohydrates

The major pathway of oxidation of carbohydrates in mammals and specifically in brain is the oxidation of pyruvate through the Krebs tricarboxylic acid cycle. The pyruvate derives from glucose and other carbohydrates by glycoysis. The acetyl-coenzyme A derived from pyruvate by the action of PDHC enters the cycle by condensation with oxaloacetate to form citrate; KGDHC catalyzes a reaction of the cycle itself (Figure 4).

Two major physiological roles are generally assigned to this pathway, in brain as in other tissues. One is the production of energy, in the form of terminal phosphate bonds of ATP, derived by the process of oxidative phosphorylation. The other is provision of biosynthetic intermediates, including intermediates of lipid, protein, and neurotransmitter metabolism. The brain has a minute-to-minute dependence on oxidative metabolism. Even mild hypoxia produces alterations in higher integrative functions; brief anoxia induces unconsciousness.[61] Except in very young animals or during prolonged starvation, the substrate for most of this oxidative activity in the brain is pyruvate derived from glucose.[61,62] Thus, even partial impairment of these pathways would be expected to impair brain function.

Thiamin deficiency severe enough to impair neurological function is associated with decreased activity of cerebral PDHC and, if severe enough, with KGDHC as well.[63-69] Sir Rudolph Peters and his co-workers reported in the 1930s that pyruvate and its metabolic derivative, lactate, accumulated in the brains of thiamin-deficient pigeons. The accumulation was particularly marked in the brain stem and occurred before frank neurological signs were detected. These and other observations led Peters to introduce the concept of a biochemical lesion which preceeds the development of an anatomical lesion.[63] Decreased activity of PDHC has been repeatedly confirmed in the brains of a variety of species of experimental animals with symptomatic thiamin deficiency.[64-69] McCandless and Schenker[66] reported that treatment with thiamin, which largely restores neurological function, partly restored PDHC activity as well in the brains of thiamin-deficient rats (from 64 to 88% of normal in the symptomatic rats to 81 to 99% of normal in the reversed deficiency group). Finally, human patients with Wernicke-Korsakoff syndrome or other manifestations of thiamin deficiency typically accumulate excessive amounts of pyruvate and lactate in physiological fluids.[1]

The physiological consequences of the impairment of pyruvate oxidation have been controversial.[65,66,70] Mild to moderate impairments of cerebral carbohydrate oxidation, like that which occurs in thiamin deficiency, have been shown in a number of laboratories to impair neurological function without significantly reducing the level of ATP or related high energy compounds. This has been shown by direct measurement of the levels of ATP and related compounds in neural tissues in thiamin deficiency and other conditions which impair pyruvate oxidation.[61,66,71,72] Furthermore, electrophysiological studies of superior cervical ganglia from rats with thiamin deficiency, or under other circumstances where carbohydrate oxidation is impaired, have demonstrated that action potentials can be maintained in response to direct stimulation even

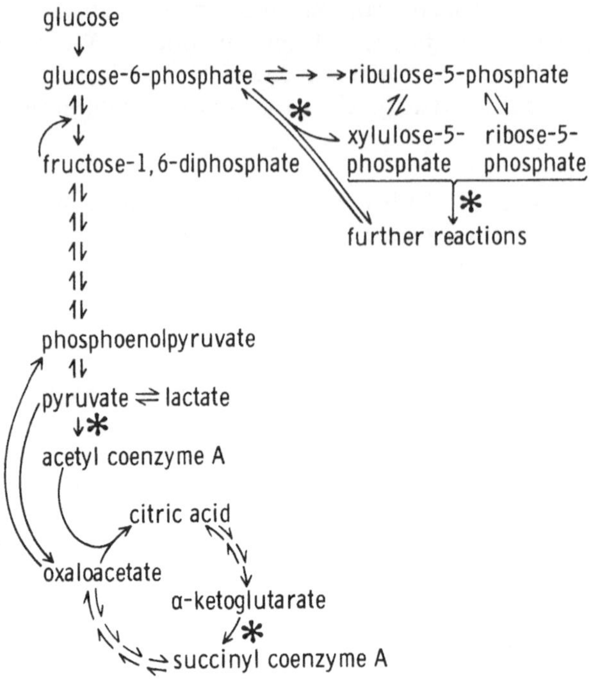

FIGURE 4. TPP-dependent reactions. The diagram shows a simplified scheme of the main pathways of mammalian carbohydrate metabolism, with the thiamin pyrophosphate dependent reactions indicated by an asterisk. The possible pathways in which transketolase participates are not shown in detail. For more extensive description of all these pathways and reactions, see White et al.[37]

strated that action potentials can be maintained in response to direct stimulation even when the biosynthesis of neurotransmitters fails.[73-76] These studies indicate that the supply of ATP is adequate to maintain ion pumps. Available evidence supports the proposal that impairment of carbohydrate oxidation in the brain impairs energy-utilizing, biosynthetic pathways before it impairs energy-producing pathways.[61] This mechanism provides a means to preserve permanent structure at the expense of transient function. The impairment of biosynthetic activity may result from a reduced provision of biosynthetic intermediates and/or alterations in potentials or other regulatory parameters.[77-79]

b. Pentose Phosphate Pathway

Transketolase participates in the pathways by which sugar phosphates are rearranged to form pentose phosphates which are intermediates in the synthesis of a number of compounds, including nucleic acids. The original formulation linked these rearrangements to the NADP-linked reactions by which glucose-6-phosphate is oxidized to ribulose-5-phosphate. A newer formulation does not require linking the rearrangement reactions to any oxidative reactions.[37]

The activity of transketolase definitely falls in thiamin deficiency, in brain, and other tissues.[65,66,80] Indeed, decreased activity of transketolase has been used to detect thiamin deficiency.[81,82] This test requires the measurement of transketolase, both with and without added exogenous TPP. In early thiamin deficiency, activity is low without added TPP but is normal with added exogenous TPP. Therefore, the stimulation by added TPP — called the TPP effect — is abnormally large. A TPP effect greater than 15% has been considered diagnostic of thiamin deficiency, and this technique has been used in extensive field studies of the adequacy of thiamin nuriture.[82-84] Recent experimental studies of rats on a thiamin-deficient diet have shown that the earliest abnormality, at 4 days, is a decrease in stores of TPP in RBC, followed at 7 days by an increased TPP-effect on transketolase from RBC, and then at 21 days an increased TPP-effect on PDHC from WBC.[85] However, very recent studies indicating genetic heterogeneity of transketolase in humans (Section IVC, below) raise questions about the interpretation of trnasketolase activity and TPP-effect in terms of thiamin nuriture.

The consequences of impaired transketolase activity are unclear. A series of studies by McCandless and co-workers have indicated that despite reductions of transketolase activity in thiamin deficiency,[66,86-88] flux through the pentose phosphate pathway is normal in brain or liver, as is synthesis of nucleic acids. Cereberal levels of reduced glutathione (a metabolite linked to the levels of NADPH and therefore presumably to the oxidation of glucose-6-phosphate to ribulose-5-phosphate) are also normal.[66]

2. Lipid Metabolism

Although there is no known TPP-dependent enzyme of lipid metabolism itself, the acetyl coenzyme A produced from pyruvate by PDHC is a key intermediate in the synthesis of fatty acids and steroids.

The reported effects of thiamin deficiency on cereberal lipid synthesis have varied. Geel and Dreyfus reported normal accumulation of lipids in the brains of thiamin-deficient rat pups born to thiamin-deficient dams.[89] However, Trostler et al. reported decreased incorporation of radioactive glucose into lipids, while acetate incorporation was normal.[90,91] Volpe and Marasa found that growth in thiamin-deficient medium reduced the ability of glial-type C_6 cells to synthesize lipids and their concentration of the lipid-synthetic enzyme fatty acid synthetase.[92] Neuronal-type cells grown under similar conditions were normal.

Loss of myelin, which is rich in lipids, is characteristic of the lesions in Wernicke-Korsakoff syndrome, and implies a defect at some level in the biosynthesis of lipids. Whether this is related to the abnormalities of lipid synthesis, which have been described in experimental systems, or is a nonspecific consequence of damage to the oligodendioglial cells which synthesize myelin, remains problematic.

3. Amino Acids and Proteins

Abnormalities in the metabolism of amino acids and in protein synthesis have been documented in thiamin deficiency.

a. Amino Acid Metabolism

Abnormalities of amino acid metabolism could be expected on two grounds in thiamin deficiency.

The 2-keto acids which are substrates for PDHC, KGDHC, and the branched chain dehydrogenases can be converted to analogous amino acids by equilibrium transamination reactions. Therefore, these amino acids — alanine, glutamate, and leucine, isoleucine, and valine — might be expected to accumulate in thiamin deficiency. However, Reynolds found no such accumulations in the urine of thiamin-deficient rats.[93] In the late stages of thiamin deficiency, there was a generalized amino aciduria, consistent with a renal tubular defect.

Transamination of tricarboxylic acid cycle intermediates is a major biosynthetic pathway for amino acids. Particularly important is the transamination of 2-ketoglutarate to glutamate and of oxaloacetate to aspartate. Indeed, much of the glucose carbon metabolized by brain is trapped in glutamate.[70,94] Particularly in brain, a portion of the glutamate is decarboxylated to form γ-aminobutyric acid (GABA), a putative inhibitory neurotransmitter.[95] Gubler and co-workers found decreased formation of glutamate, aspartate, and GABA in the brains of thiamin-deficient rats.[95] Gaitonde et al.[96-98] and others[99,100] also reported abnormalities of amino acid metabolism in thiamin-deficient brain. However, Koeppe et al.[70] did not succeed in demonstrating a significant decrease in the incorporation of glucose carbon into glutamate in thiamin-deficient brain. Their results may reflect compartmentation of glucose and amino acid metabolism in brain.[94]

b. Protein Synthesis

Defective synthesis of proteins in thiamin-deficient rat liver has been demonstrated in vivo.[101] Decreased levels of a specific protein-fatty acid synthetase have been documented in cultured cells of neural origin grown in thiamin-deficient media.[92] Yanagahira and co-workers have proposed that impaired carbohydrate catabolism leads to disaggregation of polyribosomes secondary to local changes in pH and other regulatory parameters,[102] and a similar mechanism may operate in thiamin deficiency.[101]

4. Nucleic Acids

Henderson and Schenker reported decreased incorporation of radioactive thymidine into DNA in young thiamin-deficient rat brains,[103] but Geel and Dreyfus were unable to document a difference between nucleic acid metabolism between the brains of thiamin-deficient rats and pair-fed controls.[104]

5. Neurotransmitters

Metabolism of four types of neurotransmitters have been reported to be abnormal in thiamin deficiency: acetylcholine, catecholamines, serotonin, and amino acids. Many other neurotransmitters have not been studied in this condition.

a. Acetylcholine

Since the acetyl group of acetylcholine is derived from a specific pool of pyruvate metabolized by PDHC,[79] the possibility of cholinergic deficiency in thiamin deficiency has been examined by many groups.

Earlier studies of acetylcholine levels in whole brain were contradictory: about half the groups found deficiencies and the other half did not.[93,105-109] It has subsequently become clear that accurate measurement of acetylcholine requires very rapid fixation of brain (preferably by microwave irradiation), and that measurement of levels of this compound alone may be misleading, since much of the acetylcholine is sequestered in pools which are functionally relatively inactive.[110]

Measurements of the turnover of acetylcholine in thiamin deficiency have consistently shown deficits. Cheney et al. found decreased synthesis of acetylcholine from labeled pyruvate.[107] They also reported that physostigmine, a drug which inhibits the break-down of acetylcholine, prolonged life in rats made thiamin-deficient by treatment with the thiamin antagonist pyrithiamin. Vorhees et al. measured turnover by following the levels of acetylcholine after inhibiting its synthesis.[108] They found decreases in acetylcholine metabolism in cortex, midbrain, diencephalon, and brainstem; the largest decrement (− 41%) was in the midbrain. Particularly clear results were presented by Perri et al.[73,74] who studied superior cervical ganglia from rats with symptomatic thiamin deficiency. They found that ganglia lost the ability to carry out cholinergic transsynaptic transmission in response to rapid (20/sec) dromic stimulation, while the response to slow stimulation (2/sec) and to direct axonal stimulation remained intact. They interpreted these observations as showing a decreased ability of the ganglia to synthesize acetylcholine. Subsequent studies confirmed this interpretation directly.[74] The levels of acetylcholine, the rate of its synthesis, and the rate of its release were all lower in thiamin-deficient than in normal ganglia, when the ganglia were subjected to relatively rapid stimulation (Table 1). While acetylcholine metabolism in thiamin-deficient brain still needs to be studied by modern isotopic techniques,[79] there seems little doubt that thiamin deficiency does impair cholinergic systems.

An effect of thiamin deficiency on acetylcholine synthesis agrees with a series of studies by Gibson et al. demonstrating that acetylcholine synthesis is exquisitely sensitive to impairment of pyruvate oxidation.[61,77-79,110-113] Even mild impairment of pyruvate oxidation leads to proportional impairment of acetylcholine synthesis, though less than 1% of the pyruvate oxidized is incorporated into acetylcholine. The mechanism of the linkage is not understood in detail, but appears to involve compartmentation of glucose and pyruvate metabolism.[79]

It is also notable that impairment of cholinergic function leads to impairments of memory in humans and of learning in animals.[114] There are, however, differences in detail between the psychological deficits induced by scopolamine and other anticholinergic drugs in man and those in Wernicke-Korsakoff syndrome.

b. Catecholamines

Iwata et al. have demonstrated decreased synthesis of catecholamines in severely thiamin-deficient rat brain,[115] manifested as decreased accumulation after treatment with pheniprazine to block catecholamine catabolism. Treatment with thiamin normalized catecholamine synthesis within 2 hr, although some neurological signs persisted. Blood levels of catecholamines in thiamin-deficient animals were about half of normal. These workers suggested that the slow heart beat and some of the endocrine abnormalities which have been reported in thiamin deficiency may be secondary to impaired catecholamine metabolism.

c. Serotonin

Plaitakis et al. have recently demonstrated decreased uptake of 5-hydroxytryptamine by synaptosomes from thiamin-deficient rat brains.[116-118] Furthermore, the amounts of 5-hydroxytryptamine appeared decreased by autoradiography.[118] These findings were particularly striking because the decreases were in the areas where the lesions of Wernicke-Korsakoff syndrome are most prominent.

d. Other Neurotransmitters

As noted above (Section IID3a), there is evidence for abnormalities of putative amino acid neurotransmitters such as glutamate, aspartate, and GABA in thiamin deficiency. The metabolism of peptides and other more recently discovered neurotransmitters has not yet been studied in this condition.

6. Membrane Function

The observation that electrical stimulation of isolated nerves leads to release into the media of a thiamin-like compound has been extensively confirmed.[119-123] Extensive studies have confirmed and extended this fundamental finding and led to the suggestion that this compound has a specific role in Na^+-conductance.[75,124-127] The proportion of TTP relative to other forms of thiamin increases as neural membranes are purified, suggesting that the triphosphate derivative is the membrane active form of the vitamin.[121]

However, the significance of these findings for the pathophysiology of thiamin deficiency remain unclear. The chemistry of this system is complex and still poorly

Table 1
ACETYLCHOLINE IN THIAMIN-DEFICIENT GANGLIA[a]

	Unstimulated	Stimulated
Controls		
Freely-fed	184 ± 9	149 ± 10
Pair-fed	202 ± 10	136 ± 10
Thiamin deficient	188 ± 6	63 ± 3[b]

[a] Values are pmol acetylcholine/ganglion ± SEM.
[b] Indicates P <0.001 vs. pair-fed controls. Stimulation was at 10/sec. Data are from Sacchi et al.,[74] where the experimental conditions are described in detail.

stood (see Section IIIC5).[43,44] The studies of Perri et al. and others on superior cervical ganglia indicate that membrane function including Na^+-channels are intact enough for the repetitive generation of action potentials in thiamin deficiency,[73,74] at least in the earlier stages. Finally, Pincus and Wells reported that the levels of thiamin triphosphate (TTP) remained normal in the brains of rats even in preterminal stages of thiamin deficiency.[128] Whether or not TTP is the membrane-active form of thiamin, this observation argues against a significant role of this compound in mediating the physiological effects of thiamin deficiency.

E. Experimental Thiamin Deficiency

Implicit in the studies described above is the existence of animal models of thiamin deficiency. These have been standardized in several species, using either thiamin-deficient diets or thiamin antagonists, and intensively studied pathologically.

1. Species

Standard models of thiamin deficiency have been reported for rats,[64,67] mice,[69,127] cats,[129] monkeys,[130] pigeons,[63,131] and cultured glial-type and neuronal type cell lines,[72,92] and described in ruminants as well.[132,133]

2. Induction of Thiamin Deficiency

a. *Diet*

Diets adequate in all other nutrients, but containing virtually no thiamin are commercially available. Rats fed such a diet begin to lose weight after about 2 weeks. In another 10 days to 2 weeks, they become clumsy and drag their hind legs (i.e., develop ataxia), then lose their righting reflex, and then die. Often, depending on the species and exact conditions, seizure-like shaking activity develops. The neurological signs progress through the spectrum in a few days.

The use of such diets provides the purest model of thiamin deficiency. However, the animals lose their appetites and become generally debilitated. Even using pair-fed controls, it is hard to decide what are the effects of the specific deficiency and what those of general debilitation, particularly in the late stages. One approach to this problem is using reversed controls — animals made deficient but then treated with thiamin (but not food) shortly before experimental study.

Increasing the proportion of carbohydrates relative to those of fat in the deficient diet, hastens the appearance of the symptoms,[134] emphasizing the importance of carbohydrate metabolism in thiamin deficiency. The addition of raw fish can also further the development of the syndrome, since fish contains a thiaminase which destroys thiamin.[135]

b. Antagonists

Pyrithiamin is a direct antagonist of TPP and also appears to have direct effects on membranes.[95,136,137] It crosses the blood brain barrier. Oxythiamin inhibits the pyrophosphorylation of thiamin to TPP.[95,136] It does not cross the blood brain barrier, so effects of oxythiamin poisoning on the brain must be mediated indirectly.

Other thiamin antagonists are known but less widely used.

These antagonists have the advantage that they produce neurological symptoms in animals who are not yet severely debilitated. They have the disadvantage that they may have effects beyond those on thiamin metabolism.[120,136]

3. Pathologic Lesions

The lesions induced by dietary thiamin deficiency or by thiamin antagonists have been studied in detail in several species, by both light and electron microscopy.

In general, the gross pattern of the lesions is similar to, although not identical with that of human Wernicke-Korsakoff syndrome.[1,15,137,138] Indeed, recognition of this similarity contributed to the discovery of the role of thiamin in the human disorder.[1,137,138]

By light microscopy, the pattern of the lesions — with prominent loss of myelin elements — is very similar to that of Wernicke-Korsakoff syndrome in humans.[137,138]

Electron-microscopic studies have generally confirmed the results of light-microscopy.[137-139] However, Tellez and Terry[139] found the most prominent changes in very early lesions were in presynaptic vesicles and mitochondria. They suggested that the earliest changes in thiamin deficiency were in neurones and that damage to myelin elements and glia reflected chronic or repeated insult. In this regard, it is interesting that Welch et al. recently suggested a similar mechanism of selective white matter damage in impairments of cereberal carbohydrate metabolism secondary to graded ischemia insults.[140] They concluded that

> "Selective damage to white matter may be manifest only when the ... insult is mild enough to spare ... gray matter, yet of sufficient duration ... in white matter to trigger the sequence of irreversible pathological alterations."

IV. PROBLEMS

A number of major problems remain in understanding Wernicke-Korsakoff syndrome. All patients who develop the clinical or pathological features of Wernicke-Korsakoff syndrome are not deficient in thiamin. Most patients with thiamin deficiency do not develop Wernicke-Korsakoff syndrome. Genetic variations in TPP-dependent enzymes now appear to modify the effects of dietary thiamin deficiency. All of these observations must be explained in a more complete description of the pathophysiology.

A. Wernicke-Korsakoff-Like Syndromes

Clinical or neuropathological abnormalities indistinguishable from those of Wernicke-Korsakoff syndrome have been reported in patients who did not respond to thiamin. These include both acquired disorders and the hereditary disorder first described by Leigh.

1. Other Acquired Disorders

Cole et al. described several alcoholics who, presented with typical signs and symptoms of Wernicke-Korsakoff syndrome, but did not respond to treatment with high doses of thiamin.[141] These patients appeared not to be deficient in Mg, which is important since Zieve et al.[142,143] have reported experimental evidence that Mg deficiency (which is common in alcoholics) impairs the biochemical utilization of thiamin. These patients ultimately did respond to a mixture of vitamins. None died and the neuropathology is not known. The nature of the process in these patients is unclear.

2. Leigh's Disease

In 1951, Leigh[144] described a child with progressive neurological disease in which opthalmoplegia, ataxia, and progressive mental deterioration were prominent clinical abnormalities and the neuropathological changes were very similar to those in Wernicke-Korsakoff syndrome. There was no evidence at all for dietary thiamin deficiency.

Subsequent studies have broadened the spectrum of clinical presentation of this disorder.[145-148] Even in autopsy-proven cases, the clinical picture is very variable. Respiratory abnormalities including perhaps failure of automatic respiration (Ondine's course) may be common.[149] Juvenile and adult forms, which may present as spinocerebellar ataxias, have been recognized.[147,148] Neuropathologically, the resemblance to Wernicke-Korsakoff syndrome has become, if anything, more marked. Previous distinctions between the two syndromes — including specifically such criteria as the presence or absence of involvement of the mammillary body — have not been borne out by further experience.

Although Leigh's disease appears to be inherited in an autosomal recessive pattern, extensive studies have not defined a discrete molecular defect. The patients do not have dietary thiamin deficiency. Early reports describing encouraging responses to large doses of thiamin or thiamin derivatives have not been borne out by further experience.[145,149-151] Pincus et al.[150,151] have reported deficiencies of TTP in brain from these patients and the presence of an inhibitor of the kinase which synthesizes TTP from TPP in their tissues and urine, and Murphy[152] described the inhibitor in cultured cells from the patients. Pincus et al.[150,151] have even derived a clinical test for this disorder based on these findings. However, both false positive and false negative tests occur, and the apparent specificity of the test has decreased as experience with it has increased. Also, as discussed above, the published data on this very complicated system still defy discrete interpretation (Section IIID6). Most if not all of these patients accumulate excess amounts of lactate and pyruvate in their blood, urine, or spinal fluid. Individual patients with this syndrome have been described as having defects in pyruvate carboxylase (EC6.4.1.1),[153,154] cytochrome oxidase (EC1.9.3.1),[155] and PDHC,[156] and ragged-red neuromuscular disease.[157] However, most of the patients have normal activities of those enzymes and of KGDHC and transketolase and have not been noted to have abnormal muscle on biopsy.

Whatever the molecular etiology of Leigh's disease, elucidating it would probably be a major contribution to understanding the origin of the lesions of Wernicke-Korsakoff syndrome.

B. Varying Presentations of Thiamin Deficiency

Most humans with dietary thiamin deficiency do not develop Wernicke-Korsakoff syndrome. Mild thiamin deficiency induced in clinically normal individuals leads to impairments in higher integrative functions, including specifically memory. These changes are similar qualitatively to those reported in mild hypoxia. More severe thiamin deficiency usually leads either to cardiac disease with edema formation (wet beriberi) or to neuropathy and other neurological manifestations (dry beriberi). The occurrence of frank clinical manifestation of both together is rare — in only 2 of 269 patients in the series of Victor et al.[1] Furthermore, many more patients with dry beriberi develop severe neuropathy than develop the Wernicke-Korsakoff syndrome. This is unlikely to be simply a matter of severity, since the neuropathy in patients with Wernicke-Korsakoff syndrome may be milder than in patients without frank CNS manifestations. It might relate to details of diet — e.g., the syndrome may be more common in steady drinkers who do not have intervals of sobriety and better diet and whose intestinal absorption of thiamin is not continuously inhibited with ethanol.[158-161] However, such suggestions have not been documented in controlled studies. (Indeed, it would be formidably difficult to do so in this population.)

C. Genetic Aspects

Both epidemiological and biochemical information indicate the potential importance of genetic factors in Wernicke-Korsakoff syndrome.

1. Epidemiologic Observations

There is a recurrent although poorly documented observation in the literature that Asians are more likely to develop cardiac (wet) beriberi and Europeans neurologic (dry) beriberi even on the same diet.[5,162,163] This observation was made with particular frequency during World War II in P.O.W. camps, where strict epidemiological criteria were hard to achieve. It suggests that ethnic and therefore genetic differences play a role in determining the form which clinical thiamin deficiency takes.

2. Biochemical Genetic Studies

There is now evidence for genetic variations of each of the known TPP-dependent enzymes of humans and also for the synthesis of the triphosphate, TTP.

a. Transketolase

Two lines of evidence suggest genetic variation of this enzyme in patients with Wernicke-Korsakoff syndrome.

There is one recent report of an abnormality in the binding of TPP to transketolase from four patients with well-documented Wernicke-Korsakoff syndrome.[33] The apparent K_m for TPP of transketolase for their cells was one to two orders of magnitude higher than in six controls (Figure 5). These studies were done with fibroblasts cultured from skin biopsies from the patients and controls. Persistence of the abnormality through serial cultures indicate that they are genetically determined, at least in a global sense.[2,33,164] These studies must be considered preliminary on several grounds. They have not yet been independently replicated. The molecular basis of the abnormality has not been determined. While available data are consistent with a structural mutation of the enzyme protein, they do not rule out other possibilities such as a competitive inhibitor of TPP binding. Finally, the specificity of the abnormality for Wernicke-Korsakoff syndrome rather than for alcoholism or some other abnormality has not been examined extensively.

Less detailed studies of transketolase in RBC of patients with Wernicke-Korsakoff syndrome are also consistent with at least some of them having a genetic abnormality of the apotransketolase protein.[2,165] There is a well-established observation that in some of these patients, total transketolase activity (i.e., after addition of TPP) does not return to normal even after months of therapy with parenteral thiamin.[81] Although this lack of biochemical response has been attributed to concomitant liver disease or bone marrow damage by ethanol, extensive studies by Wood et al.[166] in Australia failed

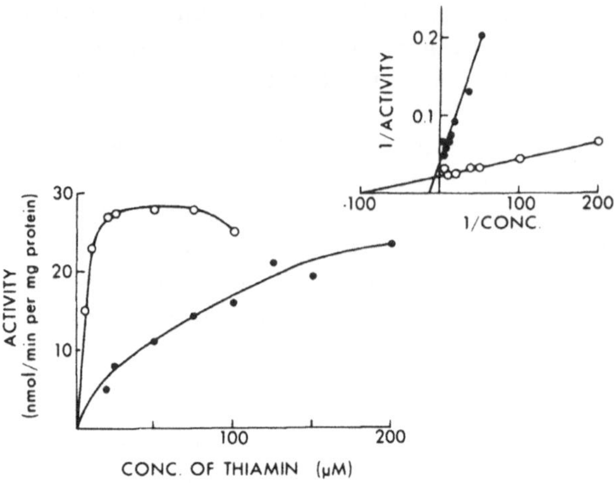

FIGURE 5. K_m for TPP for transketolase in Wernicke-Korsakoff syndrome. Transketolase activity was measured at various concentrations of TPP in fibroblast extracts from six controls and four patients with well-treated Wernicke-Korsakoff syndrome. Results for a typical patient and a typical control (O) are shown. The apparent K_m for this patient (•) was 160 μM, compared to 10 μM for the control. (From Blass, J. P. and Gibson, G. E., *N. Engl. J. Med.*, 297, 1367, 1977. With permission.)

to demonstrate an association between liver disease and persistantly low total erythrocyte transketolase activity. Of 215 alcoholics they studied, 47 (22%) had a low transketolase activity which failed to return to normal after addition of TPP. Only 10 of these 47 (21%) had liver disease. There are also anecdotal reports of patients who are not alcoholics but have abnormally low total transketolase or an elevated TPP effect.

Although further studies are clearly needed, these observations strongly suggest that there are patients in whom a hereditary, constitutional abnormality in transketolase predisposes to the development of symptomatic thiamin deficiency and specifically to Wernicke-Korsakoff syndrome. If so, then deficiency of transketolase presumably has an important role in the pathophysiology of at least this form of thiamin deficiency.

b. Pyruvate Dehydrogenase

In the last decade, patients have been identified in whom neurological disease is associated with hereditary deficiencies of PDHC activity.[167-168] Available data are consistent with the assumption that in at least some of these patients, there are mutations in PDHC peptides rather than alterations in the complex regulation of PDHC secondary to some other abnormality. However, that point has not been established definitively.[168] Impaired mentation, ataxia, and abnormal eye movements are prominent in these patients, but none of them have had Wernicke-Korsakoff syndrome. Several patients with Leigh's disease may have had PDHC deficiencies.[156,168] Several patients with abnormal elevations of pyruvate in body fluids had the clinical syndrome of a spinocerebellar degeneration and the autopsy findings of Leigh's disease.[147,169] Wick et al.[170] have described a patient with demonstrated deficiency of PDHC who responded clinically and biochemically to very large doses of thiamin. It is not clear whether this thiamin-dependence reflected deficient binding of TPP to a mutant enzyme or increased activation of PDHC due to thiamin inhibition of the inactivating PDH-kinase.

There is no reported patient with the clinical findings of Wernicke-Korsakoff syndrome in whom a constitutional abnormality of PDHC has been found. Thus, although genetic variations of PDHC in humans are reasonably well established, there is no evidence that they contribute to the development of this particular syndrome.

c. *Ketoglutarate Dehydrogenase*

Genetic deficiencies of ketoglutarate dehydrogenase activity have been reported, but only in conjunction with deficiencies of PDHC.[171-173] They appear to be secondary to deficiency of a FAD-NAD dependent enzyme, lipoamide dehydrogenase (EC1.6.4.3), which appears to be a protein common to the PDHC, KGDHC, and BCDHC complexes.[60,168,173] Deficiency of the thiamin-dependent enzyme of KGDHC has not been reported, at least as yet.

d. *Branched Chain Dehydrogenase*

Genetic variations of BDHC in humans are well documented by the existence of patients with Maple-Syrup-Urine Disease (MSUD; see Section IIIC4).[174-176] Both a severe infantile form and a milder intermittant form have been described. Although impaired mentation and ataxia commonly occur in symptomatic MSUD, the clinical syndrome in these patients is very different from Wernicke-Korsakoff syndrome. It does have some similarities to PDHC deficiency. The branched chain keto acids which accumulate in MSUD probably impair pyruvate oxidation perhaps by inhibiting KGDHC.[176]

Recently two patients with MSUD have been described in whom there was a clinical and biochemical response to large doses of thiamin.[175] These patients appear to have a mutation in BCDHC which impairs the binding of TPP to the enzyme.[60]

e. *General Genetic Considerations*

It is now widely accepted that generic variations in proteins — genetic heterogeneity — is the rule rather than the exception for proteins in human populations. This observation confirms a prediction of Garrod.[177] He pointed out that such variations could not only predispose to the development of specific diseases, but also modify, in a general way, the responses of different individuals to specific environmental stresses. Thus, in analyzing a disease process, one must consider not only what happens, but to whom it happens. This generalization applies as well on a molecular level.

These considerations are directly relevant to the role of vitamins in health and disease. Rosenberg[178] has pointed out that there are at least two kinds of disorders related to vitamins:

> "Acquired deficiency ... states affect, in varying degree, each of the reactions catalyzed by the vitamin and respond to physiological amounts of the missing vitamin. In contrast, inherited vitamin-dependency states may be defined as genetic disturbances that lead to specific biochemical abnormalities affecting only one reaction catalyzed by a vitamin and that respond only to pharmacologic amounts of the vitamin."

Studies of transketolase have led to the suggestion of a third type of vitamin-related disorder. *Vitamin insufficiency* disorders result when patients with an abnormality in a specific vitamin-dependent enzyme ingest a diet which is marginal in the specific vitamin.[33] There states fall midway in the spectrum between vitamin deficiencies and true vitamin dependency.

It is important to stress that the existence of vitamin dependency and insufficiency syndromes in no way justifies the nonspecific, faddist use of vast amounts of vitamins for nonspecific indications. Even water-soluble vitamins are toxic in large enough doses. Thiamin overdose leads to anxiety, headache, convulsions, weakness, trembling, and neuromuscular collapse.[179] What these considerations do document is the need for more research on genetic aspects of vitamin requirements.

D. Metabolic Inhibitors

Since the earliest days of the study of thiamin deficiency the suggestion has been made that a toxic metabolite accumulates which inhibits the relevant enzymes and leads to the biochemical and physiological abnormalities.[1,180,181] Not only is no such compound known, but there is no direct evidence for its existence. Were such a toxic compound to be found, it would confirm the original pathophysiologic speculations of Wernicke and Korsakoff (see Section IID1).

V. FORMULATION OF PATHOPHYSIOLOGY

The data described above can be marshaled to present the following more-or-less consistent picture of pathophysiology of Wernicke-Korsakoff syndrome. The fundamental consequence of the deficiency of thiamin is impairment of the main pathways of carbohydrate catabolism. The brain and other tissues adapt to this impairment by regulatory changes which maintain the levels of ATP and other high energy metabolites, but impair biosynthetic activities. Impairment of the synthesis of acetylcholine and other neurotransmitters leads to reversible functional changes. Chronic, low-grade metabolic impairment ultimately leads to anatomic damage which, if extensive enough, leads to the permanent deficits of Korsakoff Psychosis. Breakdown of white matter preceeds dissolution of other structures because white matter tolerates chronic low-grade impairment of oxidative metabolism less well than do other neural structures.

There are several problems with this formulation. It is unproven. It does not explain the anatomic distribution of the lesions. It does not explain why only a minority of thiamin-deficient patients develop Wernicke-Korsakoff syndrome. It does not explain why an inherited abnormality of transketolase should occur in any of these patients, and indeed does not take into account the effect of genetic variation at all.[182] It does not clarify the existence of patients with Wenicke-Korsakoff syndrome unresponsive to thiamin nor that of Leigh's disease. These problems can be solved only by further research.

VI. FUTURE DIRECTIONS

There are certain classical problems in human biology whose periodic reexamination at increasingly deep levels have repetitively led to important conceptual and practical advances. The study of glycogen metabolism and its relation to maintenance of the milieu interne and to glucagon, insulin and cyclic nucleotides is one example. The study of thiamin, with its wide implications for nutrition and metabolism, is another.

It is probably fair to say that the discovery of vitamins and other minor nutrients ranks with techniques to control infectious diseases in its implications for public health and demography. Although criteria for preventing frank vitamin deficiency diseases have been developed, recent studies have demonstrated that relatively subtle variations in diet may affect the function of the nervous system[183] and other tissues and that specific genetic variations may affect dietary requirements for specific nutrients.[33,178,182] Intensive study of thiamin deficiencies in the light of these newer approaches may again prove to have broad and unexpected implications.

REFERENCES

1. **Victor, M., Adams, R. D., and Collins, G. H.** *The Wernicke-Korsakoff Syndrome,* F. A. Davis, Philadelphia, 1971.
2. **Blass, J. P. and Gibson, G. E.,** Genetic factors in Wernicke-Korsakoff syndrome, *Alcoholism,* 3, 126, 1979.
3. **Slater, E. and Roth, M.,** *Clinical Psychiatry,* 3rd ed., Balliere, Tindall & Casell, London, 1969, chap. 7.
4. **Wernicke, C.,** *Textbook of Brain Diseases for Physicians and Students,* Fischer, Kassel, Germany, 1881, 229.
5. **DeWardener, H. E. and Lennox, B.,** Cereberal beriberi (Wernicke's encephalopathy), *Lancet,* 1, 11, 1947.
6. **Shimojyo, S., Scheinberg, P., and Reinmuth, O. M.,** Cerebral blood flow and metabolism in the Wernicke-Korsakoff syndrome, *J. Clin. Invest.,* 46, 849, 1967.
7. **Korsakoff, S. S.,** Disturbance of psychic functions in alcoholic paralysis and its relation to the disturbance of the psychic sphere in multiple neuritis of non-alcoholic origin, *Vestnik Psichiatrii,* 4, 1887.
8. **Korsakoff, S. S.,** On a special form of psychic disturbance combined with multiple neuritis, *Arch. Psychiatr.,* 21, 1890.
9. **Korsakoff, S. S.,** On memory-disturbances (pseudoreminiscences) in polyneuritic psychosis, *Allg. Zeitschr. Psychiatr.,* 47, 1891.
10. **Soukhanoff, S. and Boutenko, J.,** A study of Korsakoff's disease, *J. Ment. Pathol.,* 4, 1, 1903.
11. **Brain, W. R. and Walton, J. N.** *Diseases of the Nervous System,* Oxford University Press, New York, 1969, chap. 16.
12. **Berenberg, R. A., Pellock, J. M., DiMauro, S., Schotland, D. L., Bonilla, E., Eastwood, A., Hays, A., Vicale, C. T., Behrens, M., Chuturian, A., and Rowland, L. P.,** Lumping or splitting? Opthalmoplegia-plus or Kearns-Shy syndrome? *Ann. Neurol.,* 1, 37, 1977.
13. **Engle, W. K. and Meltzer, H.,** Histochemical abnormalities of skeletal muscle in patients with acute psychoses, *Science,* 168, 273, 1970.
14. **Talland, G. A.,** *Deranged Memory,* Academic Press, New York, 1965.
15. **Adams, R. D. and Victor, M.,** *Principles of Neurology,* McGraw-Hill-Blakiston, New York, 1977, chap. 36.
16. **Murawieff, W.,** Two cases of polioencephalitis acuta hemorrhagica superior (Wernicke), *Neurol. Centralb.,* 16, 56, 1897.
17. **Pershing, H. T.,** Alcoholic multiple neuritis with characteristic mental derangement, *Int. Med. Mag.,* 1, 803, 1892.
18. **Henderson, P. K.,** Korsakoff's psychosis occurring during pregnancy, *Johns Hopkins Hosp. Bull.,* 25, 261, 1914.
19. **Sheehan, H. L.,** Discussion of the neurologic complications of pregnancy, *Proc. R. Soc. Med.,* 32, 584, 1939.
20. **Környey, S.,** Ascending paralysis and Korsakoff Psychosis in lymphogranulomatosis, *Deutsch. Ztschr. Nervenh.,* 125, 129, 1932.
21. **Neuberger, K.,** Nonalcoholic Wernicke's disease, particularly its occurence with carcinoma, *Virchow's Arch. Pathol. Anat. Physiol.,* 298, 68, 1937.
22. **Ecker, A. D. and Woltman, H. W.,** Is nutritional deficiency the basis of Wernicke's disease? Report of a case, *JAMA,* 112, 1794, 1939.
23. **Lopez, R. I. and Collins, G. K.,** Wernicke's encephalopathy: a complication of chronic hemodialysis, *Arch. Neurol.,* 18, 248, 1968.
24. **Prickett, C. O.,** The effect of a dificiency of vitamin B_1 upon the central and peripheral nervous systems of the rat, *Am. J. Physiol.,* 107, 459, 1934.
25. **Alexander, L., Pijoan, M., and Myerson, A.,** Beriberi and scurvy, *Trans. Am. Neurol. Assoc.,* 64, 135, 1938.
26. **Alexander, L.,** Wernicke's disease. Identity of lesions produced experimentally by B_1 avitaminosis in pigeons with hemorrhagic polioencephalitis occurring in chronic alcoholism in man, *Am. J. Pathol.,* 16, 61, 1940.
27. **Bowman, K. M., Goodhart, R., and Jolliffe, N.,** Observations on the role of vitamin B_1 in the etiology and treatment of Korsakoff psychosis, *J. Neurol. Ment. Dis.,* 90, 569, 1939.
28. **Jolliffe, N., Wortis, H., and Fein, H. D.,** The Wernicke syndrome *Arch. Neurol. Psychiatr.,* 46, 569, 1941.
29. **Victor, M. and Adams, R. D.,** On the etiology of the alcoholic neurologic diseases with special reference to the role of nutrition, *Am. J. Clin. Nutr.,* 9, 379, 1961.

30. **Marcus, H.**, Korsakoff's disease (Etiologic and pathological anatomical studies), *Rev. Neurol.*, 2, 500, 1931.
31. **Rosenbaum, M. and Merritt, H. H.**, Korsakoff's syndrome, *Arch. Neurol. Psychiatr.*, 41, 978, 1939.
32. **Leigh, D.**, personal communication.
33. **Blass, J. P. and Gibson, G. E.**, Abnormality of a thiamine-requiring enzyme in patients with Wernicke-Korsakoff syndrome, *N. Engl. J. Med.*, 297, 1367, 1977.
34. **Williams, R. R.**, *Toward the Conquest of Beriberi*, Harvard University Press, Cambridge, 1961.
35. **Eijkman, C.**, A beriberi-like illness of chickens, *Virchows Arch. Pathol. Anat. Physiol.*, 148, 523, 1897.
36. **Williams, R. R.**, Structure of vitamin B_1, *J. Am. Chem. Soc.*, 58, 1063, 1936.
37. **White, A., Handler, P., Smith, E. L., Hill, R. L., and Lehman, I. R.**, *Principles of Biochemistry*, McGraw-Hill, New York, 1978, chap. 50.
38. **Krampitz, L. O.**, *Thiamine Diphosphate and Its Catalytic Functions*, Marcel Dekker, New York, 1970.
39. **Wakabayashi, Y.**, Purification of the thiamine pyrophosphokinase from pig brain, *Vitamins*, 52, 223, 1978.
40. **Inoue, A., Shim, S., and Iwata, H.**, Activation of thiamine diphosphatase by ATP in rat brain, *J. Neurochem.*, 17, 1373, 1970.
41. **Cooper, J. R. and Kini, M. M.**, Partial purification and characterization of thiamine pyrophosphatase from rabbit brain, *J. Neurochem.*, 19, 1809, 1972.
42. **Barchi, R. L. and Braun, P. E.**, Thiamine in neural membranes: enzymic hydrolysis of thiamine diphosphate, *J. Neurochem.*, 19, 1039, 1972.
43. **Ruenwongsa, O. and Cooper, J. R.**, The role of bound thiamine pyrophosphate in the synthesis of thiamine pyrophosphate in rat liver, *Biochem. Biophys. Acta*, 482, 64, 1977.
44. **Schrijver, J.**, Biotin Deficiency in the Rat as a Model for Reduced Pyruvate Carboxylase Activity — The Effect of Thiamine, Ph.D. thesis, State University of Groningen, Groningen, The Netherlands. 1978, chap. 4.
45. **Edwin, E. E. and Jackmon, R.**, Rapid radioactive method for determination of thiaminase activity and its use in the diagnosis of cerebrocortical necrosis in sheep and cattle, *J. Sci. Food Agric.*, 25, 357, 1974.
46. **Reed, L. J. and Cox, L. J.**, Macromolecular organization of enzyme systems, *Ann. Rev. Biochem.*, 35, 57, 1966.
47. **Blass, J. P. and Lewis, C. A.**, Kinetic properties of the partially purified pyruvate dehydrogenase complex of ox brain, *Biochem. J.*, 131, 31, 1973.
48. **Ngo, T. T. and Barbeau, A.**, Kinetic studies on the flavoprotein component (E_3) of the cat brain pyruvate dehydrogenase multienzyme complex, *Int. J. Biochem.*, 9, 681, 1978.
49. **Butler, J. R., Pettit, F. H., Davis, P. F., and Reed, L. J.**, Binding of thiamine thiazolone pyrophosphate to mammalian pyruvate dehydrogenase and its effects on kinase and phosphatase activities, *Biochem. Biophys. Res. Commun.*, 74, 1667, 1977.
50. **Hommes, F. A., Berger, R., and Luit-de-Haan, G.**, The effect of thiamine treatment on the activity of pyruvate dehydrogenase in relation to the treatment of Leigh's encephalomyelopathy., *Pediatr. Res.*, 7, 616, 1973.
51. **Sugden, P. H., Huston, N. J., Kerbey, A. L., and Randle, P. J.**, Phosphorylation of additional sites on pyruvate dehydrogenase inhibits its reactivation by pyruvate dehydrogenase phosphatase, *Biochem. J.*, 169, 433, 1978.
52. **Kovachich, G. B. and Haugaard, N.**, Pyruvate dehydrogenase activation in rat brain cortical slices by elevated concentrations of external potassium ions, *J. Neurochem*, 28, 923, 1977.
53. **Lynen, A., Sedlaczek, E., and Wieland, H. O.**, Partial purification and characterization of a pyruvate dehydrogenase complex inactivating enzyme from rat liver, *Biochem. J.*, 169, 321, 1978.
54. **Cate, R. L. and Roche, T. E.**, A unifying mechanism for stimulation of mammalian pyruvate dehydrogenase kinase by reduced nicotinamide adenine dinucleotide, dihydrolipoamide, acetyl coenzyme A, or pyruvate., *J. Biol. Chem.*, 253, 496, 1978.
55. **Olson, M. S., Dennis, S. C., DeBuysere, M. S., and Padma, A.**, The regulation of pyruvate dehydrogenase in the isolated, perfused rat heart, *J. Biol. Chem.*, 253, 7369, 1968.
56. **Gibson, G. E.**, personal communication.
57. **Horecker, B. L. and Snuyrniotis, P. Z.**, The coenzyme function of thiamine pyrophosphate in pentose phosphate metabolism, *J. Am. Chem. Soc.*, 75, 1009, 1953.
58. **Racker, E., de la Haba, G., and Leder, I. G.**, Thiamin pyrophosphate, a coenzyme of transketolase, *J. Am. Chem. Soc.*, 75, 1010, 1953.
59. **Srere, P., Cooper, J. R., Tabachnik, M., and Racker, R.**, The oxidative pentose phosphate cycle. I. Preparation of substrates and enzymes, *Arch. Biochem. Biophys.*, 74, 295, 1958.

60. **Pettit, F. H., Yeaman, S. J., and Reed, L. J.**, Purification and characterization of branched-chain α-keto acid dehydroganase complex of bovine kidney, *Proc. Natl. Acad. Sci. U.S.A.*, 75, 4881, 1978.
61. **Blass, J. P. and Gibson, G. E.**, Effects of mild, graded hypoxia, *Adv. Neurol.*, 26, 229, 1979.
62. **Sokoloff, L.**, Metabolism of ketone bodies by the brain, *Ann. Rev. Med.*, 24, 271, 1973.
63. **Peters, R. A.**, Biochemical lesion and its historical development, *Br. Med. Bull.*, 25, 223, 1969.
64. **Gubler, C. J.**, Studies on the physiological functions of thiamine. I. The effects of thiamine deficiency and thiamine antagonists on the oxidation of α-keto acids by rat tissues, *J. Biol. Chem.*, 236, 3112, 1961.
65. **Dreyfus, P. M. and Hauser, G.**, The effect of thiamine deficiency on the pyruvate decarboxylase system of the central nervous system, *Biochem. Biophys. Acta*, 104, 78, 1965.
66. **McCandless, D. W. and Schenker, S.**, Encephalopathy of thiamine deficiency: studies of intra-cerebral mechanisms, *J. Clin. Invest.*, 47, 2268, 1968.
67. **Murdock, D. S. and Gubler, C. J.**, Effects of thiamine deficiency and treatment with the antagonists, oxythiamine and pyrithiamine, on the levels and distribution of thiamine derivates in rat brain, *J. Nutr. Sci. Vitaminol*, 19, 237, 1973.
68. **Januschke, D., Reinauer, H., and Hollmann, S.**, Pyruvate and α-ketoglutarate respiration in mitochondria of thiamine-deficient rats, *Z. Ernaehrungswiss.*, 12, 261, 1973.
69. **Seltzer, J. L. and McDougal, D. B.**, Temporal changes of regional carboxylase levels in thiamine-depleted mouse brain, *Am. J. Physiol.*, 227, 714, 1974.
70. **Koeppe, R. E., O'Neal, R. M., and Hahn, C. H.**, Pyruvate decarboxylation in thiamine-deficient brain, *J. Neurochem.*, 11, 695, 1964.
71. **McCandless, D. W. and Cassidy, C. E.**, Adenine nucleotides in thiamine-deficient rat brain, *Res. Commun. Chem. Pathol. Pharmacol.*, 14, 579, 1976.
72. **Schwarz, J. P. and McCandless, D. W.**, Glycolytic metabolism in cultured cells of the nervous system. IV. The effects of thiamine deficiency on thiamine levels, metabolites, and thiamine-dependent enzymes of the C-6 glioma and C-1300 neuroblastoma lines, *Mol. Cell. Biochem.*, 13, 49, 1976.
73. **Perri, U., Sacchi, O., and Casella, C.**, Nervous transmission in the superior cervical ganglion of the thiamine-deficient rat, *J. Exp. Physiol.*, 55, 25, 1970.
74. **Sacchi, O., Ladinsky, H., Prigioni, I., Consolo, S., Peri, G., and Perri, U.**, Acetylcholine turnover in the thiamine—depleted superior cervical ganglion of the rat, *Brain Res.*, 151, 609, 1978.
75. **Eder, L., Hirt, L., and Dunont, Y.**, Possible involvement of thiamine in acetylcholine release, *Nature (London)*, 264, 186, 1976.
76. **Dolivo, M.**, Metabolism of mammalian sympathetic ganglia, *Fed. Proc. Fed. Am. Soc. Exp. Biol.*, 33, 1043, 1974.
77. **Gibson, G. E. and Blass, P. J.**, Impaired synthesis of acetylcholine in brain accompanying hypoglycemia and mild hypoxia, *J. Neurochem.*, 217, 37, 1976.
78. **Gibson, G. E. and Blass, J. P.**, A relation between [NAD^+]/[NADH] potentials and glucose utilization in rat brain slices, *J. Biol. Chem.*, 251, 4127, 1976.
79. **Gibson, G. E., Blass, J. P., and Jenden, D. J.**, Measurement of acetylcholine turnover using glucose as precursor. Evidence for compartmentation of glucose metabolism in brain, *J. Neurochem.*, 30, 71, 1978.
80. **Wildemann, L., Boehm, M., Pabst, W., and Hess, B.**, Relation between thiamine uptake and the content of thiamine, thiamine pyrophosphate, and the activity of transketolase in rat organs, *Enzymol. Biol. Clin.*, 10, 81, 1969.
81. **Dreyfus, P. M.**, Clinical applications of blood transketolase determinations, *N. Engl. J. Med.*, 267, 596, 1962.
82. **Warnock, L. G., Prudhomme, C. R., and Wagner, C.**, The determination of thiamine pyrophosphate in blood and other tissues, and its correlation with erythrocyte transketolase activity, *J. Nutr.*, 108, 421, 1978.
83. **Pongpanich, B., Svikrikkrich, N., Dhanamitta, S., and Valyasevi, A.**, Biochemical detection of thiamine deficiency in infants and children in Thailand, *Am. J. Clin. Nutr.*, 27, 1399, 1974.
84. **Gontzea, I., Gorcea, V., and Popescu, F.**, Biochemical assessment of thiamin status in patients with neurosis, *Nutr. Metab.*, 19, 153, 1975.
85. **Hathcock, J. N.**, Thiamin deficiency effects on rat leukocyte pyruvate decarboxylation rates, *Am. J. Clin. Nutr.*, 31, 250, 1978.
86. **McCandless, D. W., Curley, A. D., and Cassidy, C. E.**, Thiamine deficiency and the pentose phosphate cycle in rats; intracerebral mechanisms, *J. Nutr.*, 106, 1144, 1976.
87. **McCandless, D. W., Cassidy, C. E., and Curley, A. D.**, Thiamine deficiency and the hepatic pentose phosphate cycle, *Biochem. Med.*, 14, 384, 1975.
88. **McCandless, D. W., Curley, A. D., and Cassidy, C. E.**, Cardiac and renal pentose phosphate pathway activity in thiamin deficiency, *Int. J. Vitam. Nutr. Res.*, 46, 297, 1976.

89. **Geel, S. E. and Dreyfus, P. M.**, Brain lipid composition of immature thiamine-deficient and undernourished rats, *J. Neurochem.*, 24, 353, 1975.
90. **Trostler, N. and Sklan, D.**, Lipogenesis in the brain of thiamine-deficient rat pups, *J. Nutr. Sci. Vitaminol.*, 24, 105, 1978.
91. **Trostler, N., Guggenheim, K., Havivi, E., and Sklan, D.**, Effect of thiamin deficiency in pregnant and lactating rats on the brain of their offspring, *Nutr. Metab.*, 21, 294, 1977.
92. **Volpe, J. J. and Marasa, J. C.**, A role for thiamine in the regulation of fatty acid and cholesterol biosynthesis in cultured cells of neural origin, *J. Neurochem.*, 30, 975, 1978.
93. **Reynolds, S. F. and Blass, J. P.**, Normal levels of acetylcholine and of acetyl-coenzyme A in the brains of thiamine-deficient rats, *J. Neurochem.*, 24, 185, 1975; **Reynolds, S. F.**, unpublished observations, 1975.
94. **Berl, S., Clarke, D. D., and Schneider, D.**, *Metabolic Compartmentation and Neurotransmission*, Plenum Press, New York, 1974.
95. **Gubler, C. J., Adams, B. L., Hammond, B., Yuan, E. C., Guo, S. M., and Bennion, M.**, Physiological functions of thiamine. IV. Effect of thiamine deprivation and thiamine antagonists on the level of γ-aminobutiric acid and on α-oxoglutarate metabolism in rat brains, *J. Neurochem.*, 22, 831, 1974.
96. **Gaitonde, M. K. and Nixey, R. W. K.**, The effect of deficiency of thiamine on the metabolism of [U-^{14}C] glucose and [U-^{14}C] ribose and the levels of amino acids in rat brain, *J. Neurochem.*, 22, 53, 1974.
97. **Gaitonde, M. K.**, Conversion of [U-^{14}C] threonine into ^{14}C-labelled amino acids in the brain of thiamine-deficient rats, *Biochem. J.*, 150, 285, 1975.
98. **Gaitonde, M. K., Fayein, N. A., and Johnson, A. L.**, Decreased metabolism *in vivo* of glucose into amino acids of the brain of thiamine-deficient rats after treatment with pyrithiamine, *J. Neurochem.*, 24, 1215, 1975.
99. **Nesedov, L. I., Fustochenko, B. P., and Ostrovskii, Y. M.**, Levels of free amino acids in rat tissues during thiamin deficiency, *Biokhimya*, 43, 1834, 1978.
100. **Manz, H. J., Robertson, D. M., Haas, R. A., and Meyers, N.**, Cycloleucine intake in the brainstem of thiamine-deficient rats, *Acta Neuropathol.*, 36, 47, 1976.
101. **Nakagowa, H.**, Changes in protein synthesis in rat liver during the administration of a thiamine-deficient diet, *Kurume Med. J.*, 16, 101, 1969.
102. **Yanagahira, T.**, Cerebral anoxia: effect on neuron-glia fractions and polysomal protein synthesis, *J. Neurochem.*, 27, 539, 1976.
103. **Henderson, G. I. and Schenker, S.**, Reversible impairment of cerebral DNA synthesis in diet-induced thiamine deficiency, in *Thiamine*, Gubler, C. J., Fujiwara, M., and Dreyfus, P. M., Eds., Wiley, New York, 273.
104. **Geel, S. E. and Dreyfus, P. M.**, Thiamine deficiency encephalopathy in the developing rat, *Brain Res.*, 76, 435, 1974.
105. **Mann, P. J. G. and Quastel, J. H.**, Vitamin B_1 and acetylcholine formation in isolated brain, *Nature (London)*, 145, 856, 1940.
106. **Heinreich, C. P., Stadler, H., and Weiser, H.**, The effect of thiamine deficiency on the acetylcoenzyme A and acetylcholine levels in the rat brain, *J. Neurochem.*, 21, 1273, 1973.
107. **Cheney, P. L., Gubler, C. J., and Jaussi, A. W.**, Production of acetylcholine in rat brain following thiamine deprivation and treatment with thiamine antagonists, *J. Neurochem.*, 16, 1283, 1969.
108. **Vorhees, C. V., Schmidt, D. E., Barrett, R. J., and Schenker, S.**, Effects of thiamine deficiency on acetylcholine levels and utilization *in vivo* in rat brain, *J. Nutrition*, 107, 1902, 1977.
109. **Speeg, K. V., Jr., Chen, D., McCandless, D. W., and Schenker, S.**, Cerebral acetylcholine in thiamine deficiency is normal, *Proc. Soc. Exp. Biol. Med.*, 134, 1005, 1970.
110. **Jenden, D. J., Ed.**, *Cholinergic Mechanisms and Psychopharmacology*, Plenum Press, New York, 1978.
111. **Gibson, G. E. and Blass, J. P.**, Inhibition of acetylcholine synthesis and of carbohydrate utilization by metabolites from maple-syrup-urine-disease, *J. Neurochem.*, 26, 1073, 1976.
112. **Gibson, G. E. and Blass, J. P.**, Proportional inhibition of acetylcholine synthesis accompanying impairment of 3-hydroxybutyrate oxidation in rat brain slices, *Biochem. Pharmacol.*, 28, 133, 1979.
113. **Gibson, G. E., Shimada, M., and Blass, J. P.**, Alterations in acetylcholine synthesis and in cyclic nucleotides in mild cerebral hypoxia, *J. Neurochem.*, 31, 757, 1978.
114. **Drachman, D. A. and Leavitt, J.**, Human memory and the cholinergic system — relation to aging, *Arch. Neurol.*, 30, 113, 1974.
115. **Iwata H., Matsuda, T., Yamagami, S., Veda, N., and Baba, A.**, Effects of thiamin on glucose intolerance and bradycardia of thiamin-deficient rats, *Vitamins*, 50, 377, 1976.
116. **Plaitakis, A., Nicklas, W. J., and Berl, S.**, Thiamin deficiency: selective impairment of the cerebellar serotonergic system, *Neurology*, 28, 691, 1978.

117. **Chan-Palay, V.,** Indoleamine neurons and their processes in the normal brain and in chronic diet-induced thiamine deficiency demonstrated by uptake of serotonin-^{3}H, *J. Comp. Neurol.*, 176, 463, 1977.
118. **Chan-Palay, V., Plaitakis, A., Nicklas, W., and Berl, S.,** Autoradiographic demonstration of loss of labelled indoleamine axons of the cerebellum in chronic diet-induced thiamine deficiency, *Brain Res.*, 138, 380, 1977.
119. **Waldenkind, L.,** Release of thiamin and formation of a methylthiamin-like substance in the phrenic nerve-diaphragm preparation of the rat, *Acta Physiol. Scand.*, 101, 22, 1977.
120. **Armett, C. J. and Cooper, J. R.,** The role of thiamine in nervous tissue: effect of antimetabolites of the vitamin on conduction in mammalian nonmyelinated nerve fibers, *J. Pharmacol. Exp. Ther.*, 148, 137, 1965.
121. **Itokawa, Y., Schulz, R. A., and Cooper, J. R.,** Thiamine in nerve membranes, *Biochem. Biophys. Acta*, 266, 293, 1972.
122. **Barchi, R. L. and Braun, P. E.,** A membrane-associated thiamin triphosphatase from rat brain, *J. Biol. Chem.*, 247, 7668, 1972.
123. **Berman, K. and Fishman, R. A.,** Thiamine phosphate metabolism and possible coenzyme — independent functions of thiamine in brain, *J. Neurochem.*, 24, 457, 1975.
124. **Barchi, R. L. and Braun, P. E.,** Thiamine in neural membranes: enzymic hydrolysis of thiamine disphosphate, *J. Neurochem.*, 19, 1039, 1972.
125. **Hashitani, Y. and Cooper, J. R.,** The partial purification of thiamine triphosphatase from rat brain, *J. Biol. Chem.*, 247, 2117, 1972.
126. **Marzotto, A. and Galzigna, L.,** Thiamine as an artificial acetylcholine receptor. Effect of copper (II) on thiamine-acetylcholine interaction, *Int. J. Vitam. Nutr. Res.*, 41, 401, 1971.
127. **Watanabe, I.,** Pyrithiamine-induced acute thiamin-deficient encephalopathy in the mouse, *Exp. Mol. Pathol.*, 28, 381, 1978.
128. **Pincus, J. H. and Wells, K.,** Regional distribution of thiamine-dependent enzymes in normal and thiamine-deficient brain, *Exp. Neurol.*, 37, 495, 1972.
129. **Jubb, K. V., Saunders, L. Z., and Coates, H. V.,** Thiamine deficiency encephalopathy in cats, *J. Comp. Pathol.*, 66, 217, 1956.
130. **Rinehart, J. F., Friedman, M., and Greenberg, L. D.,** Effects of experimental thiamine deficiency on the nervous system of the rhesus monkey, *Arch. Pathol.*, 48, 129, 1949.
131. **Ferrari, G., Cappelli, V., and Ceriani, T.,** Liver and brain thiamine depletion and neurologic signs in pigeon antivitaminosis, *Int. J. Vitam. Nutr. Res.*, 46, 303, 1976.
132. **Roberts, G. W. and Boyd, J. W.,** Cerebrocortical necrosis in ruminants. Occurrence of thiaminase in the gut of normal and affected animals and its effect on thiamine status, *J. Comp. Pathol.*, 84, 365, 1974.
133. **Markson, L. M., Lewis, G., Terlecki, S., Edwin, E. E., and Ford, J. E.,** The aetiology of cerebrocortical necrosis; the effects of administering antimetabolites of thiamine to preruminant calves, *Br. Vet. J.*, 128, 488, 1972.
134. **Yudkin, J.,** The Vitamin B_1 sparing action of fat and protein, *Biochem. J.*, 48, 608, 1951.
135. **Everett, G. M.,** Observations on the behavior and neurophysiology of acute thiamine deficient cats, *Am. J. Physiol.*, 141, 439, 1944.
136. **Gubler, C. J.,** Biochemical changes in thiamine deficiencies, *J. Nutr. Sci. Vitaminol.*, Suppl., 22, 33, 1976.
137. **Robertson, D. M., Wasan, S. M., and Skinner, D. B.,** Ultrastructural features of early brain stem lesions of thiamine-deficient rats, *Am. J. Pathol.*, 52, 1081, 1968.
138. **Collins, G. H.,** Glial cell changes in the brain stem of thiamine-deficient rats, *Am. J. Pathol.*, 50, 791, 1967.
139. **Tellez, I. and Terry, R. D.,** Fine structure of the early changes in the vestibular nuclei of the thiamine-deficient rat, *Am. J. Pathol.*, 52, 777, 1968.
140. **Welsh, F. A., O'Connor, M. J., and Marcy, V. R.,** Effect of oligemia on regional metabolite levels in cat brain, *J. Neurochem.*, 31, 311, 1978.
141. **Cole, M., Turner, A., Frank, O., Baker, H., and Leevy, C. M.,** Extraocular palsy and thiamine therapy in Wernicke's encephalopathy, *Am. J. Clin. Nutr.*, 22, 44, 1969.
142. **Zieve, L., Doizaki, W. M., and Stenroos, L. E.,** Effect of magnesium deficiency on growth response to thiamine of thiamine-deficient rats, *J. Lab. Clin. Med.*, 72, 261, 1968.
143. **Zieve, L., Diozaki, W. M., and Stenroos, L. E.,** Effect of magnesium deficiency on blood and liver transketolase activity and on the recovery of enzyme activity in thiamine-deficiency in thiamine-deficient rats receiving thiamine, *J. Lab. Clin. Med.*, 72, 268, 1968.
144. **Leigh, D.,** Subacute necrotizing encephalomyelopathy in an infant, *J. Neurol. Neurosurg. Psychiatry* 14, 216, 1951.

145. **Pincus, J. H.,** Subacute necrotizing encephalomyelopathy (Leigh's Disease): a consideration of clinical features and etiology, *Dev. Med. Child Neurol.*, 14, 87, 1972.
146. **Montpetit, V. J. A., Anderman, F., Carpenter, S., Fawcett, J. S., Zborowsko-Sluis, D., and Giberson, H. R.,** Subacute necrotizing encephalomyelopathy, a review and study of two families, *Brain*, 94, 1, 1971.
147. **Dunn, H. G. and Dolman, C. L.,** Necrotizing encephalomyelopathy: report of a case with relapsing polyneuropathy and hyperalaninemia and with manifestations resembling Friedreich's ataxia, *Neurology*, 19, 536, 1969.
148. **Exss, R., Gulotta, F., Kallfelz, H. C., and Völpel, M.,** Wernicke's encephalopathy and cardiomyopathy in a boy with Friedreich's ataxia, *Neuropaediatrie*, 5, 162, 1974.
149. **Kissach, A. W., Currie, S., Harriman, D. G. F., Littlewood, J. M., Payne, R. B., and Walker, B. E.,** Leigh's disease and failure of automatic respiration, *Lancet*, 2, 662, 1974.
150. **Pincus, J. H., Cooper, J. R., Piros, K., and Turner, V.,** Specifity of the urine inhibitor test for Leigh's disease, *Neurology*, 24, 885, 1974.
151. **Pincus, J. H., Solitare, G. B., and Cooper, J. R.,** Thiamine triphosphate levels and histopathology. Correlation in Leigh's disease, *Arch. Neurol.*, 33, 759, 1976.
152. **Murphy, J. V.,** Subacute necrotizing encephalomyelopathy (Leigh's disease): Detection of the heterozygous carrier state, *Pediatrics*, 51, 710, 1973.
153. **Hommes, F. A., Polman, H. A., and Reerink, J. D.,** Leigh's encephalomyelopathy and inborn error of gluconeogenesis, *Arch. Dis. Child.*, 43, 423, 1968.
154. **Grover, W. D., Auerbach, V. H., and Patel, M. S.,** Biochemical studies and therapy in subacute necrotizing encephalomyelopathy (Leigh's syndrome), *J. Pediatr.*, 81, 39, 1972.
155. **Willems, J. L., Monnens, L. A. H., Trijbels, J. M. F., Veerkamp, J. H., Meyer, A. E. F. H., Van Dam, K., and Von Haelst, U.,** Leigh's encephalomyelopathy in a patient with cytochrome-c oxidase deficiency in muscle, *Pediatrics*, 60, 850, 1977.
156. **Blass, J. P., Cederbaum, S. D., and Dunn, H. G.,** Biochemical abnormalities in Leigh's disease, *Lancet*, 1, 1237, 1976.
157. **Crosby, T. W. and Chou, S. M.,** Ragged-red fibers in Leigh's disease, *Neurology*, 24, 49, 1974.
158. **Tomasulo, P. A., Kater, R. M. H., and Iber, F. L.,** Impairment of thiamine absorption in alcoholism, *Am. J. Clin. Nutr.*, 21, 1340, 1968.
159. **Baker, H., Frank, O., Zetterman, R. K., Rajan, K. S., Ten Hove, W., and Leevy, C. M.,** Inability of chronic alcoholics with liver disease to use food as a source of folates, thiamine, and vitamin B_6, *Am. J. Clin. Nutr.*, 28, 1377, 1975.
160. **Rindi, G. and Ventura, U.,** Thiamine intestinal transport, *Physiol. Rev.*, 52, 821, 1972.
161. **Abe, T. and Itokawa, Y.,** Effect of ethanol administration on thiamine metabolism and transketolase activity in rats, *Int. J. Vitam. Nutr. Res.*, 47, 307, 1977.
162. **Van Italie, T. B. and Follis, R. H.,** Thiamine deficiency ariboflavinosis, and vitamin B_6 deficiency, in *Harrison's Principles of Internal Medicine*, 7th ed., Wintrobe, M., Thorn, G. W., and Adams, R. D., Eds., McGraw-Hill-Blakiston, New York, 1974, 430.
163. **Burgess, R. C.,** Deficiency diseases in prisoners of war at Changi, Singapore, *Lancet*, 2, 411, 1946.
164. **Blass, J. P., Milne, J. A., and Rodnight, R.,** Newer concepts of psychiatric disgnosis and biochemical research on mental illness, *Lancet*, 1, 738, 1977.
165. **Yokomine, R., Kuriyama, M., Arima, H., and Igata, R.,** Biochemical analyses of vitamin B_1 metabolism in normal subjects, *Vitamins*, 52, 89, 1978.
166. **Wood, B., Breen, K. J., and Penington, D. G.,** Thiamine status in alcoholism, *Aust. N. Z. J. Med.*, 7, 475, 1977.
167. **Blass, J. P., Avigan, J., and Uhlendorf, B. W.,** A defect in pyruvate decarboxylase in a child with an intermittent movement disorder, *J. Clin. Invest.*, 49, 423, 1970.
168. **Blass, J. P.,** Disorders of pyruvate metabolism, *Neurology*, 29, 280, 1979.
169. **Guggenheim, M. A. and Stumpf, D. A.,** Familial metabolic disease with clinicopathological findings of both Leigh's disease and adult-type spinocerebellar degeneration, *Ann. Neurol.*, 2, 264, 1977.
170. **Wick, H., Schweizer, K., and Baumgartner, R.,** Thiamine dependency in a patient with congenital lacticacidemia due to pyruvate dehydrogenase deficiency, *Agents Actions*, 7, 405, 1977.
171. **Haworth, J. C., Perry, T. L., Blass, J. P., Hansen, S., and Urguhart, N.,** Lacticacidosis in three sibs due to defects in both pyruvate dehydrogenase and α-ketoglutarate dehydrogenase complexes, *Pediatrics*, 58, 564, 1976.
172. **Blass, J. P., Kark, R. A. P., Menon, N., and Harris, S. H.,** Decreased activities of the pyruvate and ketoglutarate dehydrogenase complexes in fibroblasts from five patients with Friedreich's ataxia, *N. Engl. J. Med.*, 295, 62, 1976.
173. **Robinson, B. H., Taylor, J., and Sherwood, W. G.,** Deficiency of dihydrolipoyl dehydrogenase (a component of the pyruvate and α-ketoglutarate dehydrogenase complexes): a cause of congenital chronic lactic acidosis in infancy, *Pediatr. Res.*, 11, 1198, 1977.

174. **Dancis, J. and Levitz, M.**, Abnormalities of branched chain amino acid metabolism (Hypervalinemia, Maple-Syrup-Urine-Disease, Isovaleric acidemia, and β-methylcrotonic Aciduria) in: *The Metabolic Basis of Inherited Disease*, 4th ed., Stanbury, J. B., Wyngaarden, J. B., and Fredrickson, D. S., Eds., McGraw-Hill-Blakistan, New York, chap. 20.

175. **Danner, D. J., Wheeler, F. B., Lemmon, S. K., and Elsas, L. J.**, In vivo and in vitro response of human branched-chain α-ketoacid dehydrogenase to thiamine and thiamine pyrophosphate, *Pediatr. Res.*, 12, 235, 1978.

176. **Gibson, G. E. and Blass, J. P.**, Inhibition of acetylcholine synthesis and of carbohydrate utilization by metabolites from maple-syrup-urine-disease, *J. Neuroochem.*, 26, 1073, 1976.

177. **Garrod, A.**, The place of biochemistry in medicine, *Br. Med. J.*, 1, 1099, 1928.

178. **Rosenberg, L. E.**, Inherited aminoacidopathies demonstrating vitamin dependency, *N. Engl. J. Med.*, 281, 145, 1969.

179. **DiPalma, J. R. and Ritchie, D. M.**, Vitamin toxicity, *Ann. Rev. Pharmacol. Toxicol.*, 17, 133, 1977.

180. **Liang, C. C.**, Metabolic changes in rats during developing thiamin deficiency, *Biochem. J.*, 146, 739, 1975.

181. **Liang, C. C.**, Alternative deficiency, *J. Nutr. Sci. Vitaminol.*, Suppl. 22, 47, 1976.

182. **Brock, D. J. H. and Mayo, D.**, *The Biochemical Genetics of Man*, 2nd ed., Academic Press, New York, 1978.

183. **Wurtman, R. J., Lavin, F., Mostafapour, S., and Fernstrom, J. D.**, Brain catechol synthesis: control by brain tyrosine concentration, *Science*, 185, 183, 1974.

Chapter 7

MEGAVITAMIN THERAPY AND THE CENTRAL NERVOUS SYSTEM

Reynold Spector

TABLE OF CONTENTS

I. Introduction 138
A. Definition of Vitamin 138
B. Megavitamin Therapy 138
C. Effect of Megavitamin Therapy on Brain 139

II. Anatomy of Brain Barrier Systems 139
A. Blood-Brain and Blood-Cerebrospinal Fluid Barriers 139
B. Development of the Barrier 141

III. Physiology of Transport Through Brain Barrier Systems 142
A. Loci 142
B. Mechanisms 142
C. Transport and Vitamin Homeostasis 143

IV. Ascorbic Acid 143
A. Vitamers and Function in Brain 143
B. Absorption and Requirement 143
C. Transport into Brain and Cerebrospinal Fluid 144
D. Ascorbic Acid Homeostasis in Brain 146

V. Thiamin 147
A. Vitamers and Functions in Brain 147
B. Absorption and Requirement 147
C. Transport into Brain and Cerebrospinal Fluid 147
D. Thiamin Homeostasis in Brain and Cerebrospinal Fluid 148
E. Thiamin Deficiency Encephalopathy 148

VI. Folates 148
A. Vitamers and Roles in Brain 148
B. Requirement 148
C. Transport into Cerebrospinal Fluid and Brain 148
D. Folate Homeostasis in Brain and Cerebrospinal Fluid 149
E. Folate Deficiency Encephalopathy 149

VII. Vitamin B_6 149
A. Vitamers and Roles in Brain 149
B. Transport into Cerebrospinal Fluid and Brain 150
C. Vitamin B_6 Homeostatis in Brain 150
D. Pyridoxine Deficiency Encephalopathy 150

VIII. Niacin 151
A. Vitamers and Roles in Brain 151
B. Absorption 151

C. Transport into Cerebrospinal Fluid and Brain 151
D. Total Niacin Homeostasis in Brain 152
E. Niacin Deficiency Encephalopathy 152

IX. Other Vitamins 152

X. Megavitamin Responsive Brain Disorders 152

XI. Megavitamin Therapy 153

XII. Implications of Vitamin Homeostatic Systems in the Central Nervous System for Megavitamin Therapy of Brain Disorders 153

XIII. Conclusions 154

Addendum 154

References 155

I. INTRODUCTION

A. Definition of Vitamin

Webster defines vitamins as "any number of constituents of foods in their natural state, of which very small quantities are essential for the normal nutrition of animals".[1] In mammalian nutrition, the term vitamin is generally applied to several chemicals that mammals cannot synthesize and appear in small amounts in foods. Vitamins are generally cofactors for enzymatic reactions. In normal man, the daily quantities of folates, vitamin C, vitamin B_6, thiamin, and niacin required to prevent deficiency states are about 0.1 mg, 50 mg, 2 mg, 2 mg and 15 mg, respectively. In brain, however, the term vitamin is often used in a slightly different sense. For example, in certain mammals, with the notable exceptions of the primates and guinea pig, vitamin C can be synthesized from glucose in the liver but not in the brain. Hence, vitamin C would not be considered a vitamin for these species.[2-5] Yet, vitamin C for the brain must be obtained from the blood, either from vitamin C that entered blood from the diet or was synthesized in the liver.[2] Similarly, niacin cannot be synthesized from tryptophan in mammalian brain, whereas in mammalian liver, the synthesis of niacin from tryptophan occurs.[6,7] In this sense, niacin and vitamin C could be considered vitamins for the brain since they must be obtained from outside the brain. In the rest of this review, the term vitamin will be used in that sense, i.e., a vitamin for brain will be defined as a substance the brain cannot synthesize but requires in small amounts. Generally, vitamins for the brain are cofactors in enzymatic reactions.

B. Megavitamin Therapy

Megavitamin therapy generally involves the use of the water-soluble vitamins, either singly or in combination, for the treatment of various disorders of the central nervous system or other organ systems.[8-11] Amounts of the water-soluble vitamins employed in megavitamin therapy are generally 10 to 100 times that required to prevent vitamin deficiency states; hence, the term megavitamin therapy. Megavitamin therapy has been used to treat various disorders of the central nervous system, including schizophrenia and minimal brain dysfunction as will be discussed further below.[8-11]

The rationale for the use of megavitamin therapy in certain disorders of the brain has been discussed at length by Dr. Pauling.[8] Dr. Pauling proposes that there are several plausible mechanisms by which megavitamin therapy could affect the functioning of the brain.[8] Dr. Pauling points out that the brain is more sensitive to changes in its molecular composition than other organs in the body.[8] Dr. Pauling suggests that certain apoenzymes in the brain in various individuals may have varying affinities for the (vitamin) coenzymes.[8] Dr. Pauling argues that, in the human brain, many of the enzymes may have multiple forms just as there are over 100 forms of human hemoglobin.[8] The direct consequence of this would be that the concentration of (vitamin) coenzyme needed to produce the amount of active enzyme necessary for optimal health probably varies from individual to individual.[8] Because of altered affinity for the coenzyme, some of these aopenzymes in brain may function inadequately in certain, crucial enzymatic reactions when only the usual concentrations of the various (vitamin) coenzymes are present in the brain.[8] Dr. Pauling suggests that megavitamin therapy could increase the concentration of coenzyme in brain and shift the equilibrium of the reaction of apoenzyme and coenzyme toward more of the active enzyme and, therefore, more of the required product. This theory Dr. Pauling terms orthomolecular theory. Orthomolecular psychiatry is defined as "the achievement and preservation of mental health by varing the concentration in the human body of substances that are normally present such as vitamins".[8] Alternative mechanisms by which megavitamin therapy could affect cerebral functioning would include the alteration of enzymatic activity in other organs such as the liver with secondary effects on the brain.

C. Effect of Megavitamin Therapy on Brain

In order for megavitamin therapy to work directly on the brain, one might expect that megavitamin therapy would alter the vitamin (or coenzyme) concentrations in cerebrospinal fluid and brain. If megavitamin therapy did not alter the vitamin (and coenzyme) levels in brain, then it would be difficult to postulate that megavitamin therapy had a direct effect on brain by shifting the equilibrium of the postulated abnormal apoenzyme and coenzyme toward more of the active enzyme. The purpose of the present review will be to assess the consequences of megavitamin therapy in terms of the levels of vitamins (and coenzymes) in mammalian brain. Specifically, I will deal with five water-soluble vitamins about which there is substantial information including vitamins B_1, B_6, C, niacin, and the folates. Where available, data from man will be discussed. I will also review briefly megavitamin-responsive brain disorders, the controversy about megavitamin therapy, and the implications of the vitamin homeostatic systems in the central nervous system for megavitamin therapy of brain disorders.

II. ANATOMY OF BRAIN BARRIER SYSTEMS

A. Blood-Brain and Blood-Cerebrospinal Fluid Barriers

Before water-soluble molecules in the blood can enter the brain, the molecule must enter the extracellular space of brain or the cerebrospinal fluid (CSF) which bathes the brain's surfaces.[12,13] The extracellular space of brain constitutes about 15% of the brain by weight.[12,13] There is no barrier to the movement of molecules between the extracellular space of brain and the cerebrospinal fluid.[12,13] However, barriers impede the passage into either one.[12,13] To enter the brain's extracellular space from blood by simple diffusion, the molecule must pass through the cerebral capillaries, the site of the blood-brain barrier which, unlike the capillaries in other parts of the body, are joined by tight conjunctions (Figure 1).[5,12,13] To enter the cerebrospinal fluid by simple diffusion, the molecule must pass through the choroid plexus, whose fronds float in

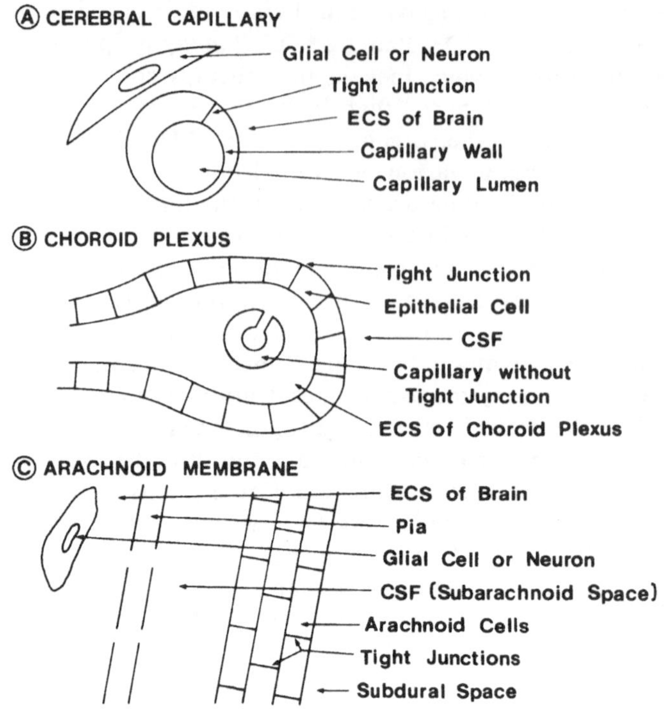

FIGURE 1. Schema of blood-brain barrier (A) and blood-CSF barriers (B and C), in which tight junctions retard simple diffusion from blood into the extracellular space (ECS) of brain and the CSF. The tight junctions that join the cerebral capillaries impede passage by simple diffusion into the ECS of brain, and those that join the choroid-plexus epithelial cells impede passage between the ECS of the choroid plexus and the CSF. The choroid-plexus capillaries are not joined by tight junctions and are "leaky". In the arachnoid membrane, tight junctions prevent passage between the subdural space, which freely communicates with the circulating blood, and the subarachnoid space. There is no barrier to movement between the ECS of brain and the CSF.

the cerebrospinal fluid, or the arachnoid membrane.[5,12,13] The epithelial cells of these blood-cerebrospinal fluid barriers are also joined by tight junctions that impede the entry of water-soluble molecules (Figure 1).[5,12,13]

These symmetrical barriers isolate the extracellular space of brain and the cerebrospinal fluid from most of the circulating blood. These barrier systems are especially effective against large water-soluble molecules such as albumin, whose concentration in normal CSF is less than that of 1% of that in plasma.[5,12,13] However, the entry of small water-soluble molecules such as mannitol and sucrose is also markedly retarded (Table 1).[5,12,13] In a few small areas of the brain, the capillaries do not have tight junctions and small amounts of water-soluble molecules enter the extracellular space of brain and CSF through these "leaky" regions.[13] Once inside the extracellular space of brain or the CSF, the molecules must traverse the plasma membrane of the brain cell itself in order to enter the brain cell.

Although the cerebral capillaries act as a barrier against many water-soluble substances, they facilitate the diffusion of nutrients not synthesized in brain such as glucose, amino acids, and fatty acids from blood into the extracellular space of the brain.[12-14] The cerebral capillaries have, therefore, two well-defined functions for

Table 1
VITAMIN CONCENTRATIONS IN RABBIT PLASMA, CSF, CHOROID PLEXUS, AND BRAIN

Substance	Molecular weight (daltons)	Plasma (μM)	CSF (μM)	CSF/ plasma ratio	Choroid plexus (μM)	Brain (μM)
Vitamins						
Ascorbic acid[2]	176	57	232	4.1	1,085	1,867
Niacinamide[6,7]	122	0.5	0.7	1.4	—	∼500
Folates[36-38]	465	0.014	0.068	4.9	7.7	∼0.6
Thiamin[26]	302[a]	0.41	0.36	0.9	15.3	10.3
B_6[32]	169[a]	0.30	0.39	1.3	14.9	9.6
Molecules transported by simple diffusion[b]						
Mannitol[12,13]	182	1.0	0.2	0.2	0.6	0.2
Sucrose[12,13]	342	1.0	0.1	0.1	0.4	0.1

[a] Nonphosphorylated moieties.
[b] Constant plasma levels were maintained by i.v. infusions. At various times, tissues were sampled to obtain steady-state levels.

transport: (1) to facilitate the transport of nutrients necessary for the brain's proper functioning, and (2) to retard the entry of many useless molecules.[12,13] Recently, evidence has been presented which suggests that the cerebral capillaries may also be able to transport unwanted substances from the extracellular space of brain into blood such as iodide.[15] Thus, although glucose and mannitol have approximately the same molecular weight and solubility characteristics, glucose readily enters the brain and CSF, but the entry of mannitol, which mammalian brain cannot utilize, is severely retarded (Table 1).[12-14]

Beside acting as a barrier, the choroid plexus has several functions. It manufactures most, if not all, of the cerebrospinal fluid.[12,13] The choroid plexus also transports waste products of brain metabolism and certain drugs such as penicillin and methotrexate from the cerebrospinal fluid into blood.[13] The epithelial cells of the choroid plexus resemble those of the renal tubule, both microscopically and functionally.[13] The role of the choroid plexus in the transport of water-soluble vitamins into CSF is discussed below.[5]

B. Development of the Barrier

The evidence, both morphologically and functionally, suggests that the blood-brain and blood-cerebrospinal fluid barriers are basically intact at birth.[13,16] Several early studies suggested, however, that the blood-brain and blood-cerebrospinal fluid barriers were immature at birth and during early development, based on studies of the penetration of substances like inulin (a nonmetabolized sugar polymer with a molecular weight of about 5000) into brain and cerebrospinal fluid.[13,16] In the newborn animal, the concentration of inulin in brain and cerebrospinal fluid was much higher than in adult animals after the parenteral injection of inulin.[13,16] The interpretation of these studies as showing that the barriers are immature in newborn animals is probably in error for several reasons.[13,16] First, the production of cerebrospinal fluid in immature animals is much less than in mature animals.[13,16] This would tend to increase the concentration of a substance like inulin in the cerebrospinal fluid, even if the barriers were perfectly intact.[13,16] Second, the extracellular space of brain is much larger in immature animals.[13,16] Third, there is substantial morphological evidence that both the blood-brain

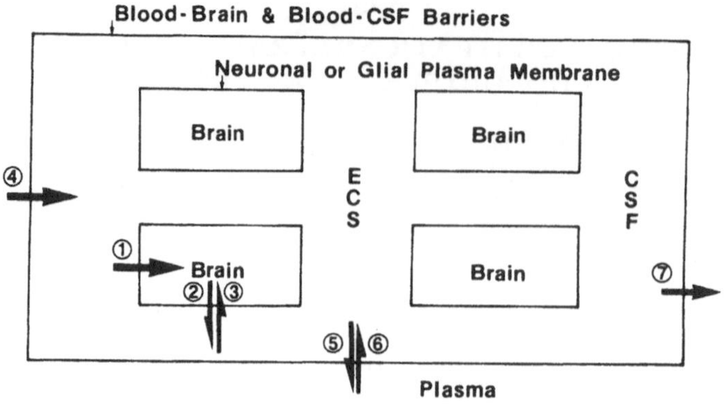

FIGURE 2. Model of ascorbic acid-transport processes. Arrow 4 indicates active transport into the CSF and extracellular space (ECS) of brain, and arrow 1 indicates active transport from there into the brain. These processes can be described by a Michaelis-Menten transport model.[5] Arrows 2, 3, 5, and 6 represent transport by simple diffusion, and 7 unidirectional transport from CSF to plasma by bulk flow of CSF through the arachnoid granulations.[5] The constants that describe these processes have been determined for ascorbic acid transport in the rabbit.[5] With knowledge of the plasma ascorbic acid concentration as a function of time, the initial concentrations in brain and CSF and the constants for processes 1 through 7, the nonlinear differential equations that describe these processes can be solved on a ditital computer. In this way, the CSF and brain ascorbic acid concentrations as a function of time can be determined.[5] The data in Table 3 were obtained from this model.

and blood-cerebrospinal fluid barriers are anatomically intact at birth.[13,16] Consistent with this view is the fact that the protein concentration in the cerebrospinal fluid of newborn children (human) is approximately twice that of adults, but still only approximately 1% of the concentration in plasma.[16]

III. PHYSIOLOGY OF TRANSPORT THROUGH BRAIN BARRIER SYSTEMS

A. Loci

For a vitamer molecule to pass from the blood into the cerebrospinal fluid or extracellular space of brain, the molecule would first have to pass through the cerebral capillaries or the choroid plexus or the arachnoid membrane (Figure 1).[5,13] Then the molecule would have to pass through the cellular membrane of the glial or neuronal cells. In a sense, the brain exists within a compartment that is anatomically separated from the plasma. This is schematically shown in Figure 2.

B. Mechanisms

Beside passing through the anatomical blood-brain and blood-cerebrospinal fluid barriers described above, a vitamin's ability to enter brain cells from blood by passive mechanisms would also depend on several other factors.[13,17] These would include the plasma concentration of the vitamin, its physical-chemical properties, particularly with regard to lipid solubility, and the degree of binding of the vitamin to plasma proteins.[13,17] Other important factors would include cerebral blood flow and the general permeability of the cerebral blood vessels in the brain and choroid plexus.[13,17] On a regional basis, important factors would include the regional vascularity of the brain, the regional blood flow of the brain, and the rate of regional diffusion through the interstitial spaces and cerebrospinal fluid.[13,17] Also, the rate of efflux or diffusion from brain into cerebrospinal fluid, or from the extracellular space of brain back into blood as well as the bulk flow of cerebrospinal fluid into blood would be other factors.[13,17]

Passive mechanisms, however, cannot explain the distribution of the various vitamins in the central nervous system. In the discussion of the individual vitamins, examples of facilitated diffusion, which does not require an energy source, or active transport of unchanged vitamin which does, will be documented. Also, more complex, energy-requiring processes which involve the transfer of certain vitamins between blood and cerebrospinal fluid and/or into brain cells will be documented. These mechanisms are generally called carrier-mediated transport mechanisms and are theoretical constructs that need to be postulated to explain the vitamin pharmacokinetics in the central nervous system.[17]

C. Transport and Vitamin Homeostasis

In all the water-soluble vitamin transport systems studied to date, the transport of the vitamin between blood and cerebrospinal fluid (and the extracellular space of brain), and/or from the extracellular space or cerebrospinal fluid into brain cells occurs by saturable transport mechanisms.[5-7,18-21] Simple diffusion of the vitamins listed in Table 1 through the blood-brain and blood-cerebrospinal fluid barriers, and thence into brain cells, is relatively unimportant.[5,7,19] Almost all of these transport systems can be described by Michaelis-Menten transport kinetics with a half-saturation concentration (K_T) and V_{max} (maximal transport velocity).[5-7,18-21] Since these transport systems are in series; that is, the molecule must pass through either the blood-brain or blood-cerebrospinal fluid barrier first and then through the brain cell membrane itself, there is a powerful regulatory system possible. In fact, many, although not all, of the vitamins traverse saturable transport systems at both the blood-cerebrospinal fluid and extracellular space of brain-brain interface.[5-7,18,19] Moreover, many of these carrier-mediated transport processes are approximately half or more saturated at normal plasma and cerebrospinal fluid vitamin concentrations.[5] Since passive transport mechanisms (such as simple diffusion) are relatively unimportant in transferring vitamins from blood into brain cells, the saturable mechanisms allow vitamin entry into brain cells to be rather tightly controlled.[5] Although less is known about efflux of vitamins from brain cells and the central nervous system, there is substantial evidence that control of efflux is also another important component of the vitamin homeostatic systems in the central nervous system.[18,19] Specific examples of these concepts will be illustrated below under the various vitamins.

IV. ASCORBIC ACID

A. Vitamers and Function in Brain

Ascorbic acid and dehydroascorbic acid (which can be readily reduced back to ascorbic acid) are present in high concentrations in brain (Table 1).[3] The only clear function of ascorbic acid in brain is to serve as a cofactor for the enzyme dopamine-β-hydroxylase which converts dopamine to noradrenaline.[3] However, there is very strong evidence that ascorbic acid also plays a protective role in brain and protects brain molecules against oxidation, e.g., lipid peroxidation.[22]

B. Absorption and Requirement

In man, ascorbic acid is absorbed in the gut by a specific saturable transport mechanism.[2,4] Thus, the ingestion of large amounts of ascorbic acid by mouth, does not increase the amount absorbed in a linear way. Approximately 50 mg of vitamin C in

Table 2
VITAMIN CONCENTRATIONS IN HUMAN PLASMA AND CSF

Vitamin	Plasma (μM)	CSF (μM)	CSF/ plasma ratio
At low or normal plasma concentrations			
Ascorbic acid[2,23]	24	78	3.3
Folates[5,36,37]	0.013	0.032	2.5
At higher plasma concentrations			
Ascorbic acid[2,23]	52	106	2.0[a]
Folates[5,36,37]	0.129	0.090	0.7

[a] At higher plasma concentrations, the CSF/plasma ratio becomes < 1.[23]

the diet per day in patients with normal gastrointestinal function are required to prevent the development of scurvy.

C. Transport into Brain and Cerebrospinal Fluid

Table 1 shows the plasma, CSF, choroid-plexus, and brain concentrations of ascorbic acid in overnight fasted rabbits. Table 2 shows the plasma and CSF concentration of ascorbic acid, at low or normal, or higher plasma concentratons in man. In rabbits and in man, CSF levels of ascorbic acid are higher than normal plasma levels (Tables 1 and 2). These levels in CSF are not the result of vitamin binding and are much higher than would be predicted on the basis of passive diffusion from blood.[3,5] As mentioned above, substances such as mannitol and sucrose enter the CSF from blood by diffusion (Table 1). The solubility characteristics and molecular weight of mannitol are comparable to those of ascorbic acid (Table 1).[3] Yet, the level of ascorbic acid in CSF is much higher than the steady-state levels of mannitol. (The concentrations of substances in the extracellular space of brain cannot be measured directly, but it is assumed that these concentrations are about equal to those in the CSF, since there is no barrier to transport between these fluids.)[5,13]

The CSF concentrations of ascorbic acid is 2 to 4 times higher than the normal plasma concentrations in both rabbit and man (Tables 1 and 2). Since no ascorbic acid is synthesized in brain or CSF, the inescapable conclusion is that vitamin-transport systems exist in the central nervous system and maintain these concentration gradients.[5]

The transport of ascorbic acid into CSF and brain is well understood and can be used as an example of vitamin transport since ascorbic acid is not synthesized or me tabolized to inactive products in the central nervous system.[2-5] (However, the available evidence suggests that the other vitamins in Table 1 may enter the CSF and brain by separate, although similar, mechanisms and is discussed below.) Ascorbic acid is homogeneously distributed throughout the mammalian brain.[2-4] In both man and rabbit, the vitamin enters the brain and CSF from blood. In the rabbit, some ascorbic acid is synthesized in the liver, whereas in man, all ascorbic acid comes from the diet.[2-4]

In an early investigation of ascorbic acid transport from blood into the central nervous system, Hammerström showed autoradiographically that after intravenous injection of (^{14}C) ascorbic acid in guinea pigs, tracer amounts appeared in the nervous system, at first within the choroid plexus.[24] The (^{14}C) ascorbic acid next appeared in the CSF and then entered the brain over a period of several days.[24] On the basis of these studies, Hammerström suggested that the choroid plexus was the locus of the process that transports ascorbic acid from blood into CSF. My colleagues and I obtained similar results when we sampled the CSF and brain of rabbits at various times after intravenous injections of (^{14}C) ascorbic acid.[3] Thus, unlike glucose and essential amino acids which enter the brain predominantly through the cerebral capillaries, ascorbic acid appeared to enter guinea pig and rabbit brain, in large part, from CSF via the choroid plexus.[14]

The entry of ascorbic acid into the CSF and brain of rabbits can be saturated by increasing the plasma ascorbic acid concentration; in other words, when the plasma concentration is high, a lesser increase in ascorbic acid occurs in the CSF.[2] Similarly, the data in man in Table 2 show that the CSF-plasma ratios of ascorbic acid decrease when the plasma levels increase. Thus in man and rabbit, entry of these vitamins from plasma into CSF occurs by a saturable mechanism.[2,4,23]

In the rabbit, the isolated choroid plexus is able to concentrate ascorbic acid (and the other vitamins in Table 1) by separate, active, saturable, transport processes.[2] These processes are specific in that only the individual vitamin or closely related congeners have affinity for the transport system. For example, the ascorbic acid system is stereospecific.[3] The stereoisomer, isoascorbic acid, has only 5% as much affinity as ascorbic acid itself for the ascorbic-acid transport system in the choroid plexus.[3] In this regard, the choroid plexus behaves like the kidney; substances that are actively transported through the renal tubules in vivo are concentrated in kidney slices by specific, active transport processes in vitro. The half-saturation concentration for ascorbic acid entry into rabbit CSF in vivo is 44 μM — comparable to the value (50 μM) for entry into isolated choroid plexus in vitro.[2] The exact half-saturation concentration for ascorbic acid entry into CSF is unknown in man, but the data in Table 2 suggest that it is similar to that in rabbits.

Thus, the experimental evidence shows that ascorbic acid is concentrated in vivo by the choroid plexus and then enters the CSF. In vitro, the choroid plexus concentrates ascorbic acid by an active-transport process with kinetic properties similar to those of the process that transports ascorbic acid into CSF in vivo. These findings support the hypothesis that the choroid plexus is a locus for the transfer of ascorbic acid from blood into CSF.[2-5] Although some of this vitamin must enter the extracellular space of brain from blood by simple diffusion, the exact amount is unknown.[4] There is no evidence of carrier-mediated transport of ascorbic acid across the cerebral capillaries.[25]

Once it has entered the CSF and the extracellular space of brain, ascorbic acid is concentrated by brain cells by an active transport system.[4] The half-saturation concentration for active transport of the vitamin into rabbit brain slices is 33 μM.[4] Since brain slices do not contain an intact blood-brain barrier, the transfer of ascorbic acid into these slices indicates that brain cells are themselves capable of transporting ascorbic acid intracellularly from the extracellular space of brain (and the CSF) in vivo.[4] Thus, in normal rabbits and probably man, the brain is surrounded by ascorbic acid at a concentration several times higher than the half-saturation concentration for entry into brain.[2,4] This high concentration in the CSF (and extracellular space of brain) depends on the active transport of ascorbic acid into CSF by the choroid plexus, and on the blood-brain and blood-CSF barriers which retard simple diffusion of ascorbic acid back down the concentration gradient into blood (Table 1).[4]

An estimate of the turnover of ascorbic acid in rabbit choroid plexus, CSF, and brain provides an indication of how much of the vitamin enters and leaves these tissues per hour. At the normal ascorbic acid concentration in rabbit plasma, 12% of the ascorbic acid in the CSF, 18% of that in the choroid plexus, and 2% of that in the brain are replaced per hour by ascorbic acid from the blood.[2] Under steady-state conditions, the ascorbic acid that enters the CSF, choroid plexus, and brain from blood

Table 3
ASCORBIC ACID CONCENTRATIONS IN CSF AND BRAIN AS A FUNCTION OF THE PLASMA CONCENTRATION[a]

Plasma (μM)	CSF (μM)	CSF/plasma ratio	Brain (μM)	Brain/plasma ratio
5.7	40	7.0	1,017	178
28.0	148	5.3	1,591	57
56.0	210	3.7	1,767	31
114	290	2.5	1,886	17
568	506	0.9	2,170	4

[a] Values are derived from the model in Figure 2 and the experimentally determined constants for the processes in Figure 2. The validity of this model has been verified experimentally.[4,5]

replaces what is lost by bulk flow of CSF into blood and simple diffusion back into blood.[2,4]

D. Ascorbic Acid Homeostasis in Brain

A simplified model of ascrobic acid transport is shown in Figure 2.[5] This model makes it possible to compare the importance of simple diffusion through the brain-barrier systems with that of active transport into CSF and brain. At the normal plasma ascorbic acid concentration, the amount transferred by simple diffusion into the extracellular space-CSF region (process 6) and the brain (process 3) is less than 10% of the amount transferred by carrier-mediated active transport (processes 4 and 1).[5] Table 3 shows predictions of steady-state plasma, CSF, and brain concentrations (rabbit) based on this model and experimentally determined kinetic constants for transport from blood into CSF (44 μM), choroid plexus (50 μM), and brain (33 μM).[2-4] These predictions have been verified experimentally at a plasma concentration of 568 μM.[4] An increase in the ascorbic acid concentration in rabbit plasma from 5.7 μM to 568 μM would increase CSF levels almost 13 times, but brain levels only twice.[4] The normal rabbit plasma ascorbic acid concentration is 57 μM.[4]

On the other hand, even at very low plasma levels of ascorbic acid, the brain levels are well maintained by the active-transport systems shown in Figure 2. When plasma ascorbic acid is low, these systems pump relatively more of the vitamin into the CSF and brain, to replace what is lost by diffusion (processes 2 and 5) and bulk flow of CSF (process 7) into blood. In other words, although diffusion into CSF and brain does not provide an adequate new supply of ascorbic acid except at very high plasma levels, the active-transport processes help control brain levels by regulating entry. The system is ideally set for homeostasis, since entry of ascorbic acid into CSF is half-saturated at the normal plasma level, whereas entry into brain is almost completely saturated at the normal CSF level.[2-4] Even in vitamin-deficiency states, the relative maintenance of brain levels of ascorbic acid can be verified experimentally.[2]

Another striking aspect of the ascorbic acid transport system is the analogous physiologic behavior of the choroid plexus and renal tubule epithelial cells. Both tissues bidirectionally transport similar substances. As noted above, penicillin is transported from blood to urine by kidney tubules and from CSF to blood by the choroid plexus.[5,17] Similarly, ascorbic acid and other water-soluble vitamins are transported from urine to blood by kidney tubules and from blood to CSF by the choroid plexus.[5,17] Thus, the choroid plexus behaves like a kidney interposed between blood and CSF, and it helps regulate the molecular environment of the brain.[4,5]

V. THIAMIN

A. Vitamers and Functions in Brain

Thiamin is found in brain as well as other tissues as free thiamin and the monophosphate, diphosphate, and triphosphate. Thiamin diphosphate is the predominant moiety.[26] In brain as well as other tissues, there are several enzymes which are able to convert thiamin to its various phosphorylated forms.[26] There are also several phosphatases in brain and other tissues which can dephosphorylate the various thiamin phosphates back to free thiamin.[26] In brain, thiamin, in the form of thiamin diphosphate, is a coenzyme for several important enzymatic reactions, including the conversion of pyruvate to acetylCoA.[27] A second probable role for thiamin in brain is its neurophysiological function in nerve conduction.[28,29] Thiamin triphosphate is thought to be the specific form of thiamin that is involved in this noncoenzymatic function of thiamin in the central nervous system.[28,29] However, the evidence for this noncoenzymatic, neurophysiological function for thiamin triphosphate, although suggestive, is still incomplete.

B. Absorption and Requirement

The thiamin requirement in the diet is approximately 2 mg/day.[30] Dietary thiamin is absorbed in the intestine by a carrier-mediated process.[30] The process is saturable and has an affinity constant in isolated preparations of about 0.2 m*M*.[30] The transport system in the intestine requires energy and is sodium-dependent.[30]The evidence suggests that the moiety actually transported is thiamin itself.[30,31] The transport of thiamin from the lumen to the serosal side of the gut does not appear to be associated with phosphorylation and dephosphorylation intracellularly.[30,31] Because of this active carrier-mediated transport system in the intestine, an oral intake of thiamin greater than 10 mg does not generally increase blood levels of the vitamin significantly.[30,31] This is because the carrier-mediated transport system is essentially completely saturated at this dose.[30,31]

C. Transport into Brain and Cerebrospinal Fluid

The concentrations of total thiamin (all thiamin forms) in brain, cerebrospinal fluid, choroid plexus, and plasma of rabbits are shown in Table 1. Like ascorbic acid, the entry of thiamin into rabbit cerebrospinal fluid and brain is by a saturable process in vivo.[26] This process is one half saturated at a concentration of approximately 0.5 μM total thiamin in the plasma.[26] At the normal plasma total thiamin concentration, less than 5% of the total thiamin entry into cerebrospinal fluid, choroid plexus, and brain is by simple diffusion in rabbits.[26] The relative turnover of total thiamin in rabbit choroid plexus, brain, and cerebrospinal fluid was 5, 2, and 14%/hr, respectively, when measured by the penetration of ^{35}S labeled thiamin injected into blood.[26] In vivo, the clearance of ^{35}S thiamin in tracer concentrations relative to mannitol from CSF was not saturable after the intracerebroventricular injection of various concentrations of thiamin.[26] However, a portion of the (^{35}S) thiamin cleared from the CSF entered rabbit brain by a saturable mechanism that depended upon phosphorylation of the (^{35}S) thiamin.[26] In vitro, choroid plexuses and brain slices accumulate (^{35}S) thiamin against a concentration gradient by an active saturable process.[26] As in the case of ascorbic acid, these results with thiamin have shown that the entry of total thiamin

thiamin into brain and CSF from blood is regulated by saturable transport systems.[26] The loci of these system are, in part, in the choroid plexus and brain cells themselves.[26] Such systems exist in man.[26] However, the biochemical details of the mechanism by which thiamin and/or thiamin phosphates enter cerebrospinal fluid and brain, and the quantitative aspects about the route of entry remain uncertain.[26]

D. Thiamin Homeostasis in Brain and Cerebrospinal Fluid

That the total thiamin levels in brain are homeostatically regulated has been known for many years. In thiamin-depleted rats, brain levels of total thiamin are much better maintained than total thiamin levels in other organs.[26] On the other hand, in intact animals, there is a barrier to the entry of free thiamin and even more so to the phosphorylated forms of thiamin after intravenous injections.[26] The studies discussed above suggest that several mechanisms in the mammalian body regulate thiamin levels in brain. First, the gut and kidney tend to regulate the plasma levels of total thiamin by a group of saturable transport systems and metabolism.[26,30,31] Second, there are mechanisms (discussed above) at the blood-cerebrospinal fluid barrier (choroid plexus and/or blood-brain barrier) that regulate the concentration of total thiamin in the extracellular space of brain and cerebrospinal fluid.[26] Third, the cells of the brain themselves have saturable uptake and release systems that further tend to regulate the total thiamin concentration inside the cell.[32] These mechanisms make a three-tiered system and tend to protect the brain from fluctuations in dietary ingestion of thiamin.[26]

E. Thiamin Deficiency Encephalopathy

Thiamin deficiency encephalopathy will be considered in great detail by Dr. Blass in his discussion of thiamin in the Wernicke-Korsakoff syndrome in this book.

VI. FOLATES

A. Vitamers and Roles in Brain

In brain, the folates exist predominantly as polyglutamate forms of dihydrofolate and tetrahydrofolate.[33] The roles of the various folates in brain are uncertain, but may involve the synthesis of purines, thymidine, glycine, and the recycling of homocysteine back to methionine. However, all these substances may enter brain from blood and the relative importance of the folate-dependent pathways in human brain for the synthesis of these essential substances is unknown.[14] We have recently been able to show unequivocally that, in adult rats, folates are normally involved in the recycling of methionine from homocysteine via methyltetrahydrofolate.[34] Thus, at least in rats, there is one unequivocal role for folate in brain.

B. Requirement

A human requires approximately 0.1 mg of folate per day. The polyglumate forms are hydrolyzed in the gut and are almost as effective nutritionally as the oxidized or reduced folate monoglutamates.[35]

C. Transport into Cerebrospinal Fluid and Brain

The concentrations of folates in plasma, CSF, choroid plexus, and brain in rabbits and in plasma and CSF of man are shown in Tables 1 and 2. The predominant form of folate in both man and rabbits in cerebrospinal fluid and plasma is N-5-methyltetrahydrofolate.[36-38] As noted in Tables 1 and 2, the concentration of folate in cerebrospinal fluid is generally 3 to 5 times higher than in plasma. Methytetrahydrofolic acid, but not folic acid, readily enters the choroid plexus and cerebrospinal fluid and probably brain by a saturable transport system.[37] In both rabbits and man, the entry of methyltetrahydrofolic acid from blood into CSF is approximately one half saturated at the normal plasma concentration of approximately 0.02 μM methyltetrahydrofolate (Tables 1 and 2).[37,38] In contrast, folic acid, but not methyltetrahydrofolic acid, was readily cleared from cerebrospinal fluid to blood by a saturable system after the intracerebroventricular injection of radiolabeled folic acid or methyltetrahydrofolic acid into rabbits.[37] As in the case of ascorbic acid, the evidence is very strong that the choroid plexus is involved in the transport of methyltetrahydrofolic acid up a concentration gradient (Tables 1 and 2) from plasma into cerebrospinal fluid. In vitro, rabbit choroid plexus contains a vigorous transport system for N-5-methytetrahydrofolate that is one half saturated at about 0.02 μM which is approximately the normal plasma concentration.[36] Rabbit choroid plexus is also able to concentrate oxidized folates like folic acid by an extremely high affinity concentrating mechanism.[36] Within the choroid plexus, a folate-binding protein has been partially purified and characterized.[39] However, the exact relationship between the choroid plexus folate binding protein and folate transport in vivo between plasma and cerebrospinal fluid is only incompletely understood.[39] It should be recognized that without the folate transport system in the choroid plexus, the concentration of folate (due to its size and charge) in cerebrospinal fluid would be expected to be 1 or 2% of that in plasma and not 400% (Tables 1 and 2).[12,13] In summary, methyltetrahydrofolate is readily transported from plasma into cerebrospinal fluid through the choroid plexus by a saturable transport mechanism, whereas folic acid is readily transported out of cerebrospinal fluid into plasma.[37] The exact mechanism by which methyltetrahydrofolate enters brain cells from the extracellular space of brain and the cerebrospinal fluid is at present unknown.

D. Folate Homeostasis in Brain and Cerebrospinal Fluid

The above-mentioned studies help explain the finding that it is much more difficult to cause folate depletion in brain than in other organs, such as the liver and bone marrow.[37] On the other hand, it is difficult to raise the concentration of methyltetrahydrofolate in cerebrospinal fluid and brain by raising the plasma levels of methyltetrahydrofolate.[37,38] As discussed above in the case of ascorbic acid, the entry of a vitamin (methyltetrahydrofolate in this case) from plasma into cerebrospinal fluid by an active carrier-mediated process explains, in large part, the relationship of the cerebrospinal fluid to plasma concentrations (when the system is normally approximately one half saturated). Presumably, it is difficult to deplete the brain of folates since the brain cells are surrounded by fluid whose concentration of folates is maintained relatively constant.

E. Folate Deficiency Encephalopathy

In recent years, it has been established in man that there is an association between folate deficiency and certain neuropsychiatric and neurologic abnormalities.[40,41] These abnormalities are quite rare (presumably) because of the difficulty in depleting folates in the central nervous system.[37,40,41] However, definite neurological and neuropsychiatric symptoms have been described which are related to a folate deficiency encephalopathy.[40,41] This process may resolve, although slowly, after the ingestion of large amounts of folic acid.[40,41]

VII. VITAMIN B_6

A. Vitamers and Roles in Brain

In brain, as in other tissues, there are six vitamers of vitamin B_6 normally found including pyridoxine, pyridoxal, pyridoxamine, pyridoxine-5′-phosphate, pyridoxal-5′-phosphate, and pyridoxamine-5′-phosphate.[18-21] The nonphosphorylated vitamin B_6 vitamers can all be phosphorylated by the enzyme pyridoxal kinase to their respective phosphates.[18,19] Pyridoxine-5′-phosphate and pyridoxamine-5′-phosphate can be oxidized to pyridoxal-5′-phosphate by the enzyme pyridoxine phosphate oxidase.[18,19,21] Phosphorylated forms of the vitamer can be dephosphorylated by phosphatases present in brain.[20,21] The active forms of vitamin B_6 are pyridoxal-5′-phosphate and pyridoxamine-5′-phosphate which serve as enzymatic cofactors.[18] There are more than 50 vitamin B_6-dependent enzymes in mammalian tissues.[42] Most of these enzymes have been shown to be present in brain.[42] Many of these enzymatic reactions are involved in amino acid or carbohydrate metabolism.[42] However, vitamin B_6 also plays an essential role in other pathways in brain, including catecholamine synthesis.[42]

B. Transport into Cerebrospinal Fluid and Brain

The concentration of total vitamin B_6 (all six vitamers) in rabbit plasma, CSF, choroid plexus, and brain are shown in Table 1. The nonphosphorylated forms of vitamin B_6 enter cerebrospinal fluid, choroid plexus, and brain in vivo by a process that is one half saturated at a plasma concentration of approximately 1 to 2 μM.[18] The phosphorylated forms of vitamin B_6 in plasma enter brain and cerebrospinal fluid poorly compared to the nonphosphorylated forms.[18,20] When injected directly into cerebrospinal fluid, the nonphosphorylated forms of vitamin B_6 enter brain by a saturable accumulation system.[18] The phosphorylated forms do not seem to be able to pass through brain cell membranes intact.[20] The probable mechanism by which brain cells accumulate vitamin B_6 intracellularly depends on faciltated diffusion of the nonphosphorylated vitamin B_6 forms through the plasma membrane with intracellular trapping by phosphorylation.[18-20] Brain slices release only dephosphorylated vitamers, whereas choroid plexus releases both phosphorylated and nonphosphorylated vitamers.[19] The choroid plexus is probably the source of the phosphorylated vitamers in cerebrospinal fluid.[19] As with vitamin C, there is a saturable transport system for vitamin B_6 (presumably, in part, within the choroid plexus) that regulates the entry of vitamin B_6 from plasma into the cerebrospinal fluid and the extracellular space of brain.[18,19] Then, from the extracellular space of brain and cerebrospinal fluid, vitamin B_6 enters brain cells by a high-affinity saturable accumulation system.[18,19] This latter system (in brain cells) depends, in part, on pyridoxal kinase and the phosphorylation of the nonphosphorylated (vitamin B_6) vitamers which pass through the plasma membrane.[18-21]

C. Vitamin B_6 Homeostasis in Brain

The studies reported above help explain why it is difficult to elevate the total levels of vitamin B_6 in brain, even after the parenteral administration of massive doses of pyridoxine for several days.[21] The injection of 200 mg/kg of pyridoxine hydrochloride for 3 days to rabbits led only to an increase of pyridoxal phosphate in brain by 30%.[21] There was no effect on the concentration of pyridoxal itself in brain.[21] The transport studies of vitamin B_6 between plasma and cerebrospinal fluid and brain have shown that the regulation of entry and exit of vitamin B_6 by saturable transport systems is one important factor in determining the total vitamin B_6 concentration in brain.[19,21] Other important factors (besides the actual transfer of vitamin B_6 through the cell membranes) would include the concentration of vitamin B_6 interconverting enzymes and tissue binding (for example, to enzymes).[18,19] In vivo, there appears to be no metabolism of vitamin B_6 to inactive forms in brain.[18,19] Metabolism appears to occur predominantly in the liver with the formation of 4-pyridoxic acid from pyridoxal.[18]

D. Pyridoxine Deficiency Encephalopathy

In children, a clear cut syndrome of pyridoxine deficiency has been reported.[42] This deficiency syndrome primarily affects the central nervous system.[42] A well-documented epidemic of vitamin B_6 deficiency encephalopathy with convulsive seizures occurred in infants receiving a proprietary milk formula containing almost no vitamin B_6.[42] The seizures were relieved immediately after the parenteral injection of pyridoxine.[42] In adults, central nervous system symptoms including ataxia, hyperacusis, and hyperirritability as well as convulsions have been ascribed to severe vitamin B_6 deficiency states.[42] Thus, as in the case of thiamin deficiency, when severe, the homeostatic systems for vitamin B_6 in the central nervous system can ultimately fail with the evolution of a pyridoxine-deficiency encephalopathy.

VIII. NIACIN

A. Vitamers and Roles in Brain

The concentration of total niacin in brain, cerebrospinal fluid, and serum is shown in Table 1.[6,7] Total niacin is defined as the amount of niacin in niacin-containing vitamers plus niacin itself in the tissues. The following niacin-containing compounds (vitamers) are found in the tissues including brain: niacin riboside, niacin mononucleotide, niacin adenine dinucleotide, niacinamide adenine dinucleotide (NAD), niacinamide adenine dinucleotide phosphate, and the reduced forms of the latter two vitamers. The niacinamide-containing coenzymes are involved in many essential enzymatic reactions in brain.[6,7] In brain, niacin (and niacinamide) cannot be synthesized from tryptophan because, unlike liver and kidney, brain lacks the final enzyme in the synthetic pathway-quinolinate phosphoribosyltransferase.[43] However, brain homogenates, like liver, have the capacity to synthesize niacinamide adenine dinucleotide from niacinamide via niacinamide mononucleotide and from niacin via niacin mononucleotide and niacin adenine dinucleotide.[6,7] Also, like liver, brain contains NADase which exchanges niacinamide in NAD with unbound niacinamide.[6,7] Thus, as discussed above, niacin (or niacinamide) are requiements for brain, although niacin can be synthesized from tryptophan in mammalian liver if adequate tryptophan is provided in the diet.

B. Absorption

Niacin and niacinamide are readily absorbed from the gut. A daily intake of 15 mg/day of niacin or niacinamide are recommended to prevent deficiency. Approximately 1 mg of niacin can be formed from 60 mg of tryptophan.

C. Transport into Cerebrospinal Fluid and Brain

Niacinamide and niacin behave differently insofar as their transport into the cerebrospinal fluid, choroid plexus, and brain are concerned.[6,7] Niacinamide readily enters the cerebrospinal fluid of rabbits from plasma by facilitated diffusion.[7] The half-saturation concentration for that process is greater than 2.0 mM, but has not been determined.[7] Niacinamide enters brain and choroid plexus from plasma by saturable process with a half-saturation concentration of approximately 2 to 4 μM.[7] When tracer concentrations of ^{14}C niacinamide are injected intracerebroventricularly into rabbits, some of the ^{14}C niacinamide in brain is incorporated directly into NAD probably via NADase.[7] The remaining (^{14}C) niacinamide is rapidly removed from cerebrospinal fluid by facilitated diffusion.[7] In vitro, rabbit brain slices contain a high affinity accumulation system for (^{14}C) niacinamide.[6] This system rapidly accumulates (^{14}C) niacinamide by incorporating it into NAD presumably by way of NADase.[6]

Unlike niacinamde, niacin itself poorly penetrates into the cerebrospinal fluid (and presumably the extracellular space of brain) unless it is converted to niacinamide.[7] However, once within the cerebrospinal fluid, the ^{14}C niacin can be converted by brain cells into niacin mononucleotide, niacin adenine dinucleotide, and finally to niacinamide adenine dinucleotide (NAD) itself.[7,44] The uptake of niacin from the extracellular space of brain by brain slices is a very high-affinity, low-capacity system.[6] Niacin riboside and niacinamide riboside can probably also be transported from the extracellular space of brain into brain cells.[6,44] However, as in the case of the phosphorylated vitamin B_6 vitamers, the phosphorylated forms of niacin and niacinamide cannot penetrate brain cells without first being dephosphorylated.[44]

D. Total Niacin Homeostasis in Brain

It has been known for many years that total niacin and NAD levels in brain are homeostatically regulated. Although it is relatively easy to produce symptomatic niacin deficiencies in animals, total niacin and NAD levels in brain are much better maintained than in liver in deficient animals.[45,46] At extremely high plasma niacinamide levels, brain NAD levels, unlike liver, increase only slightly.[45] Even the injection of massive amounts of niacinamide or NAD (500 mg/kg) results in only about a 50% increase in NAD levels in brain. The results of the transport studies reported above suggest that the control of entry and exit of niacinamide and/or niacin is the mechanism, at least in part, by which total niacin and NAD levels in brain cells are regulated. In the case of niacinamide which readily passes between the cerebrospinal fluid and plasma (most probably by a low-affinity facilitated diffusion system), the regulation of entry of niacinamide into brain cells by a high affinity accumulation system is an integral part of the homeostatic system.[7] In the case of niacin, penetration into CSF and the extracellular space of brain from plasma (which is poor) as well as regulation of entry into brain cells themselves by a saturable accumulation system, are two distinct parts of the homeostatic system which occur in series.[6,7] In vivo, any niacin that enters the central nervous system by simple diffusion is converted to niacinamide.[7]

E. Niacin Deficiency Encephalopathy

Niacin and tryptophan deficiency is well known as Pellagra.[47]

IX. OTHER VITAMINS

Megavitamin doses of riboflavin and vitamin B_{12} have been used for the therapy of several brain disorders.[8,11] However, there is inadequate information available in both animals and man to further discuss, at length, the possible transport and/or homeostasis of these vitamins in the central nervous system. In the case of riboflavin as in the case of vitamin B_6, it does appear that riboflavin (and not riboflavin mononucleotide or flavin adenine dinucleotide) is the moiety transported from plasma into brain.[48]

X. MEGAVITAMIN RESPONSIVE BRAIN DISORDERS

Approximately 14 vitamin-responsive inherited diseases of the nervous system have been described in the last 30 years.[49,50] These patients need more than the usual daily requirement of the vitamin to prevent vitamin deficiency states.[49,50] In some cases, 10 or 100 times the usual daily requirement are necessary to prevent the disease process. Perhaps the best known example of a vitamin responsive brain disorder is that of pyridoxine dependent seizures.[49,50] In 1954, several children were found who had episodes of jitteriness that progressed to grand mal seizures.[42,49,50] It was discovered that these seizures could be contolled with pharmacologic (but not physiologic) doses of vitamin B_6. Moreover, it was found that, if the diagnosis was made early and the treatment of pyridoxine dependent seizures instituted immediately upon diagnosis, normal mental development occurred.[49,50] The pathophysiology of this disorder (pyridoxine dependent seizures) has not been proven, but some evidence suggests that glutamic acid decarboxylase activity may be inadequate in these patients.[42,49]

Beside pyridoxine responsive seizures, other vitamin responsive disorders (affecting the central nervous system) have been reported that can be reversed by thiamin, vitamin B_6, vitamin B_{12}, folic acid, biotin, and niacin.[49,50] These disorders have recently been reviewed in-depth by several authors. It is worth noting that Dr. Leon Rosenberg, one of the pioneers in the field of vitamin responsive neurologic diseases, feels that although these conditions are quite rare, they must be considered in the differential diagnosis of patients with various types of neurological deficits, because their recognition leads to effective long-term treatment.[49] Dr. Rosenberg is "sympathetic to the idea that a short course of vitamins in large amounts represents a valuable therapeutic trial in patients with neurological or psychiatric problems of unknown nature. This, after all, is the way this field began with a description of pyridoxine dependent seizures."[49] Dr. Rosenberg, however, is opposed to the uncritical use of pharmacologic doses of vitamins for two reasons.[49] First, he feels that megavitamin therapy may lead to false hopes on the part of the patient or the patient's family, and second, may also run counter to the dictum "Above all do no harm."[49]

XI. MEGAVITAMIN THERAPY

The use of pharmacologic doses of the various water-soluble vitamins, either singly or in combination, has been suggested for a wide variety of neurologic or neuropsychiatric disorders.[8-10] These disorders include schizophrenia and minimal brain dysfunction in hyperkinetic children.[8-10] Dr. Pauling and Dr. Hoffer have argued forcefully for the use of megavitamin therapy in the treatment of schizophrenia.[8,9] Dr. Cott feels that megavitamin therapy is a useful approach to behavioral disorders and learning disabilities in children.[10] However, the medical establishment, in general, has frowned upon the use of megavitamin therapy for these conditions because of the lack of well-controlled, reproducible studies that have shown the effectiveness of megavitamin therapy in these conditions.[8] A critical review of this literature suggests to the author that, if megavitamin therapy is, in fact, useful for the various conditions for which it has been advocated, its utility is minimal and can only be observed in a relatively small numbers of patients. In the author's view, the claims for the efficacy of megavitamin therapy are excessive in view of the lack of clear evidence for these claims.[51] The reader is invited to review some of the evidence on this subject before making up his mind about the utility or lack of utility of megavitamin therapy.[8-11,51]

XII. IMPLICATIONS OF THE VITAMIN HOMEOSTATIC SYSTEMS IN THE CENTRAL NERVOUS SYSTEM FOR MEGAVITAMIN THERAPY OF BRAIN DISORDERS

It is perhaps worthwhile to discuss first the implications of the ascorbic acid homeostatic systems in the central nervous system for megavitamin therapy with ascorbic acid. This would appear reasonable since we know more about the ascorbic acid systems in both animals and man. Except at very high or very low plasma ascorbic acid concentrations, the homeostatic systems discussed above would tend to maintain relatively constant brain concentrations of ascorbic acid irrespective of the plasma levels (Tables 2 and 3). In addition, very high plasma levels of vitamin C are unlikely because of the regulatory mechanisms in the gut and kidneys. As noted above, a saturable transport system limits ascorbic acid absorption in the gut.[4,5] Tubular ascorbic acid reabsorption in the kidney is exceeded at a plasma concentration of 90 μM.[4,5] Because of these two mechanisms, it is difficult to achieve plasma levels greater than 100 μM unless the vitamin is injected intravenously. Moreover, even if the plasma level were increased from 57 μM to 114 μM, the brain level would increase by less than 10% (Table 3).[4,5] On the other hand, only in extreme vitamin deficiency states would the brain ascorbic acid level fall below half the normal value 1770 μM (Table 3).[4,5] Therefore, unless a small change in the brain concentration of ascorbic acid (Tables 1 and 3) makes a large functional difference, it is difficult to see how megavitamin therapy would alter brain function in a patient who was not markedly vitamin deficient.[4,5]

It is possible that megavitamin therapy could alter brain function indirectly by changing the concentrations of other substances which secondarily enter brain from blood. However, if there are patients who respond to megavitamin therapy, it is also possible that such patients have deficient or altered vitamin transport systems with high saturation concentrations or low maximal transport velocities or both. Several retarded children who are unable to transfer folates from blood into cerebrospinal fluid have been described.[52] Almost certainly, the folate transport systems in the central nervous system of these patients are abnormal.[36-39] Finally, in diseases such as meningitis, the normal cerebrospinal fluid to plasma vitamin gradients are diminished or abolished because of interference with transport. Whether some of the unexplained signs or symptoms of meningitis result from focal vitamin deficiencies is uncertain.

Similar considerations apply to thiamin, the folates, niacin, and vitamin B_6. However, the homeostatic systems of these vitamins are probably not quite as effective as the homeostatic systems for ascorbic acid, in that no clear ascorbic acid deficiency encephalopathy has been described. An alternative possibility is that the brain is much less sensitive to the lowering of the vitamin C concentration (in brain) than for the other vitamins.

Thus, unless small increases in coenzyme levels in brain are able to shift the postulated equilibrium between apoenzyme and coenzyme toward holoenzyme (as suggested by Dr. Pauling),[8] it is difficult to see how megavitamin therapy could increase altered apoenzyme activity. However, the author cautions that such considerations do not necessarily justify a rejection of orthomolecular methods as adjunctive therapy in the treatment of schizophrenia or other central nervous system diseases. Instead, direct clinical evidence from controlled trials must be the ultimate measure of the efficacy or worthlessness of megavitamin therapy in these conditions.

XIII. CONCLUSIONS

The central nervous system has many specialized processes that normally regulate the concentrations of water-soluble vitamins in the brain. These processes tend to provide an ultrastable environment for the brain. These homeostatic mechanisms can break down in extreme nutritional deficiency states, when they are disrupted by diseases such as meningitis, and, probably, in rare genetic case. These systems have important implications for megavitamin therapy and for those who attempt to increase the levels of these vitamins in brain.

ADDENDUM

Since the submission of this chapter, we have made substantial progress in our understanding of riboflavin homeostasis in the central nervous system.[53-56]

REFERENCES

1. *Webster's New Collegiate Dictionary,* G. and C. Merriam Co., Springfield, Ill., 1956.
2. **Spector, R. and Lorenzo, A. V.,** Ascorbic acid homeostasis in the central nervous system, *Am. J. Physiol.,* 225, 757, 1973.
3. **Spector, R. and Lorenzo, A. V.,** The specificity of the ascorbic acid transport system of the central nervous system, *Am. J. Physiol.,* 226, 1468, 1974.
4. **Spector, R., Spector, A. Z., and Snodgrass, S. R.,** A model for transport in the central nervous system, *Am. J. Physiol.,* 232, R73, 1977.
5. **Spector, R.,** Vitamin homeostasis in the central nervous system; seminar in medicine, *N. Engl. J. Med.,* 296, 1393, 1977.
6. **Spector, R. and Kelley, P.,** Niacin and niacinamide accumulation by brain slices and choroid plexus in vitro, *J. Neurochem.,* 33, 291, 1979.
7. **Spector, R.,** Niacin and niacinamide transport in the central nervous system, *J. Neurochem.,* 33, 1285, 1979.
8. **Pauling, L.,** On the orthomolecular environment of the mind: orthomolecular theory, *Am. J. Psychiatry,* 131, 1251, 1974.
9. **Hoffer, A.,** An examination of the double-blind method as it has been applied to megavitamin therapy, *J. Orthop. Psych.,* 2, 107, 1973.
10. **Cott, A.,** Megavitamins: the orthomolecular approach to behavioral disorders and learning disabilities, *Acad. Ther.,* 7, 245, 1972.
11. **Arnold, L. E., Christopher, J., Huestis, R. D., and Smeltzer, D. J.,** Megavitamins for minimal brain dysfunction, *JAMA,* 240, 2642, 1978.
12. **Davson, H.,** *Physiology of the Cerebrospinal Fluid,* J. and A. Churchill, London, 1970.
13. **Rapaport, S. I.,** *Blood-Brain Barrier in Physiology and Medicine,* Raven Press, New York, 1976.
14. **Pardridge, W. M. and Oldendorf, W. H.** Transport of metabolic substrates through the blood-brain barrier,. *J. Neurochem.,* 28, 5, 1977.
15. **McComb, J. G., Davson, H., and Hollingsworth, J. R.,** Attempted separation of blood-brain and blood-cerebrospinal fluid barriers in the rabbit, in *The Ocular and Cerebrospinal Fluids,* Bito, L. Z., Davson, H., and Fenstermacher, J. D., Eds., Academic Press, New York, 1977, 333.
16. **Saunders, N. R.,** Ontogeny of the blood-brain barrier, in *The Ocular and Cerebrospinal Fluids,* Bito, L. Z., Davson, H., and Fenstermacher, J. D., Eds., Academic Press, New York, 1977, 523.
17. **Lorenzo, A. V. and Spector, R.** Transport phenomena in the nervous system: physiological and pathological aspects, in *Advances in Experimental Medicine and Biology,* Vol. 69, Levi, G., Battistin, L., and Lajtha, A., Eds., Plenum Press, New York, 1976, 447.
18. **Spector, R.,** Vitamin B_6 transport in the central nervous system, *in vivo* studies, *J. Neurochem.,* 30, 881, 1978.
19. **Spector, R.,** Vitamin B_6 transport in the central nervous system, *in vitro* studies, *J. Neurochem.,* 30, 889, 1978.
20. **Spector, R. and Greenwald, L.,** Transport and metabolism of vitamin B_6 in rabbit brain and choroid plexus, *J. Biol. Chem.,* 253, 2373, 1978.
21. **Spector, R. and Shikuma, S.,** The stability of vitamin B_6 accumulation and pyridoxal kinase activity in rabbit brain and choroid plexus, *J. Neurochem.,* 31, 1403, 1978.
22. **Seugi, A., Schaefer, A., and Komlos, M.,** Protective role of brain ascorbic acid content against lipid peroxidation; *Sep. Experientia,* 34, 1056, 1978.
23. **Ridge, B. D., Fairhurst, E., Chadwick, D., and Reynolds, E. H.,** Ascorbic acid concentrations in human plasma and cerebrospinal fluid, *Proc. Nutr. Soc.,* 35, 57A, 1976.
24. **Hammerström, L.,** Autoradiographic studies on the distribution of C^{14}-labelled ascorbic acid and dehydroascorbic acid, *Acta Physiol. Scand. (Suppl.),* 289, 1, 1966.
25. **Oldendorf, W. H.,** Distribution of drugs to the brain, in *Psychopharmacology in the Practice of Medicine,* Jarvik, M. E., Ed., Appleton-Century-Crofts, New York, 1977, 167.
26. **Spector, R.,** Thiamine transport in the central nervous system, *Am. J. Physiol.,* 230, 1101, 1976.
27. **Siess, E., Wittman, J., and Weiland, O.,** Interconversion and kinetic properties of pyrovate dehydroxylase from brain, *Hoppe Seylers Z. Physiol. Chem.,* 352, 447, 1971.
28. **Von Muralt, A.,** The role of thiamine in neurophysiology, *Ann. N.Y. Acad. Sci.,* 98, 499, 1962.
29. **Copper, J. R. and Pincus, J. H.,** *CIBA Foundations Study Group No. 28,* Woklenholme, G. E. W. and O'Connor, M., Eds., Churchill, London, 1967, 112.
30. **Rindi, G. and Ventura, V.,** Thiamine intestinal transport, *Physiol. Rev.,* 52, 821, 1972.
31. **Schaller, K. and Holler, H.,** Thiamine absorption in the rat, *Int. J. Vitam. Nutr. Res.,* 45, 31, 1975.
32. **Gubler, C. J., Fujiwara, M., and Dreyfus, P. M., Eds.,** *Thiamine,* John Wiley & Sons, New York, 1976, 157.

33. **McClain, L. D., Carl, G. F., and Bridgers, W. F.,** Distribution of folic acid coenzymes and folate dependent enzymes in mouse brain, *J. Neurochem.*, 24, 719, 1975.
34. **Spector, R., Coakley, G., and Blakely, R.,** Methionine recycling in brain: a role for folates and B-12, *J. Neurochem.*, 34, 132, 1979.
35. **McIntyre, L. J., Dow, J. W., McIntyre, R. J., and Harding, N. G. L.,** Cellular acquisition of folate: a compartmental model, *J. Mol. Med.*, 1, 3, 1975.
36. **Spector, R. and Lorenzo, A. V.,** Folate transport by the choroid plexus *in vitro, Science,* 187, 540, 1975.
37. **Spector, R. and Lorenzo, A. V.,** Folate transport in the central nervous system, *Am. J. Physiol.*, 229, 777, 1975.
38. **Cramer, J. A., Mattson, R. H., and Brillman, J.,** Folinic acid. Therapy in Epilepsy, *Fed. Proc. Fed. Am. Soc. Exp. Biol.*, 35, 582, 1976.
39. **Spector, R.,** Identification of a folate binding macromolecule in rabbit choroid plexus, *J. Biol. Chem.*, 252, 3364, 1977.
40. **Reynolds, E. H.,** Neurological aspects of folate and vitamin B_{12} metabolism, *Clin. Haematol.*, 5, 661, 1976.
41. **Botez, M. I., Fontaine, F., Botez, T., and Bachevalier, J.,** Folate-responsive neurological and mental disorders: report of 16 cases, *Eur. Neurol.*, 16, 230, 1977.
42. **Weiner, W.,** Vitamin B_6 in the pathogenesis and treatment of diseases of the central nervous system, in *Clinical Neuropharmacology,* Vol. 1, Klawans, H. L., Ed., Raven Press, New York, 1976, 107.
43. **Ikeda, M., Isuji, H., Nakamura, S., Ichiyama, A., Nishizuka, Y., and Hayaishi, O.,** Studies on the brosynthesis of nicotinamide adenine dinucleotide from tryptophan in mammals, *J. Biol. Chem.*, 240, 1395, 1965.
44. **Deguchi, T., Arata, J., Nishizuka, Y., and Hayaishi, O.,** Studies on the biosynthesis of nicotinamide adenine dinucleotide in the brain, *Biochim. Biophys. Acta,* 158, 382, 1968.
45. **Kaplan, N. O., Goldin, A., and Humphreys, S. R.,** Pyridine nucleotide synthesis in the mouse, *J. Biol. Chem.*, 219, 287, 1956.
46. **Garcia-Bunuel, L., McDougal, P. B., and Burch, H. B.,** Oxidized and reduced pyridine nucleotide levels and enzyme activities in brain and liver of niacin deficient rats, *J. Neurochem.*, 9, 589, 1962.
47. **Goldsmith, G. A.,** Niacin-tryptophan relationships in man and niacin requirements, *Am. J. Clin. Nutr.*, 6, 479, 1968.
48. **Nagatsu, T., Nagatsu-Ishibashi, I., Okuda, J., and Yagi, K.,** Incorporation of peripherally administered riboflavin into flavin nucleotides in the brain, *J. Neurochem.*, 14, 207, 1967.
49. **Rosenberg, L. E.,** Vitamin-responsive inherited diseases affecting the nervous system, in *Brain Dysfunction in Metabolic Disorders,* Vol. 53, Plum, F., Ed., Raven Press, New York, 1974, 263.
50. **Hillman, R. E.,** Megavitamin responsive aminoacidopathies, *Pediatr. Clin. North Am.*, 23, 557, 1976.
51. **Cheraskin, E. and Ringsdorf, W. M.,** *Psychodietetics,* Bantam Books, New York, 1974.
52. **Lanzkowsky, P.,** Congenital malabsorption of folate, *Am. J. Med.*, 48, 580, 1970.
53. **Spector, R. and Boose, B.,** Transport of riboflavin by the choroid plexus *in vitro, J. Biol. Chem.*, 254, 10286, 1979.
54. **Spector, R.,** Riboflavin Transport by Brain Slices *in vitro, J. Neurochem.*, in press.
55. **Spector, R.,** Riboflavin homeostasis in the central nervous system, *J. Neurochem.*, in press.
56. **Spector, R.,** Riboflavin transport in the central nervous system: characterization and effects of drugs, *J. Clin. Invest.*, 66, 821, 1980.

Chapter 8

VITAMIN A IN DERMATOLOGY

W. J. Cunliffe

TABLE OF CONTENTS

I. Introduction 158

II. Cutaneous Physiology 159
- A. Summary 159

III. Pharmacology of Vitamin A Acid 160
- A. Topical Absorption 160
- B. Excretion and Metabolism 160
- C. Formulation 160
- D. Mechanism of Action 162
- E. Side Effects 163
- F. Preparations Available in the U.K. and U.S. 165
- G. Summary 165

IV. Clinical Studies with Vitamin A Acid 165
- A. Acne Vulgaris 165
  - 1. Oral Therapy 165
  - 2. Topical Therapy 166
- B. Summary 169

V. Vitamin A Acid Therapy in Other Skin Diseases 169
- A. Psoriasis 169
  - 1. Summary 170
- B. Malignant and Premalignant Conditions 170
  - 1. Summary 171
- C. Ichthyosis 171
  - 1. Summary 172
- D. Keratosis Pilaris 176
- E. Keratosis Follicularis (Darier's Disease) 176
  - 1. Summary 176
- F. Palmoplantar Keratodermas 176
- G. Verrucous Nevus 177
- H. Trichostasis Spinulosa 177
- I. Lichen Planus 180
  - 1. Summary 180
- J. Subcorneal Pustular Dermatoses 180
- K. Molluscum Contagiosum 180
- L. Warts 181
- M. Atopic Eczema 181
  - 1. Summary 181

Acknowledgments 181

References 183

Table 1
THE CHEMICAL STRUCTURE OF RETINOL, RETINAL, AND RETINOIC ACID

Vitamin A

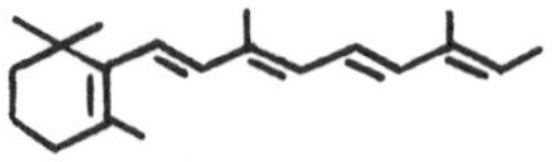	CH_2OH	Retinol, Vitamin A alcohol
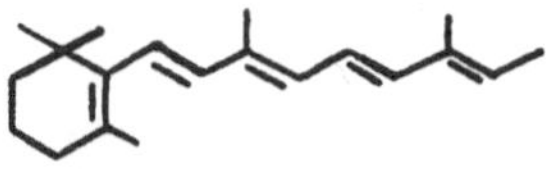	CHO	Retinal, Vitamin A aldehyde
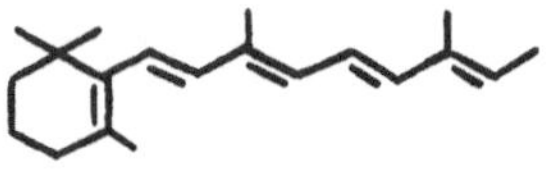	COOH	Retinoic acid, Vitamin A acid

I. INTRODUCTION

Vitamin A is well known for its importance in the differentiation of epithelial tissues. The term vitamin A is used when reference is made to the biological activity of one or more vitamin A active substances. Table 1 shows the structural formulas of three of the more important vitamin A compounds. These are: retinol (vitamin A alcohol), retinal (vitamin A aldehyde), and retinoic acid (vitamin A acid). All three compounds contain as common structural units a trimethyl-cyclohexenyl group and all-transconfigurated polyene chain with four double bonds. They are crystalline substances of limited stability, and their more detailed chemistry has been discussed in a previous chapter.

Oxidation of the alcohol or aldehyde appears to be an irreversible process, which means that reduction of the acid to the alcohol or aldehyde in the organisms is not possible. Vitamin A acid appears to be that form which is tissue (skin) active; it can function for all functions of vitamin A except in the eye and testis.

More recently the synthesis of retinoic acid analogues has been initially aiming at compounds which hopefully posses higher activity with less toxicity. From a large series of compounds (Table 2) synthesized, several retinoids for oral administration are now being studied in patients with skin problems.

Vitamin A preparations, both systemic and topical, have been administered for years for hyperkeratotic skin disorders. In the early 1960s, topical preparations of tretinoin (retinoic acid, vitamin A acid) were first reported to decrease scale formation in the ichthyosiform dermatoses. However, no further publications concerning their topical application appeared until 1968, when Frost and Weinstein[1] described the effects of various forms of vitamin A acid cream in strengths ranging from 0.1% to 0.3% in patients with ichthyosiform dermatoses and psoriasis. Subsequently, occasional papers, usually case reports, have appeared on the efficacy of vitamin A acid in the treatment of such chronic dermatoses as palmar-plantar hyperkeratosis, lamellar ichtyosis, ichthyosis vulgaris, Darier's disease, and psoriasis. Considerable differences in formulation of medication, concentration of vitamin A acid, method and frequency of application, and duration of treatment can be noted in these reports. Adequate clinical trials are often lacking, except for studies on the treatment of acne vulgaris and studies with the newer oral retinoids.

Table 2
SOME OF THE NEW DERIVATIVES OF RETINOIC ACID

New derivatives		
13-cis Retinoic acid	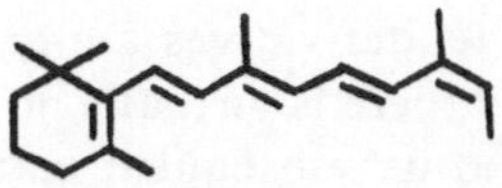	COOH
Retinoic acid ethylamide	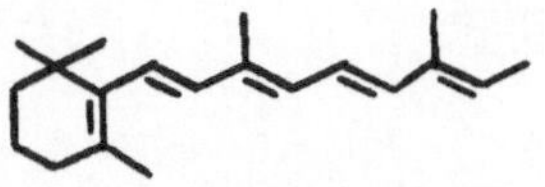	$CONHC_2H_5$
Anhydroretinon acid	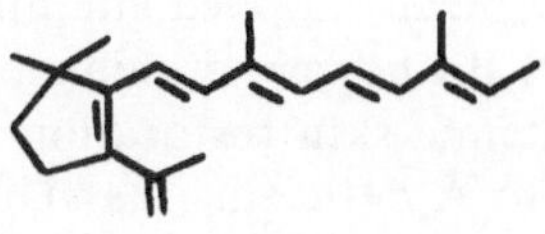	COOH
Aromatic retinoid ethyl ester, RO 10-9369	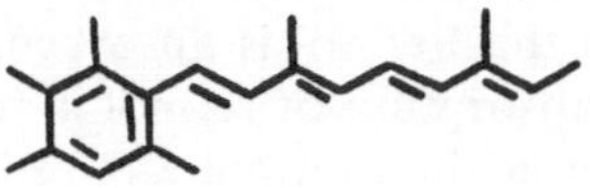	$COOC_2H_5$
Aromatic retinoid ethylamide	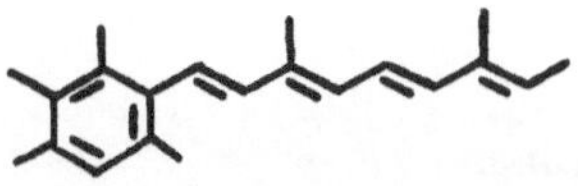	$CONHC_2H_5$

II. CUTANEOUS PHYSIOLOGY

The percutaneous absorption of vitamin A is poorly understood, but once in the skin, one of its major functions is the maintenance of normal epithelial structure. Hyperkeratosis of the skin (hypertrophy of the corneous layer of the skin) as a result of vitamin A deficiency has been observed in man and experimentally produced in animals.[2]

In rats, the effects of vitamin A deficiency in the skin lags behind other physical symptoms of hypovitaminosis A. This seems to indicate that obvious cutaneous histological changes characterizing vitamin A deficiency develop less rapidly than other physical symptoms and chemical changes. It also appears that restoration of normal structures of the A-depleted skin is slower than the resumption of growth, remission of xerophthalmia, and so forth. Although we know much about the pharmacological cutaneous effects of orally and topically applied vitamin A, we know too little about the sites of localization and the precise role of vitamin A in cutaneous physiology in man or animals. Vitamin A is undoubtedly needed for normal keratinization, and in skin vitamin A acid is a major metabolite. It may also be required for the adequate function of zinc and in prostaglandin synthesis.[4] There is evidence that zinc and prostaglandins may be involved in the mediation of cutaneous inflammation.[5,6] Furthermore, it could possibly control sebaceous gland secretion as, a vitamin A acid derivative, 13-cis-retinic acid can significantly decrease sebaceous gland production,[7] and Vitamin A acid is involved in the metabolism of steroid hormones in other organs such as the placenta, testes, and ovary.[8]

A. Summary

1. Many gaps exist in our knowledge.
2. Vitamin A acid is required for normal keratinization.

III. PHARMACOLOGY OF VITAMIN A ACID

In this section we ought to consider the cutaneous pharmacology of both topically applied and orally administered vitamin A acid. The pharmacology of orally administered vitamin A acid and its derivatives are described in detail in another chapter in this volume. Unfortunately there is virtually no information on the amount of vitamin A that reaches the skin and its subsequent cutaneous metabolism when taken orally. There is, however, more information on the pharmacology of vitamin A acid when applied topically.

A. Topical Absorption

Percutaneous absorption of vitamin A has been shown to occur in man and animals.[2] In the albino rat topically applied vitamin A is effective at the site of application only and does not restore the normal architecture of the skin at neighboring sites. This is in contrast to the systemic skin restoration when adequate amounts of vitamin A are orally administered to A-deficient rats. These results suggest that the skin is able to directly utilize topically applied vitamin A at the site of application.

Although it is known that retinol is absorbed through the intact animal skin,[9] evidence of percutaneous absorption of retinol in man is less convincing.[10] Retinoic acid is absorbed topically better than retinol and retinal being an aldehyde is more firmly bound to skin proteins in a Schiff-base linkage to a free amino group and is less well absorbed.[11]

B. Excretion and Metabolism

There is little information on the metabolism and excretion of topically applied vitamin A. The reason for the lack of information is that radioactive techniques are required to measure the levels of vitamin A in the different parts of the skin. Hopefully, the more recently developed technique of high pressure liquid chromatography (HPLC) which allows the measurement of 1 ng will prove helpful.[12] Schaefer and Zesch[13] (Figure 1) have shown that the concentration of vitamin A acid in the skin, after local application of a 0.1% preparation is 10 μM or 3 $\mu g/g$ tissue. The clinical effect resulting from such an application corresponds to that achieved by oral administation of 100 mg vitamin A acid.

Systemic absorption of topically applied vitamin A must occur for three reasons. First, in one patient, it has been shown that when radioactive vitamin A acid was applied to the skin, evidence of the radioactivity appeared in the urine.[13] Second, in hypo-vitamin A induced animals, there is a rise in serum liver and vitamin A after topical application. Third, in a few patients (7.8% of 405 patients) abnormalities of liver enzymes occurred after external application of vitamin A acid,[14] but such an event is uncommon.[15]

C. Formulation

It has been known for 30 years that the vehicle in which the vitamin A is topically applied influences the percutaneous absorption and perhaps also the utilization of the vitamin. Vitamin A is relatively unstable and antioxidants are added to all forms of vehicle. It is best, particularly when in solution, to store in dark containers to protect it from light. The most extensive study was by Plewig, Wolff, and Braun-Falco.[15] These authors showed that the effects of vitamin A acid depended on the isomer, concentration, the vehicle, duration of treatment, the skin area, and the age of the patient. They measured the effect of percutaneous absorption by measuring the induction of redness and scaling and compared the effects of the variables so mentioned.

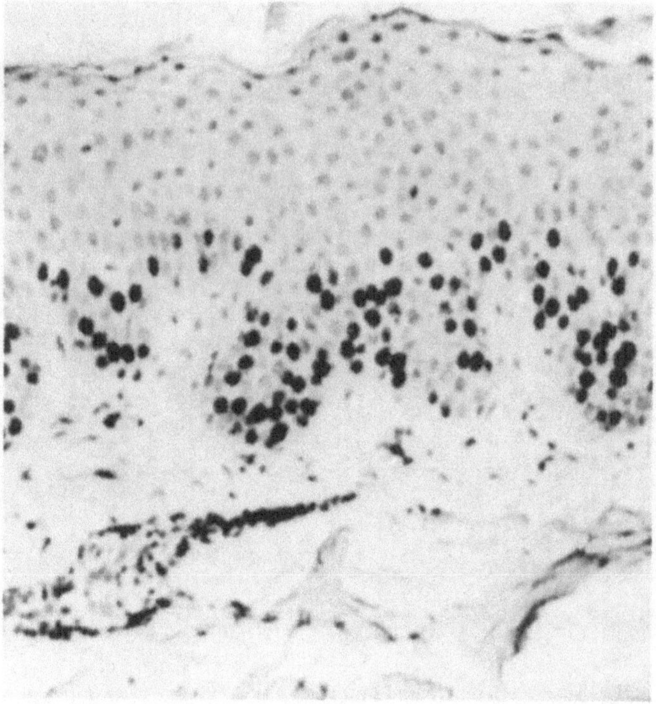

FIGURE 1. The penetration of vitamin A acid 0.1% in i-propanol is shown 100 min after topical application in vivo. (From Schaefer, H. and Zesch, A., *Acta Derm. Venereol.*, Suppl. 74, 50, 1974. With permission.)

They showed that vitamin A acid was the most effective and that concentrations greater than 0.1% virtually always produced some form of redness and scaling. With respect to the vehicle, the preparation was more active in a gel base than in a cream base, and this in turn more active than in a fatty cream base.

Clinically, an initial irritation occurred with reddening and scaling, edema, and moderate burning. After long-term treatment peeling and glistening appearance of the stratum corneum predominated.

Clinical evidence of percutaneous effect was seen within 2 to 5 days of application. This was quicker the higher the initial concentration. Also an increased reaction occurred the younger the patient. There were regional differences; the back and the extensor sides of the extremities tolerated higher concentrations better than the face, neck, and flexor side of the extremities. The intertriginous regions were specially susceptible to irritation. In the facial region, the forehead and the upper part of the lip were less sensitive than the lower part of the lips; the periorbital, perinasal and perioral areas, and submental region are especially sensitive. Experimentally, formulations can be investigated by evaluating their skin irritation potential on rabbit skin.[16] Using such techniques, it was found that relatively high concentrations of retinoic acid (0.1% to 0.2%) were necessary in lipophilic vehicles (in which retinoic acid is relatively insoluble) to achieve a desirable irritation level. Such formulations should be useful for treating ichthyosiform dermatoses. Conversely, it was found that relatively low concentrations of retinoic acid (0.0001% to 0.05%) were preferable in hydrophilic vehicles (in which retinoic acid is relatively soluble) to achieve a desirable irritation level. Such formulations should be useful for treating acne.

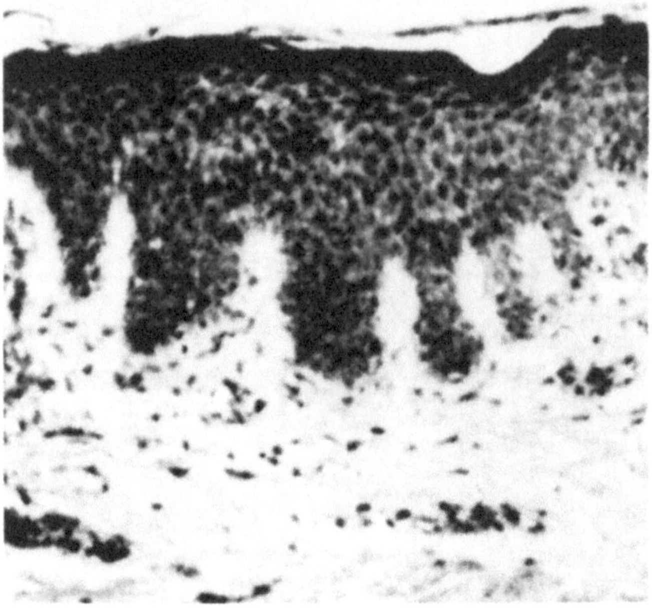

FIGURE 2. The increased uptake of ^{3}H thymidine after application of 0.1% vitamin A acid is shown. There is high proliferative activity. (From Plewig, G., et al., *Acta Derm. Venereol.*, Suppl. 74, 107, 1975. With permission.)

D. Mechanism of Action

Histological investigations following the topical application of vitamin A in man, reveals a pronounced thickening of the middle part of the epidermis (acanthosis) within 3 days. The electron microscope shows that the proliferating epidermis cells also exhibit notable evidence of increased cell function: large nucleoli within distinct nucleolenemia, profuse ribosomes, mitochondria, and Golgi's complexes. The changes in the epidermal cell nuclei, which appear even after short-term vitamin A acid application, suggest that vitamin A acid acts primarily on the nucleic acid metabolism.[15]

Christophers and Braun-Falco[17] also demonstrated a rapid increase of H^3-thymidine labeled cells of the epidermis and a threefold increase of its thickness. Biochemically, 1% retinoic acid caused an increase of RNA-polymerase activity.[18]

Besides the increase in proliferation, an important and particularly noticeable effect of vitamin A acid is its action on epidermal differentiation. Histological evidence of this action is found, in particular, when high concentrations of vitamin A acid are given, in complete loss of the stratum granulosum, and in a parakeratotic mode of cornification (retention of nuclei in the stratum corneum). Histochemical investigations show that even the apparently histologically normal stratum corneum displays a qualitatively abnormal behavior. It is observed electron microscopically that the disturbance of differentiation begins in the deeper layers of the epidermis. Despite the increased cell activity, hardly any structural protein is formed as revealed by a substantial reduction in tonofibrils and desmosomes in comparison with normal conditions. This may also correlate with a reduction in membrane coated granules, whose function is possibly linked with binding or separating the fibrous protein-containing cells. Additional support comes from Karaseck[19]who studied the growth and keratinization of human skin in cell culture in presence of retinoic acid. At low concentrations (0.1 $\mu g/m\ell$), the growth rate was increased and keratinization depressed and was accompanied

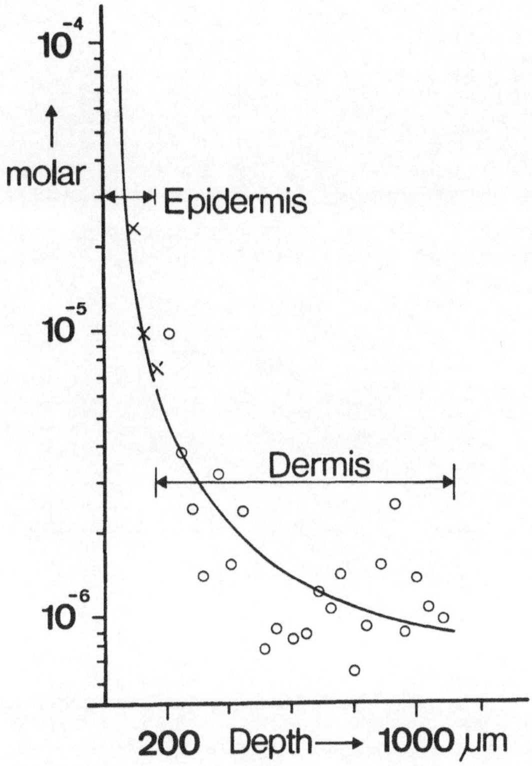

FIGURE 3. The thickening of the epidermis is shown with a prominent granule layer after a long-term (50 days) application of 0.1% vitamin A acid solution. (From Plewig, G., et al., *Acta Derm. Venereol.*, Suppl. 74, 107, 1975. With permission.)

by the formation of an amorphous substance. At more than 10-fold higher concentration, growth was inhibited.

The observed effects occur within a few days of application and become less obvious with increasing duration of application.[20]

To what extent these effects are produced by liberation of catabolic enzymes from lyosomes, remains obscure. In the concentrations employed, at any rate, vitamin A acid does not lead to major toxic effects with necrosis.

Oral anatomic retinoids influence the immunological system by stimulating delayed hypersensitivity and depressing humoral mechanisms.

Pharmacologically, the effects of vitamin A acid on epidermal differentiation are, no doubt, significant and probably explain the favorable effects of vitamin A acid in dermatoses with disturbed keratinization (Figures 2 and 3).

Optimal concentration of vitamin A acid depends on the individual patient tolerance, on the nature and site of the dermatosis, and on the formulation.

E. Side Effects

The major side effects of topically applied vitamin A acid are erythema and scaling, and in more sensitive patients the skin may become edematous and blistered. As indicated earlier, the side effects are related to the type of vitamin A acid derivative, its

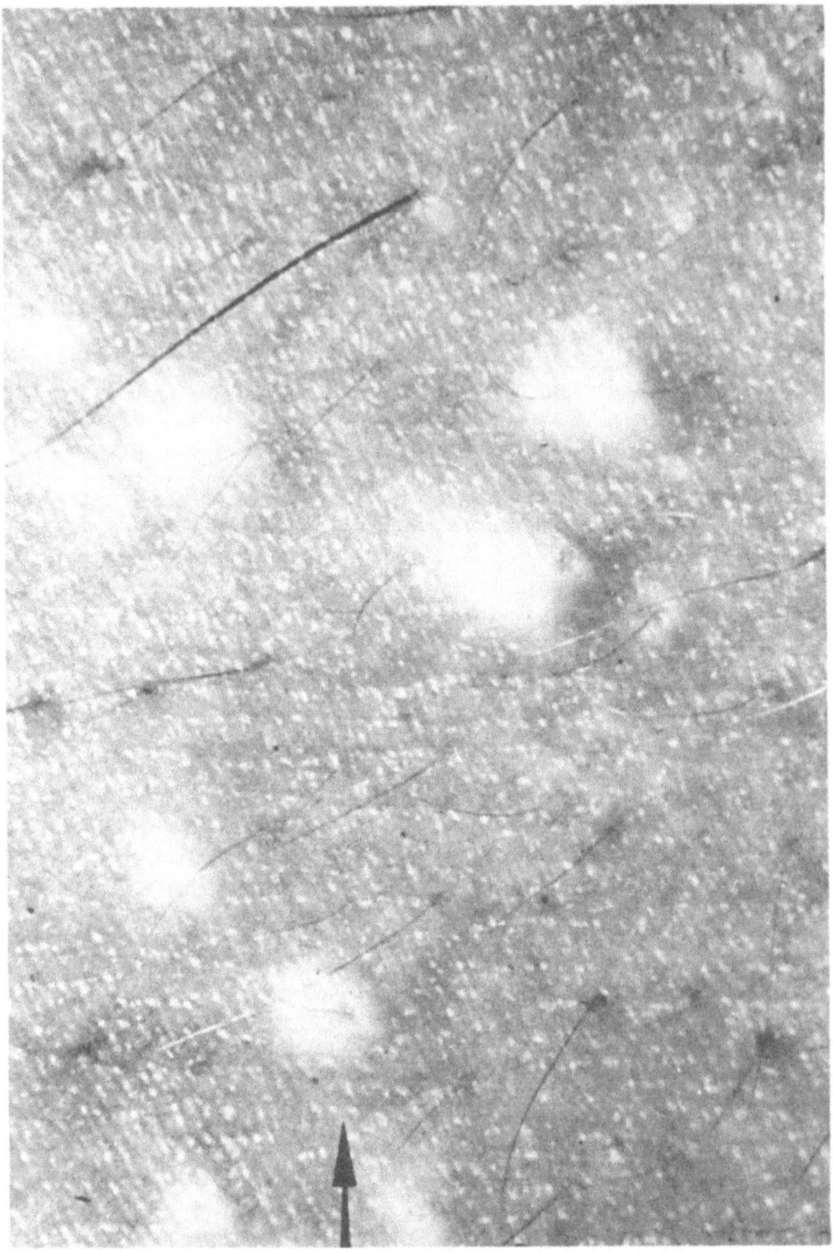

FIGURE 4. Patchy hypopigmentation following treatment with topical vitamin A acid.

vehicle, age of subject, the site of application, and the type of disease being treated. By reducing the frequency of application each day, or by reducing the application to a few times a week, the degree of erythema and scaling will settle, although cessation of therapy may be necessary. There may be an increased sensitivity to ultraviolet light and after prolonged treatment hypopigmentation (Figure 4) occurs. Other topical keratolytic agents and the excessive use of soap should be avoided.

Acne may initially be worsened because of edema in the pilosebaceous duct, producing increased obstruction at that site. However, with continuous treatment, such as in acne, there is some evidence that the erythema and scaling will settle with continued use.[21] Vitamin A acid thus exhibits the pharmacological effect known as tachyphylaxis.

Table 3
THE SIDE EFFECTS OF ORAL VITAMIN A

Side effects	Percentage
Dryness of lips	80%
Headache	77%
Flushing	73%
Dryness of skin	50%
Feeling of tiredness	33%
Lack of appetite	30%
Increase of thirst	30%
Nausea	27%

The side effects of orally administered vitamin A acid are described in detail elsewhere in this volume, but it is worth recording the side effects of oral vitamin A acid used in the treatment of dermatological conditions. Stüttgen[22] investigated 30 patients with a variety of abnormal keratinizing disorders and the major side effects are listed in Table 3.[22] Most side effects were seen at a dose of 100 mg/day, and by reducing the dosage as indicated in the appropriate clinical section, the side effects were greatly reduced, particularly when the maintenance dose is 20 to 40 mg/day. Side effects of the new retinoids are also dose dependent and contraception is essential while on therapy and for 6 months afterwards because of possible teratogenecity.

F. Preparations Available in the U.K. and U.S.

Preparations differ in different parts of the world. Some preparations are available on prescription and commercially available while other formulations, including the aromatic retinoids, can be obtained (sometimes) on request from relevant drug companies.

G. Summary

1. Vitamin A acid orally and topically affects keratinization, but little is known about the detailed cutaneous pharmacology or orally administered vitamin A acid or the related retinoids.
2. Vitamin A acid produces epidermal thickening and reduces formation of fibrous proteins (keratin).
3. The primary effect is unknown, but its action on nuclear metabolism may determine subsequent events.
4. Side effects of oral vitamin A acid are minimal at 20 to 40 mg/day; 100 mg/day can be toxic. Side effects of the new retinoids are also dose related.
5. Side effects of topical vitamin A acid are erythema and scaling, but some degree of erythema and scaling may be needed to produce adequate clinical efficacy.

IV. CLINICAL STUDIES WITH VITAMIN A ACID

A. Acne Vulgaris

1. Oral Therapy

There are no convincing reports on the value of oral retinoic acid in acne,[23] but recently 13-cis retinoic acid has been tried.[7] Although clinical reports with this drug are limited. the results appear convincing, particularly since clinical improvement was associated with a significant reduction (greater than 80%) in sebum production (a variable that usually correlates with acne severity)[24] and an alteration in surface lipid composition.[7] (Figure 5). This drug could revolutionize the treatment of acne in the 1980s.

2. Topical Therapy

A wide variety of different vehicles have been employed with retinoic acid and there are little published data on the availability and stability of retinoic acid in these various vehicles. This makes interpretation of clinical studies somewhat difficult. Kligman, Fulton, and Plewig[25] were the first authors to claim that topical retinoic acid was more effective than benzoyl peroxide, sulfur, or resorcinol. They postulated that this agent worked by increasing the turnover of the epithelium in the pilosebaceous units and comedones, on the one hand, while at the same time promoting dehiscence of keratin which would, in normal circumstances, have lead to follicular obstruction. Thus, follicular keratin becomes more loosely bound together.

More recently, the beneficial effect of topical retinoic acid in acne has been confirmed by many other authors (Pedace and Stoughton,[26] Report from the General Practitioner Research Group,[27] and Christianson et al.[28]). For instance, Pedace and Stoughton[26] studied 61 patients in a double-blind trial. Nearly 90% of the patients improved with retinoic acid, whereas only 25% of the placebo group showed improvement. These workers used a concentration of retinoic acid of 0.05% in a propylethylene glycol-ethanol vehicle. These authors described photoirritant reactions in two patients and increased sensitivity to ultraviolet light has subsequently been reported in some patients receiving this local therapy. Retinoic acid should not be used in conjunction with therapeutic exposure to either natural or artificial ultraviolet light because of this effect. Some patients with an initially darkly pigmented skin developed some depigmentation and some of their private patients found the therapy cosmetically unacceptable. Peachey and Connor,[29] in a study using 0.1% retinoic acid lotion in 22 patients with acne vulgaris, were unenthusiastic about this mode of therapy. Although 13 patients improved, there was an unacceptable level of erythema and peeling. This may have been because the concentration of retinoic acid employed in this trial was too high.

More recently, studies using retinoic acid in a concentration of 0.025% (Retin-A) have been made. For instance, Simpson et al.[30] studied 57 patients suffering from acne vulgaris and treated them for 12 weeks with a 0.025% retinoic acid lotion. Of these subjects, 95% showed some overall benefit, which was particularly marked in patients with early lesions consisting mainly of comedones, papules, and pustules. Retinoic acid at this concentration did not produce irritant side effects and only two patients withdrew from this trial because of such side effects. Retinoic acid seems to be effective in preventing the formation of comedones and in removing those already present. Furthermore, Mills et al.,[31] Leyden et al.,[32] and Christianson et al.[28] have claimed that a combination of retinoic acid and tetracycline was superior to either therapy alone after 8 weeks of treatment. The patients treated had a combination of both comedones and inflammatory lesions. Gould et al.[21] were unable to demonstrate that vitamin A acid and tetracycline was superior to tetracycline alone in an 8-week study, but after the 8 weeks when the tetracycline was stopped, therapy with vitamin A acid maintained satisfactory improvement in contrast to those patients who received placebo only.

The preparation most frequently used as either a gel or a lotion is 0.025% concentration. Retin A seems to be a useful preparation for patients particularly affected by comedones and early inflammatory lesions. Retinoic acid has been successfully used to a treat steroid acne,[33] occupational acne,[34] and senile acne[34] (Figures 6 and 7).

If the clinician decides to use this preparation, patient cooperation is vital. The patient should be told to avoid excessive exposure to ultraviolet light, excessive face wash-

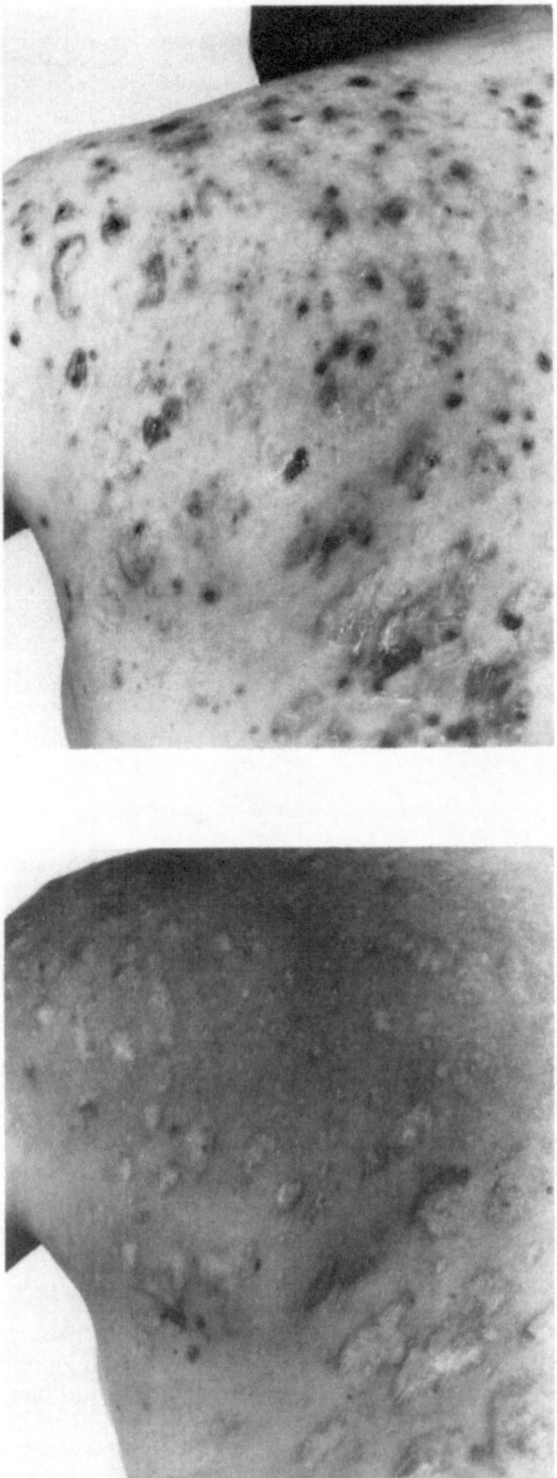

FIGURE 5. The effect of treatment on nodular cystic acne with 13-cis retinoic acid. (A) Before treatment; (B) the result of treatment. (From Peck, G., et al., *N. Engl. J. Med.*, 300, 329, 1979. With permission.)

FIGURE 6. Acne induced by oral steroids.

ing, and excessive use of the product. Care must be taken to ensure that retinoic acid does not come in contact with the corners of the mouth, nose, eyes, and mucous membranes. It must be stressed to the patient that the acne may appear to get worse before it gets better and that at least 6 to 8 weeks of therapy could be necessary before significant clinical gains occur. The patient must be told that as with all acne therapies, treatment must be continued for several months.

Opinions differ on how to use topical vitamin A acid. Some clinicians are more aggressive than others. For example, Plewig et al[15] start intensively with high concentrations and follow this with a prolonged treatment of lower concentrations. With hospitalized patients under close supervision, it is possible to use much higher concentrations than in the case of outpatients. For example, a 0.1% vitamin A acid solution twice daily for 4 to 5 days can be used until the appearance of a pronounced dermatitis. From this point a 0.05% to 0.25% vitamin A acid solution can be used.

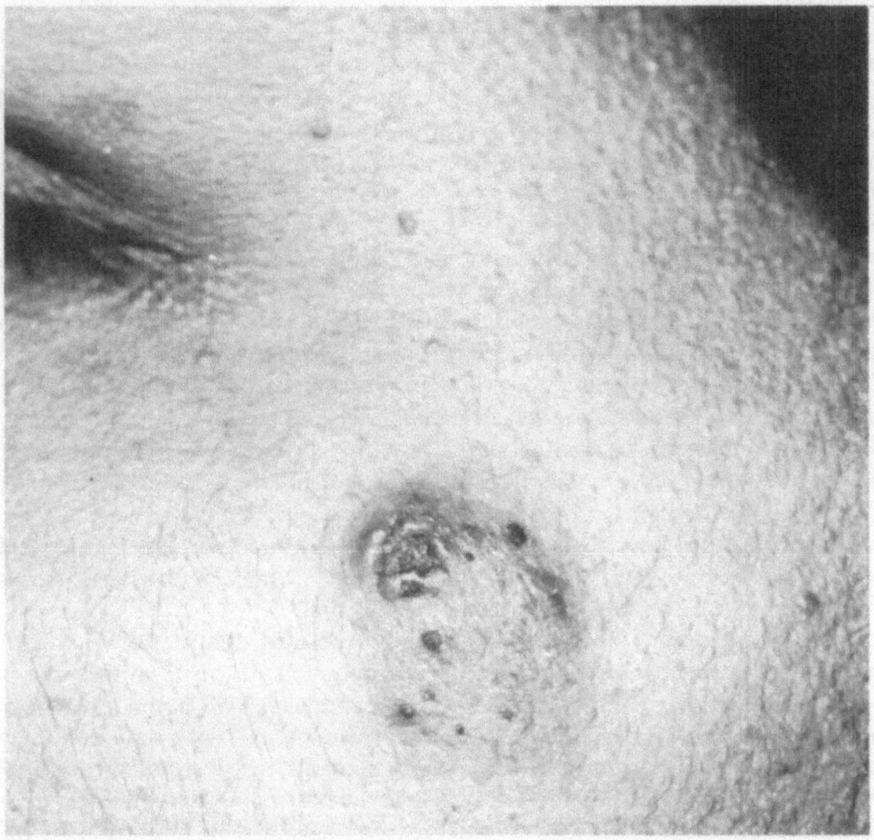

FIGURE 7. Acne induced by topical steroids. The patient had a plaque of sarcoid treated with a potent fluorinated steroid.

Other clinicians only aim to obtain mild erythema and scaling and adjust the application (once or twice daily) to obtain minimal erythema and/or scaling. Often with continued usage these controllable side effects tend to become less evident and the product remains effective.

B. Summary

1. Oral retinoic acid is of doubtful value in acne.
2. Oral 13-cis retinoic acid derivatives appear promising for patients with severe acne, but more detailed studies are required.
3. Topical retinoic acid is an established, proven, commercially available treatment for acne vulgaris.
4. In milder cases retinoic acid can be used alone, but in patients with moderate and severe acne, to obtain beneficial results, long-term oral antibiotics (such as erythromycin, tetracycline, or cotriamoxizole) are also required.[34]
5. Topical vitamin A acid is also useful as a maintainance therapy after stopping long-term oral antibiotics.

V. VITAMIN A ACID THERAPY IN OTHER SKIN DISEASES

A. Psoriasis

The first doctors to indicate a possible role for retinoic acid in psoriasis were Frost and Weinstein.[1] Fredriksson[35] found some improvement in a single-blind study of 40 psoriatics. Histologically, the effect of retinoic acid on psoriasis is characterized, somewhat surprisingly, by a decrease of epidermal thickness, formation of a granular layer, and a significant fall of the mitotic count. These authors, although showing a significantly favorable response in terms of keratin formation and reduced mitotic activity, showed, when the study was extended to 3 months, that many patients were unable to tolerate the ointment in its original concentration (0.1%) because of skin irritation. On theoretical grounds it seemed reasonable that the irritant properties of the retinoic acid could be counteracted by topical steroid preparation. Fry et al.[36] also conducted a double-blind study on 33 patients in which retinoic acid ointment combined with betamethasone-17-valerate was compared with a placebo ointment plus the steroid alone.[37] Objective assessment of the treated areas showed a significant superiority of the retinoic acid and steroid combination over the placebo. Unfortunately, the retinoic acid and steroid combinations are relatively unstable.

In contrast, Günther[38] treated 25 patients with retinoic acid in various bases and only in four was there an obvious improvement. Local skin irritation was a problem in this study. More recently, 26 patients have been treated with two derivatives of retinoic acid (13-cis retinoic acid and retinoic acid ethylamide orally).[39] In 3 weeks, 15 of the 26 showed improvement, but none completely. Many of the patients, particularly those using the 13-cis derivatives, showed chelitis.

In the past decade many dermatological centers are now treating some patients with psoriasis by a treatment referred to as PUVA treatment. In this treatment the patient takes orally, on a body weight basis, 8-methoxy psoralens. Two hours later the patient is exposed to UVA radiation. There is now convincing evidence that this therapy is helpful in psoriasis, although its overall safety remains to be established. In a study of 91 patients, Fritsch et al.[40] demonstrated that the oral administration of a derivative of an all-trans retinoic acid significantly decreased the total time and exposure of UVA needed to clear plaque psoriasis and palmar-plantar psoriasis.

At a recent international meeting in Berlin (November 1980) several eminent dermatologists (Orfanos, Polano, Grupper, Guilhou, and Jablanska) demonstrated the clinical benefits of oral aromatic retinoids in psoriasis.

1. Summary

1. Topical vitamin A acid alone is of no value in psoriasis because of local irritation.
2. When applied with a local steroid topical vitamin A may help but because of stability it cannot, as yet, be made by drug companies.
3. The new aromatic derivatives with or without PUVA treatment look promising.

B. Malignant and Premalignant Conditions

Retinoic acid, although unrelated to known groups of cytotoxic agents, affects of the growth of some tumors. Experimentally, vitamin A acid accelerates the regression of DMBA-induced keratoacanthomata and papillomata.[41-43]

Actinic keratoses are premalignant skin lesions occurring especially on exposed parts of the body in patients 50 years and beyond (Figure 8). Treatment of 60 patients with actinic keratoses with vitamin A acid (0.1% to 0.3%) in an ointment base, produced the following results: 24 responded with a complete and 27 with a partial regression of the lesion; of 16 patients with basal cell epitheliomata, 5 regressed clinically completely, 10 partially (Bollag and Ott[44]). Similar results were obtained by Schumacher and Stüttgen[45] and Belisario.[46] The efficacy of tumor treatment with retinoic acid is clearly below that obtained with the usual methods, such as radiotherapy and surgical removal. In older patients, some dermatologists use the topical antimitotic agent 5-fluorouracil cream and there is evidence that the combined application of vitamin A acid and 5-fluorouracil will satisfactorily clear actinic keratosis.[47] A decisive effect of vitamin A acid in this combination is to enable 5-fluorouracil to reach the target tissue. Oral retinoids have been shown to reduce skin tumors in patients with multiple basal cell carcinoma and xeroderma pigments.

1. Summary

1. Vitamin A acid is of limited value in the treatment of malignant and premalignant skin tumors.
2. It may possibly help patients with senile keratosis, if combined with 5-fluourouracil.
3. Vitamin A acid is not the main treatment for these conditions; surgical excision and radiotherapy are more effective, but oral retinoids may be helpful.

C. Ichthyosis

The ichthyoses refer to a group of scaly dermatoses of which there are many clinical and genetically distinct types.[48] Most appear at, or shortly after, birth and may in some patients cause little or no trouble, whereas in some subjects the defect in scale formation and separation can produce, cosmetically, a most unacceptable appearance.

The management of ichthyotic skin diseases has always been a frustrating therapeutic problem for dermatologists. Bath emollients and topical salicylic acid and urea based ointments have been, and still are, the mainstay of therapy because of their supposedly keratolytic activity. Vitamin A acid possibly is directed at the basic problem — persistent scale formation. Since the early 1960s topical retinoic acid[49] showed promising results. However, inconvenience of application and frequent skin irritation, even if mild, have limited its acceptability for many patients[50] — especially if the patient has atopic eczema.

The results in treating ichthyotic conditions with vitamin A acid are variable, but generally point towards a beneficial effect. Stüttgen[51] was the first to describe the keratolytic action of retinoic acid topically applied in patients with ichthyosis vulgaris. Beer[52] also reported good results with retinoic acid (0.1% in a greasy ointment) in ichthyosis vulgaris, as did Grice et al.[53] in eight patients with sex-linked autosomal dominant ichthyosis and six patients with autosomal dominant ichthyosis. In contrast, Frost and Weinstein[1] saw no improvement in ichthyosis vulgaris, but these authors found topical retinoic acid of value in 16 of 17 patients with lamellar ichthyosis.

Muller et al.[54] treated 23 patients with various types of ichthyosis in a double-blind study comparing 0.1% vitamin A acid cream with 2% salicylic acid ointment. Patients with ichthyosis vulgaris and lamellar ichthyosis responded well. Local adverse reactions, especially burning and excoriation, were not severe and could be controlled by modification of the treatment regime.

Stüttgen[55] used a combination of vitamin A acid (0.1% in an ointment base) combined with oral vitamin A. They preferred to use a topical preparation for about 2 weeks until the treatment became irritant. When this happened the patient was given 20 mg vitamin A acid orally, b.d., and this was decreased to 10 mg b.d. after about 4 weeks. Previous to this type of combined therapy, 100 mg had to be given orally to produce a reasonable response — a level which approaches toxic level. As the disorder improved, these authors also decreased the local concentration of vitamin A acid to either a 0.05% or a 0.01% concentration. After about 2 months the oral therapy can stop, but often has to be reinstated about 2 months later because of relapse. The dose

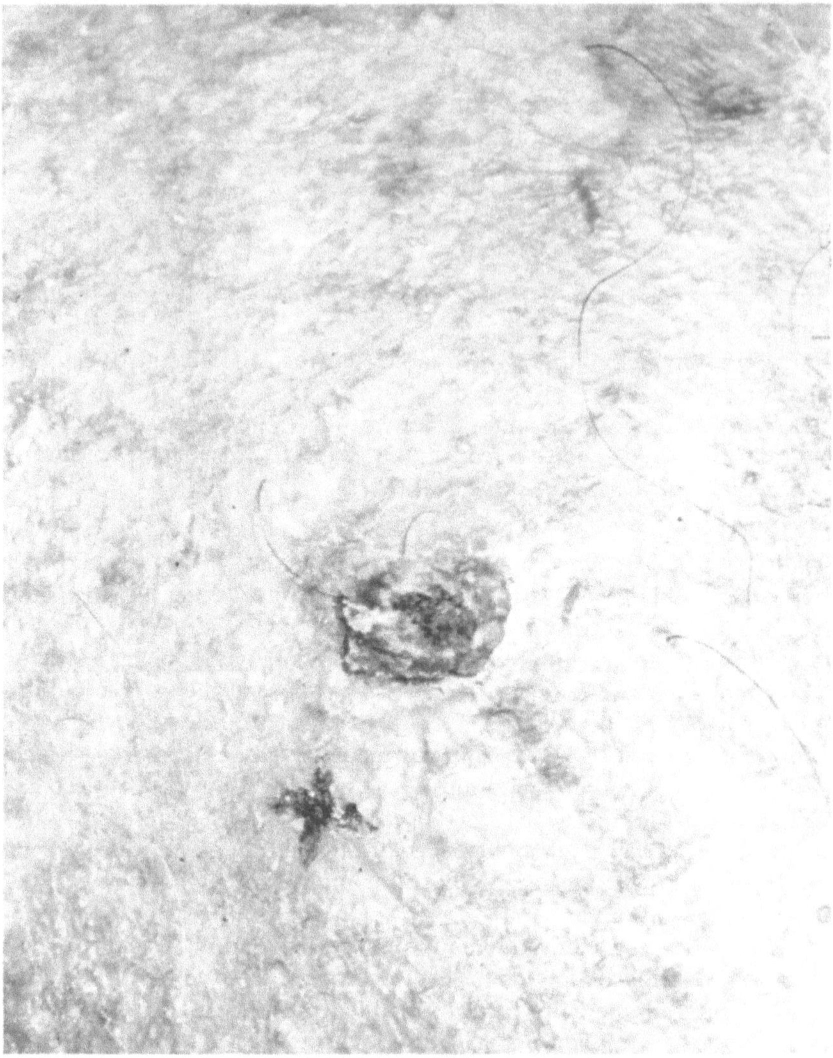

FIGURE 8. An actinic keratosis.

must be tailored to the patient, adjusting both the dose and oral therapy so that the abnormal keratin formation is just suppressed (Figures 9 and 10).

An exceedingly rare type of ichthyosis is congenital ichthyosiform erythroderma and this has been shown to respond to combined oral and topical Vitamin A acid.[11] More recently, Peck and Yoder[56] reported the successful use of oral 13-cis retinoic acid in the treatment of five out of five patients with lamellar ichthyosis (Figure 11) two of whom had nonbullous congenital ichthyosiform erythroderma; X-linked ichthyosis does not respond. At the international meeting in Berlin (1980) several of the authors (Marks, Tamay, Orfanos, and Peck) reported considerable of the aromatic retinoids in several types of ecthyoses.

1. Many patients with less severe forms of ichthyosis respond reasonably well to emollients and urea based preparations, but in the more severe cases, vitamin A acid could be tried in all forms of ichthyoses, especially in an ointment base.
2. Large doses of oral vitamin A acid are, because of side effects, contra indicated.

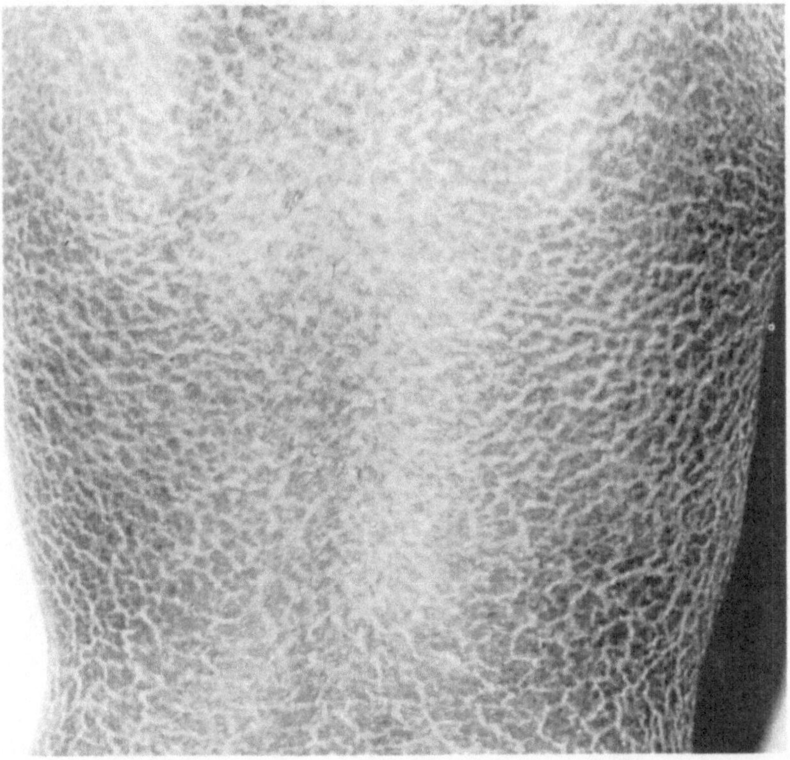

A

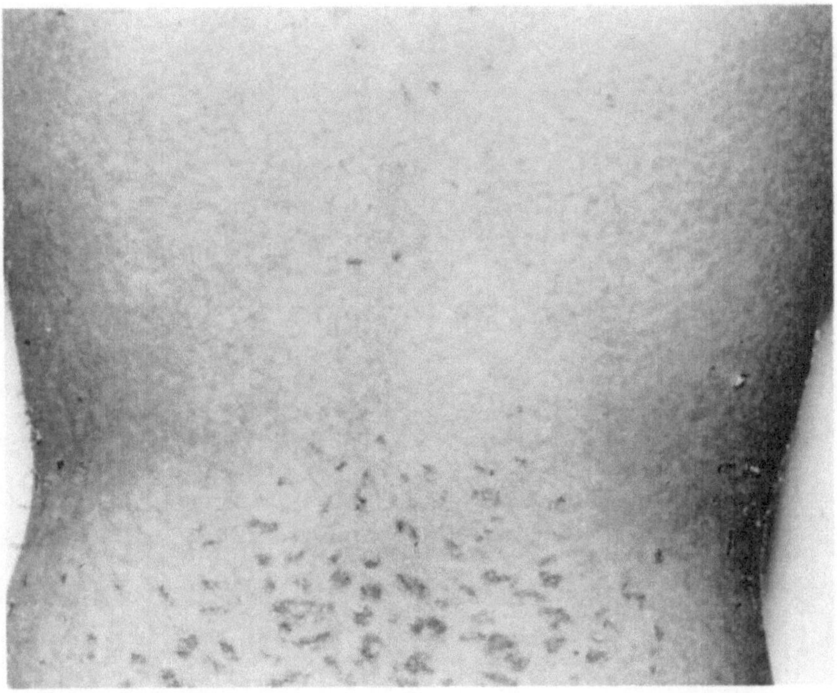

B

FIGURE 9. Lamellar ichthyosis (A) before treatment with oral aromatic retinoic acid and (B) after treatment with oral aromatic retinoic acid. (Courtesy of H. Pehamberger.).

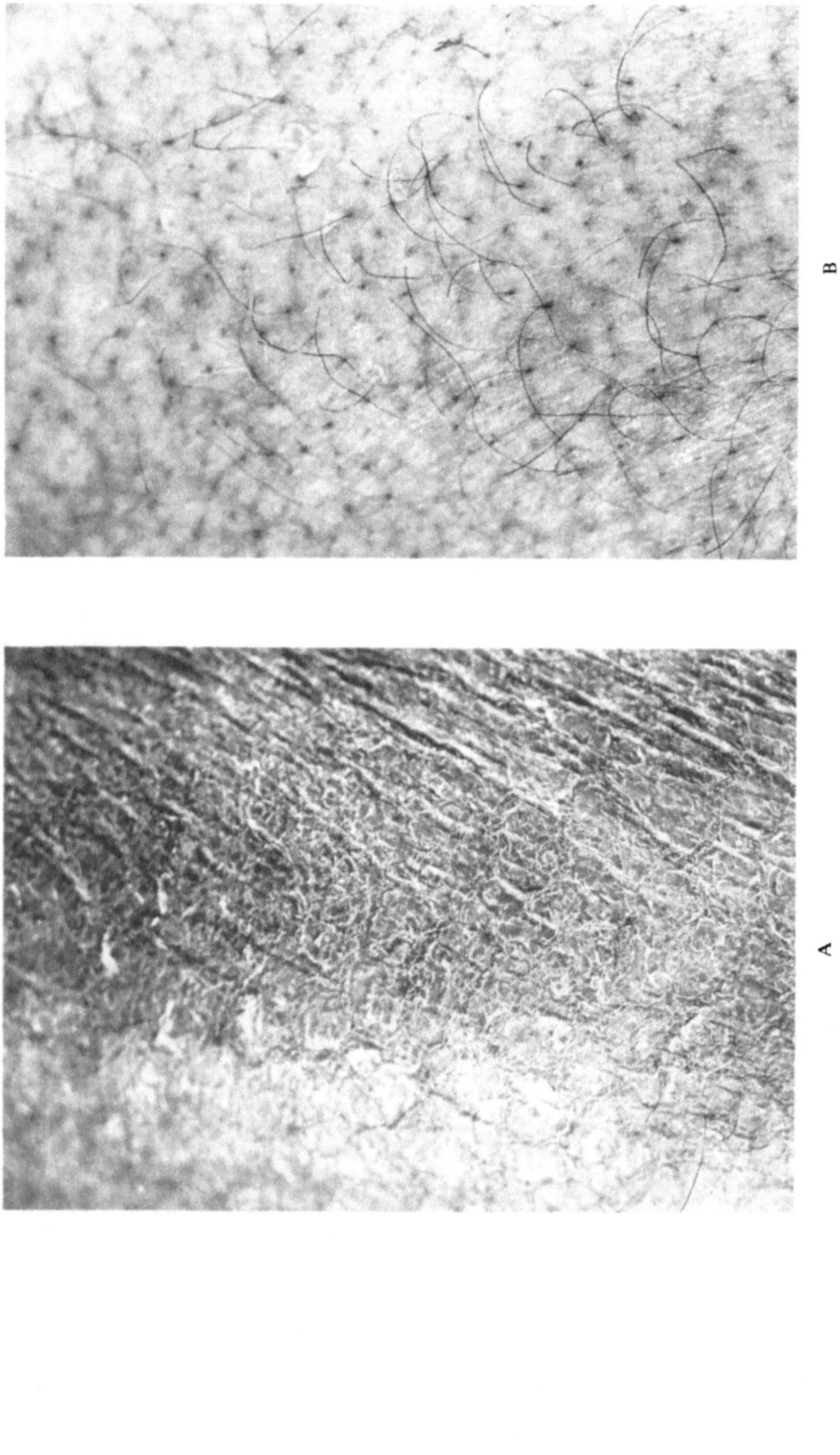

FIGURE 10. Congenital ichthyosiform erythroderma (A) before treatment with oral vitamin A acid and (B) after treatment with oral vitamin A acid. (From Stuttgen, G., *Acta Derm. Venereol.*, Suppl. 74, 174, 1975. With permission.)

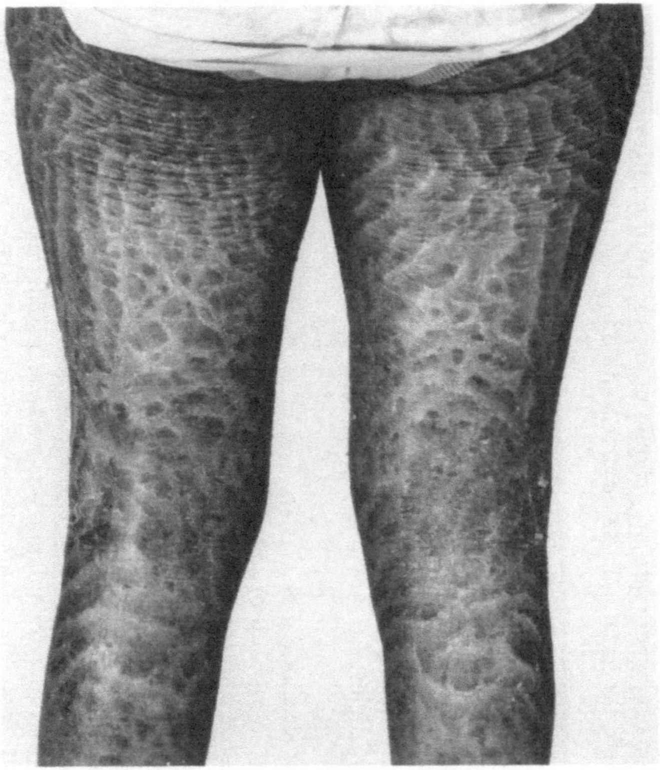

A

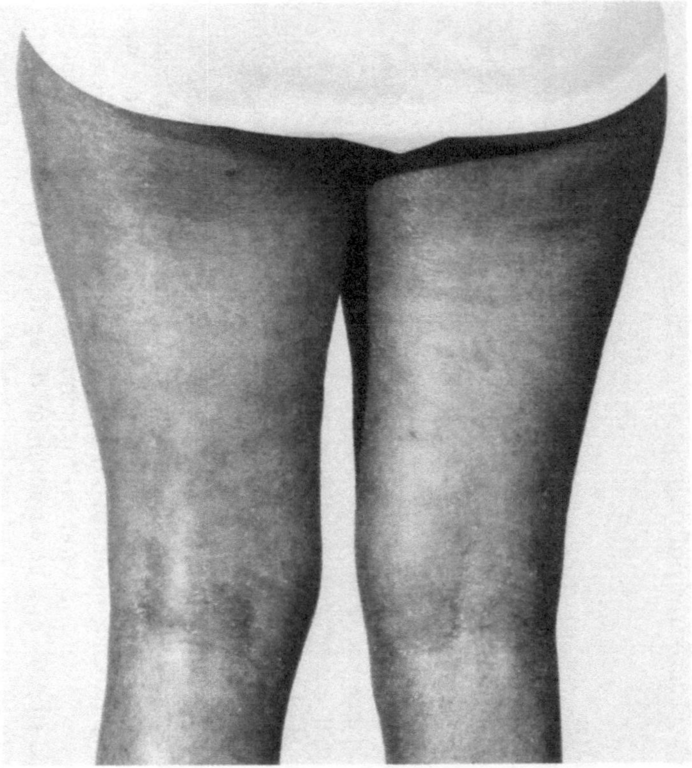

B

FIGURE 11. Lamellar ichthyosis (A) before treatment with 13-cis retinoic acid and (B) After treatment with 13-cis retinoic acid. (From Peck, G. L. and Yoder, F. W., *Lancet*, 2, 1172, 1976. With permission.)

3. Combined topical vitamin A acid treatment and oral vitamin A acid (20 to 40 mg b.d.) can also be tried; the combined therapy reduces the oral requirement.
4. The regimes, either topical or oral and topical, must be tailored to the individual.
5. The new 13-cis derivatives of retinoic acid offer considerable hope to such patients.

D. Keratosis Pilaris

This is a common disorder in which keratotic follicular papules occur especially on the extensor aspects of the arms and legs. Rarely is treatment required; a 0.1% retinoic acid in soft yellow paraffin has been reported, in an uncontrolled study, to be of help.

E. Keratosis Follicularis (Darier's Disease)

This is an autosomal dominant hereditary geno-dermatoses that occurs with the same frequency in men and women.

Beer[52] found that topical vitamin A acid was of little value in Darier's disease. In contrast, Fulton et al.[57] successfully treated two cases using 1% retinoic acid under occlusion in a base composed of ethyl alcohol (95%), ethylene glycol monomethyl ether, and propylene glycol (1:1:2).

In a similar base, Goette[58] treated, in a well-controlled study, a zosteriform type of keratosis follicularis with success. The untreated half showed no change, whereas the treated half showed a marked reduction in size and inflammation. There was some redness and scaling associated with the therapy and remission was for several months.

Four patients with extensive keratosis follicularis showed excellent clinical response to the oral administration of a new transaromatic derivative of retinoic acid.[59] Initial oral treatment with 50 to 75 mg/day was followed by substantial improvement in 4 to 7 days and the lesions cleared completely after 3 to 4 weeks. Long-term treatment with 25 to 30 mg/day was sufficient to prevent recurrence. No serious side effects were seen with this dosage after several months. However, it is recommended not to be used in children, and females, where relevant, should be on the contraceptive pill. Some dryness of the lips and the nasal mucosa occurred and one patient experienced slight nausea.

1. Summary

1. Topical vitamin A acid may help, particularly in a base containing ethyl alcohol, ethylene glycol monomethyl ether, and propylene glycol (1:1:2).
2. The new aromatic derivates appear helpful, but need more detailed clinical and laboratory studies.

F. Palmoplantar Keratodermas

This is not an easy group of disorders to classify.[48] Many are genetic in origin and the patients present with thickened palmar and plantar skin for which there is no specific treatment. The results of Beer[52] and Heiss and Gross[60] are contradictory, but in a controlled study of 9 patients, Günther[61] found that 0.1% vitamin A acid in petroleum jelly was significantly better than the base only, and that the disorder could subsequently be maintained by twice weekly application (Figure 12). Using 0.1% lotion or cream, Touraine and Revuz[62] could not detect much improvement in six patients with differing types of palmoplantar keratodermas. This data suggests that an ointment base is preferable. Oral aromatic retinoids have also been reported to help[62] (Figure 13).

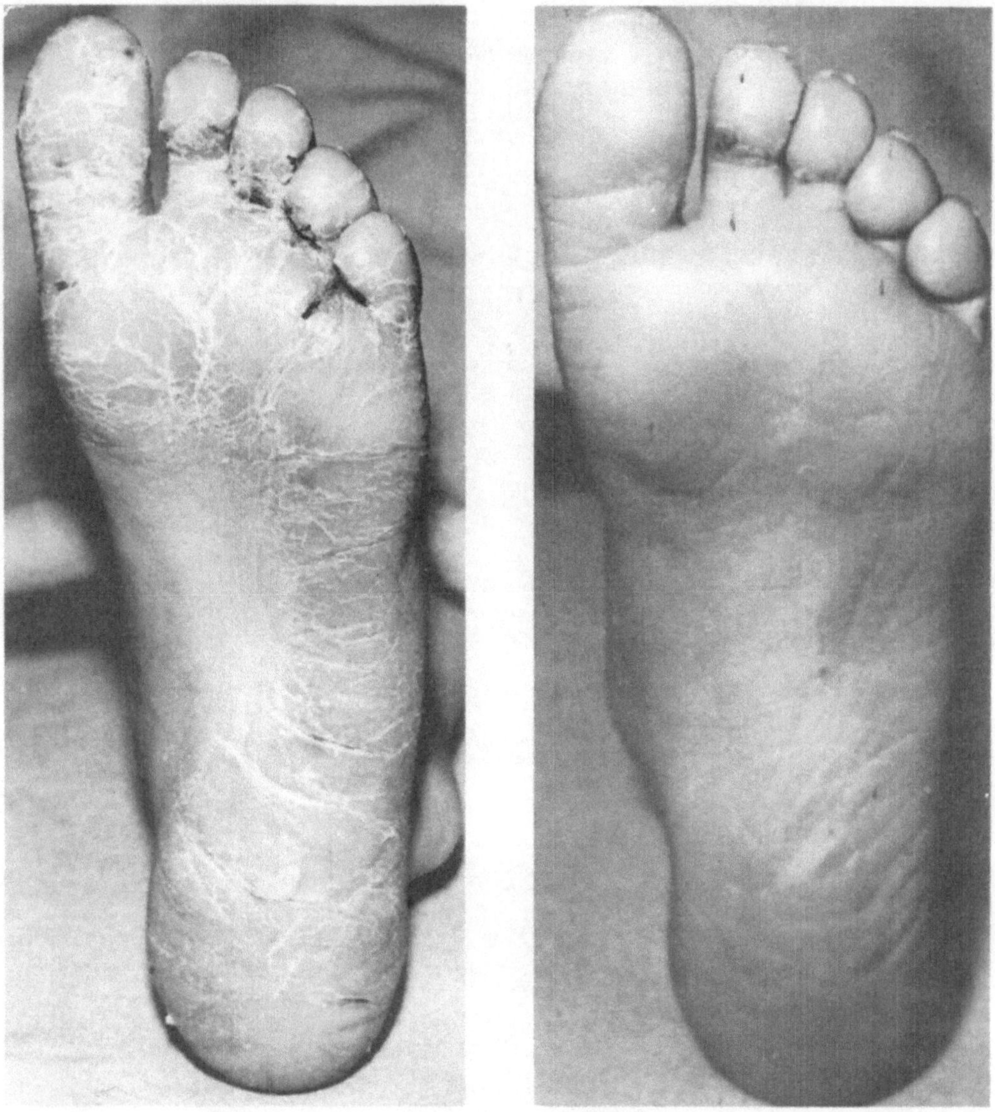

A B

FIGURE 12. Plantar keratoderma (A) before treatment with topical vitamin A acid in an ointment base and (B) after treatment with topical vitamin A acid in an ointment base. (From Günther, S., *Dermatol. Monatsschr.*, 159(7), 721, 1973. With permission.)

G. Verrucous Nevus

These type of nevi, of which there are two types, are uncommon and are often present at birth as linear hyperkeratotic lesions. They are difficult to treat and may require surgical dermabrasion. Günther[63] treated five cases with 0.1% retinoic acid in petroleum jelly. After 4 months treatment the lesions were much improved and any irritation was reduced by decreasing frequency of application to every second or third day. After 4 months, only weekly treatment was required, and if left untreated for 6 weeks, the scaling became much more obvious (Figure 14).

H. Trichostasis Spinulosa

Trichostasis spinulosa is a curious disorder of sebaceous follicles which results from

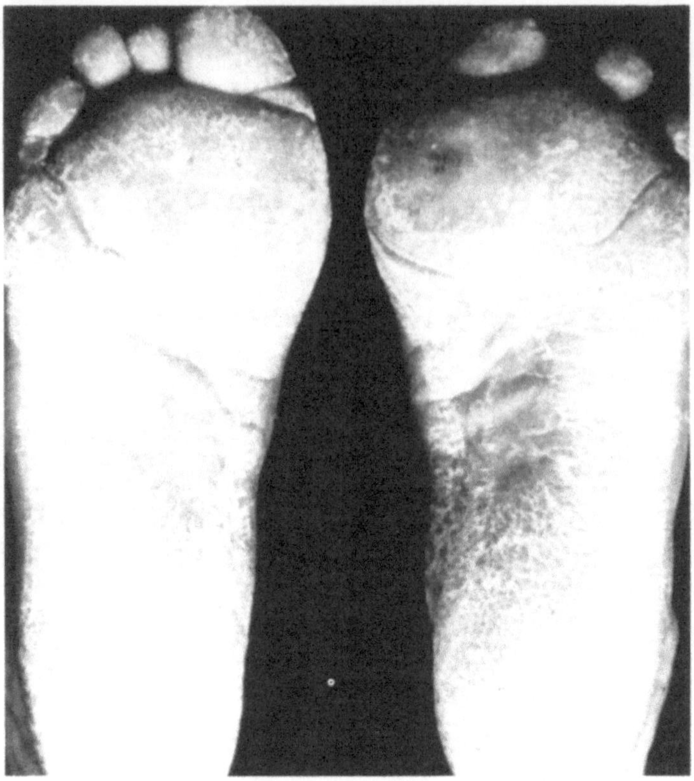

A

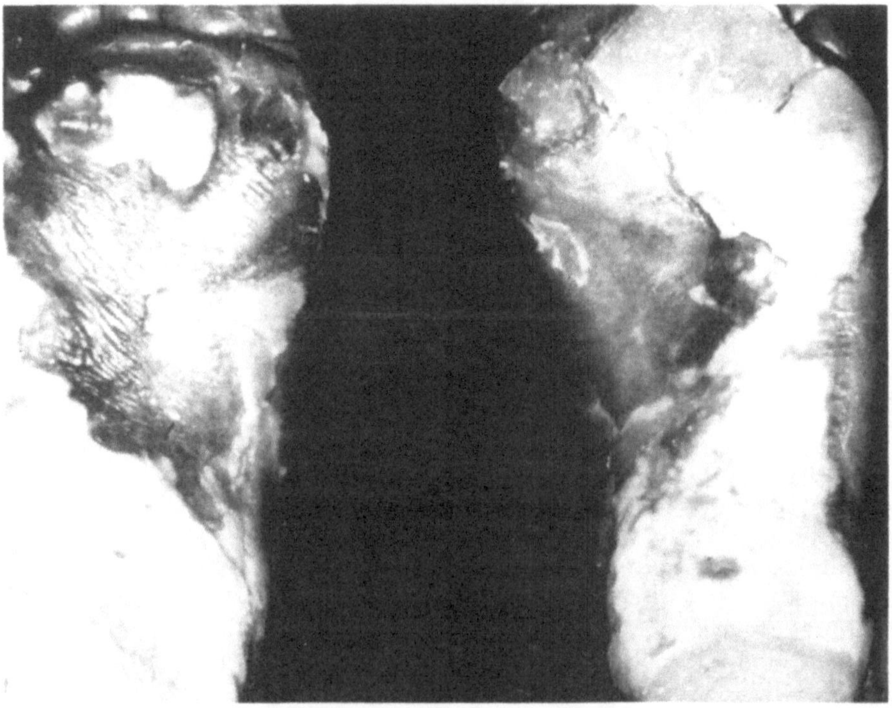

B

FIGURE 13. Plantar keratoderma (A) before treatment with oral aromatic retinoic acid and (B) after treatment with oral aromatic retinoic acid. (Courtesy of Dr. P. Fritsch.)

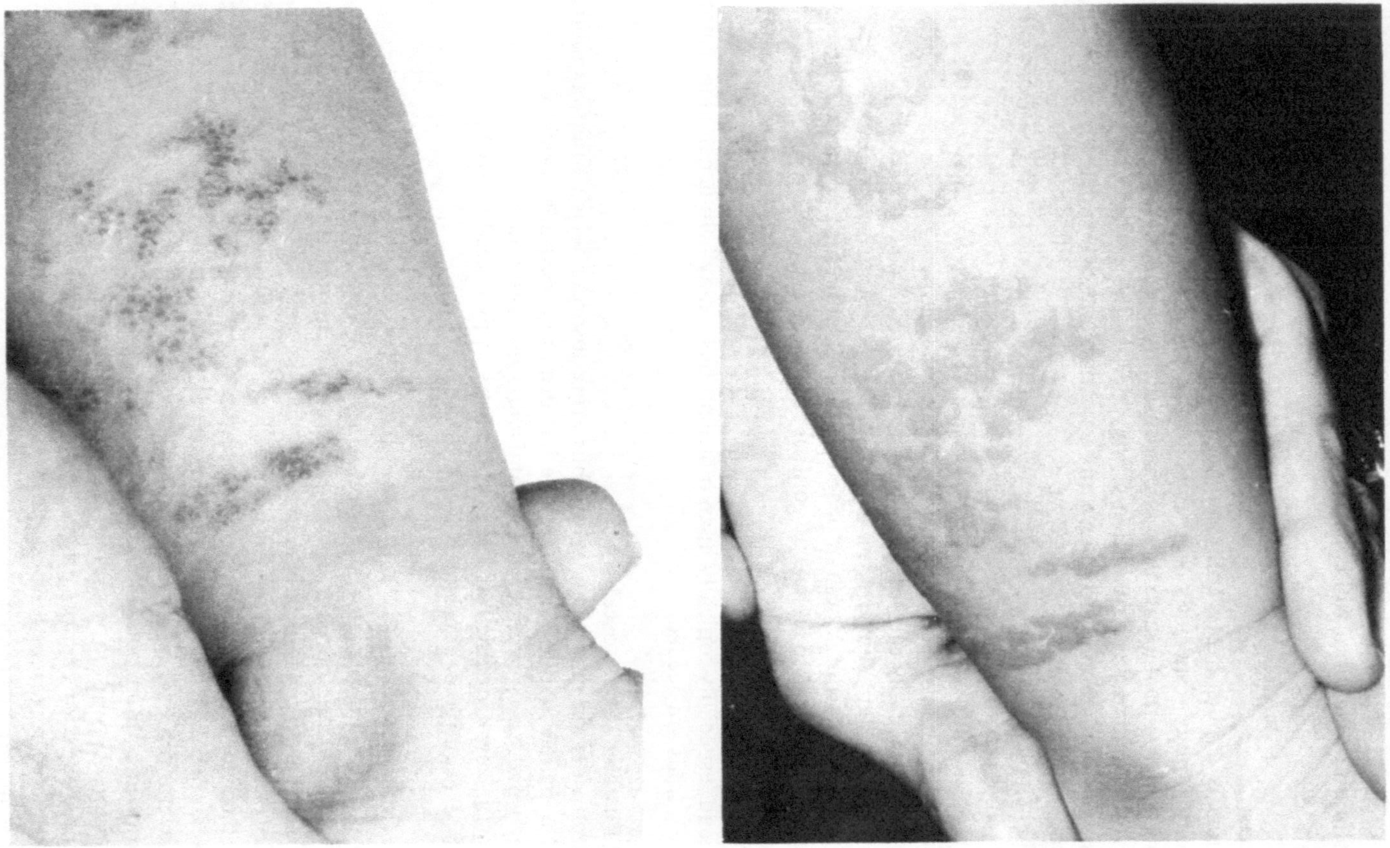

A

B

FIGURE 14. A linear verrucous naevus (A) before treatment with topical retinoic acid in an ointment base and (B) after treatment with topical retinoic acid in an ointment base. (From Gunther, S., *Dermatol. Monatsschr.*, 159(7), 721, 1973. With permission.)

the failure of telogen hairs to be shed. These are retained in neat fascicles of 5 to 40 fine hairs and are the result of follicular hyperkeratosis. The condition is mainly an symptomatic cosmetic nuisance. It is exceedingly common on the face, especially the nose, of the elderly, most often in a marginal to subclinical form.

There is one report[64] demonstrating quite conclusively that daily topical application of a 0.05% solution of tretinoin will expel the hair plugs and eradicate the manifestations of the disorder.

I. Lichen Planus

Lichen planus is a reasonably common dermatological disorder presenting with moderately itchy papules, especially on the extensor aspects of the limbs and trunk. A white lacy patterning on the buccal mucous diagnostically occurs in most patients. The cause is uncertain, and in 30% of patients, may persist for 18 months. Treatment usually consists of antihistamines to relieve any irritation and, when necessary, topical (and rarely oral) corticosteroids.

Studies on lichen planus are completely uncontrolled. In one study[65] 40 patients with typical lichen planus lesions of the oral mucosa and skin were treated with either oral administration of vitamin A acid (30 mg/day), local application (0.1% vitamin A acid or 0.05% in an ointment base), or a combination of both (Figure 15).

The skin lesions, especially the verrucous lesions, improved within a few weeks with topical therapy. A similar study by the same author[66] demonstrated the efficacy of topical application of vitamin A acid in lichen planus. Oral therapy appears to be less satisfactory, but a dose of 20 to 80 mg can control the disease within 8 weeks.[67]

1. Summary

Vitamin A acid, especially topically in an ointment base, may help lichen planus but adequate clinical trials are needed.

J. Subcorneal Pustular Dermatoses

Subcorneal pustular dermatoses is a rare blistering disorder which often responds to dapsone. One patient with a fatal outcome has been reported unresponsive to corticosteroids, dapsone, sulfapyridine, and levamisole. Only systemic treatment with retinoic acid and the new aromatic retinoic acid derivatives produced a satisfactory clinical response, but a complete remission was not obtained.[68]

K. Molluscum Contagiosum

Mollusca contagiosa are benign skin lesions due to a filterable virus which belongs to the group of poxviruses because of its shape resembling a rectangular prism. The clinical picture is characterized by disseminated wax-like grey to yellowish-white hemispherical efflorescences; these occur on normal looking skin and have a diameter of 2 to 5 mm. Characteristic are the pitted center that sometimes has a small opening from which pressure will release a cheese-like material.

There are several reports, unfortunately uncontrolled, on the successful treatment of topically applied retinoic acid in molluscum contagiosum.[69-71] In the majority of patients, the lesions cleared up with an average of 9.9 days. The treatment is simple and claimed to be superior to most other therapies in common use, especially in children with multiple or recurrent lesions. To prevent possible irritation around the lesions a little vaseline can be applied.

A direct antiviral effect may account for the therapeutic benefit of tretinoin in the molluscum. Vitamin A decreases the stability of lipoprotein membranes[72-74] and alters the morphology of viral envelope. Disruption of the virus has been observed with tretinoin.[75]

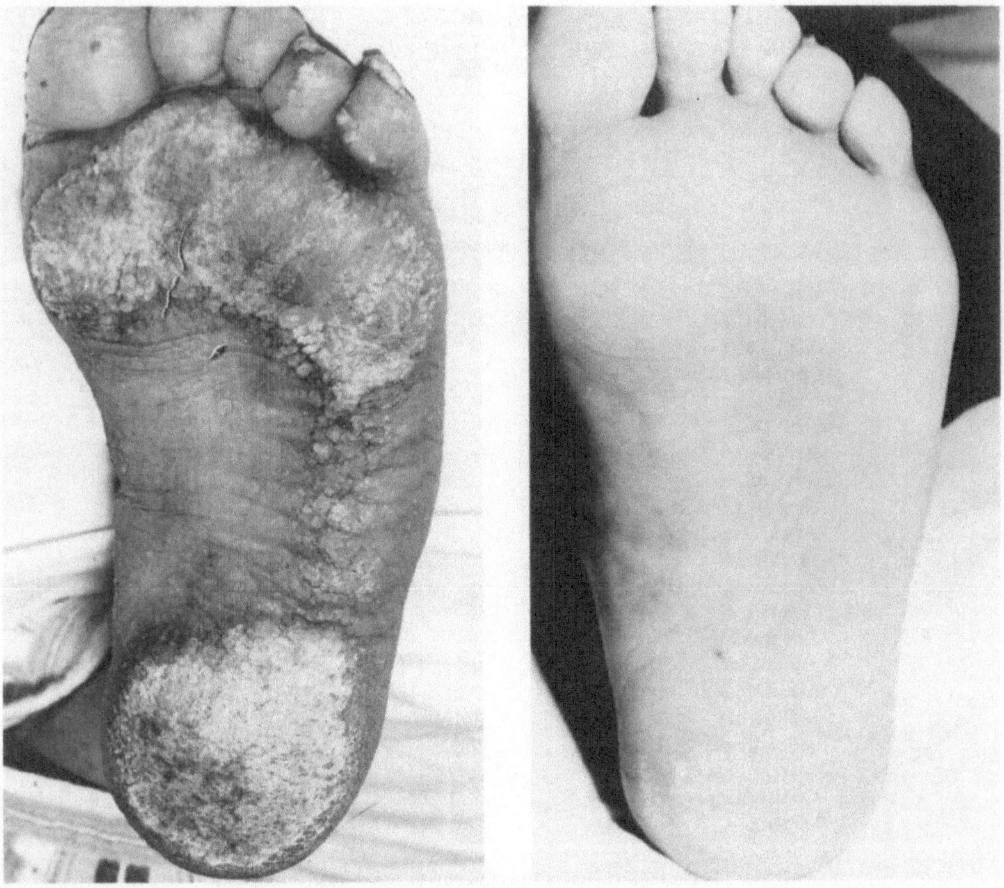

A B

FIGURE 15. Lichen planus (A) before treatment with topical retinoic acid in an ointment base beneath polythene occlusion and (B) after treatment with topical retinoic acid in an ointment base beneath polythene occulusion. (From Gunther, S., *Dermatol. Monatsschr.*, 159(7), 721, 1973. With Permission.)

L. Warts

Vitamin A acid topically probably has no effect on virus warts[70] but oral aromatic retinoids may help in severe, persistent cases (I.D.S. 1980).

M. Atopic Eczema

Several papers indicate that vitamin A acid will irritate excessively the skin of patients with constitutional eczema and a study of its effect on six patients, with constitutional eczema, showed it to have no effect whatsoever.[63,76]

1. Summary

Vitamin A acid is of some value in the treatment of dermatoses other than acne. Of particular value is the combination of low dose vitamin A with topical treatment in a dose appropriate to the condition. Also of interest, are the newer aromatic retinoids. It is possible that they could in the 1980s markedly influence the treatment of certain of these dermatoses-especially psoriasis and icthyosea.Table 4 summarizes the overall clinical role of vitamin A acid which has a place in the treatment of skin diseases.

Table 4
SUMMARY OF EFFECTIVENESS OF VITAMIN A

	Topical vitamin A acid	Oral vitamin A acid	Combined oral and topical vitamin A acid	New retinoids *
Acne	+ + + +	±	?	+ + +
Psoriasis	+ 1	?	?	+ + + 2
Malignant and pre-malignant conditions	+ 3	−	−	+ +
Ichthyosis	+ + +	+ + +	+ + +	+ + +
Keratosis follicularis	+ +	±	?	+ + +
Keratosis pilaris	+ +	?	?	?
Keratosis palmo plantaris	+	−	+	+ +
Verrucous naevus	+ +	?	?	?
Trichostasis spinulosa	+ +	?	?	?
Lichen planus	+	+	+	+
Molluscum contagiosum	+ + +	?	?	?
Warts	−	?	?	+
Constitutional eczema	−	?	?	?

Note: This table indicates the efficacy or otherwise, of vitamin A acid.
1, combined with local steroid;
2, combined with or without PUVA;
3, combined with 5-fluorouracil;
−, no good at all;
*, not yet commercially available;
?, possibly not even tried;
±, doubtful benefit;
+, possible benefit — further work needed;
+ +, some benefit — further work needed;
+ + +, reasonable benefit — further work needed;
+ + + +, proven and effective.

ACKNOWLEDGMENTS

I wish to thank my dermatological colleagues for allowing me permission to publish and giving me the following figures: (1) Schaefer, H. and Zesch, A; (2 and 3) Plewig, G., Wolff, H. H., and Braun-Falco, O.; (5) Peck, G. L., Olsen, T. G., Yoder, F. W., Strauss, J. S., Downing, D., Pandya, M., Butrus, D., and Arnaud-Battander, A.; (9) Pehamberger, H.; (10) Stüttgen, G.; (11) Peck, G. L. and Yoder, F. W.; (12, 14, and 15) Günther, S.; and (13) Fritsch, P.

I also wish to thank Mrs. P. Hick, Mrs. K. Hewitt, and Mrs. M. Storey for secretarial help.

REFERENCES

1. **Frost, P. and Weinstein, G. D.**, Topical administration of Vitamin A acid for ichthyosiform dermatoses and psoriasis, *JAMA*, 207, 1853, 1959.
2. **Sobel, A. E., Parnall, J. P., Sherman, B. S., and Bradley, D. K.**, Percutaneous absorption of Vitamin A, *J. Invest. Dermatol.*, 30, 315, 1958.
3. **DeLuca, H. F. and Zile, M.**, Aspects of teratology of Vitamin A acid (β-All-Trans retinoic acid), *Acta Derm. Venerol.* Suppl. 74, 13, 1975.
4. **Ziboh, V. A.**, Regulation of prostaglandin E_2 biosynthesis in guinea pig skin by retinoic acid, *Acta Derm. Venerol.*, Suppl. 74, 56, 1975.
5. **Michaëlsson, G., Vahlquist, A., and Juhlin, L.**, Serum zinc and retinol-binding protein in acne, *Br. J. Dermatol.*, 96, 283, 1977.
6. **Ziboh, V. A.**, Regulation of prostaglandin E_1 biosynthesis in guinea pig skin by retinoic acid, *Acta Derm. Venerol.*, Suppl., 74, 56, 1975.
7. **Peck, G. L., Olsen, T. G., Yoder, F. W., Strauss, J. S., Downing, D., Pandya, M., Butrus, D., and Arnaud-Battander, A.**, Prolonged remissions of cystic and conglobate acne with 13-cis retinoic acid, *N. Engl. J. Med.*, 300, 329, 1979.
8. **LeTreut, A. and Grangaud, R.**, Action of Vitamin A on the in-vitro transformation of 16α hydroxydehydroepiandroterone into 16α hydroxyandrostenedione by placental tissue, *C. R. Acad. Sci. Ser. D.*, 274, 951, 1972.
9. **Sobel, A. E.**, Percutaneous and oral absorption of Vitamin A, *Arch. Dermatol.*, 73, 388, 1956.
10. **Flesch, P.**, Inhibition with unsaturated compounds, *J. Invest. Dermatol.* 19, 353, 1952.
11. **Thompson, J. and Milne, J. A.**, The use of retinoic acid in congenital ichthyosiform erythroderma, *Br. J. Dermatol.*, 81, 452, 1969.
12. **McCormick, A. M., Napoli, J. L., and Deluca, H. F.**, High pressure liquid chromatographic resolution of Vitamin A compounds, *Anal. Biochem.*, 86, 25, 1978.
13. **Schaefer, H. and Zesch, A.**, Penetration of Vitamin A acid into human skin, *Acta Derm. Venerol.*, Suppl. 74, 50, 1974.
14. **Günther, S.**, Vitamin A acid. Clinical investigations with 405 patients, *Cutis*, 17, 287, 1976.
15. **Plewig, G., Wolff, H. H., and Braun-Falco O.**, Topical treatment of normal and pathological human skin with Vitamin A acid. Histological and electron microscopic studies, *Arch. Klin. Exp. Dermatol.*, 237, 390, 1971.
16. **Schlichting, D. A., Wooding, W. M., and Brown, M.**, Optimising dermatological formulations of retinoic acid, *J. Pharm. Sci.*, 62, 388, 1973.
17. **Christophers, E. and Braun-Falco, O.**, Stimulation der epidermalen DNS-synthese durch Vitamin A — Säure, *Arch. Klin. Exp. Dermatol.*, 232, 427, 1968.
18. **Lukacs, I., Christophers, E., and Braun-Falco, O.**, Zur wikungsweise der Vitamin A Säure, *Arch. Dermatol. Forsch.*, 240, 375, 1971.
19. **Karaseck, M. A.**, Effect of all-trans-retinoic acid on human skin epithelial cells in vitro, *J. Soc. Cosmet. Chem.*, 21, 925, 1970.
20. **Wolff, H. H., Christophers, E., and Braun-Falco, O.**, Beeinflussung der epidermalen aus-differenzierung durch Vitamin A — Säure, *Arch. Klin. Exp. Dermatol.*, 237, 774, 1970.
21. **Gould, D. J., Ead, R., and Cunliffe, W. J.**, Oral tetracycline and retinoic acid gel in acne, *The Practitioner*, 221, 268, 1978.
22. **Stüttgen, G.**, Oral Vitamin A acid therapy, *Acta Derm. Venerol.*, Suppl. 74, 174, 1975.
23. **Cunliffe, W. J. and Cotterill, J. A.**, *The Acnes*, W. B. Saunders, Philadelphia, 1975.
24. **Cunliffe, W. J. and Shuster, S.**, The pathogenesis of acne, *Lancet*, 1, 685, 1969.
25. **Kligman, A. M., Fulton, J. E., and Plewig G.**, Topical Vitamin A acid in acne vulgaris, *Arch. Dermatol.*, 99, 469, 1969.
26. **Pedace, F. and Stoughton, R.**, Topical retinoic acid in the treatment of acne vulgaris, *Br. J. Dermatol.*, 84, 465, 1971.
27. Report from the General Practitioner Research Group, 1974. Retinoic acid in the treatment of acne, *The Practitioner*, 213, 387, 1974.
28. **Christianson, J. V., Gadborg, E., Ludvigsen, K., Meier, C. H., Norholm, A., Osmundren, P. E., Pederson, D., Rasmussen, K. A., Reiter, H., Reymann, F., Rosman, N., Sylvert, B., Unna, P., Wehnert, R., Aastrup, B., Anderson, B., and Holm, P.**, Topical Vitamin A acid (Airol) and systemic oxytetracycline in the treatment of acne vulgaris, *Dermatologica*, 149, 121, 1974.
29. **Peachey, R. D. G. and Conner, B. L.**, Topical retinoic acid in the treatment of acne vulgaris, *Br. J. Dermatol.*, 85, 462, 1971.
30. **Simpson, J. R., Trusted, H. W., Watts, A. B., and Wilkinson, D. S.**, Clinical experience with 0.025% retinoic acid solution (Retin A) in the topical treatment of acne, *Curr. Med. Res. Opin.*, 1, 485, 1973.

31. **Mills, O. H., Marples, R. R., and Kligman, A. M.**, Acne Vulgaris. Oral therapy with tetracycline and topical therapy with Vitamin A acid, *Arch. Dermatol.*, 106, 200, 1972.
32. **Leyden, J. J., Marples, R. R., Mills, O. H., and Kligman, A. M.**, Tretinoin and antibiotic therapy in acne vulgaris, *South. Med. J.*, 67, 20, 1974.
33. **Mills, O. H., Leyden, J. J., and Kligman, A. M.**, Tretinoin treatment of steroid acne, *Arch. Dermatol.*, 108, 381, 1973.
34. **Plewig, G. and Kligman, A. M.**, Vitamin A acid in acneiform dermatoses, *Acta Derm. Venerol.*, Suppl. 74, 119, 1975.
35. **Fredriksson, T.**, Antipsoriatic activity of retinoic acid Vitamin A acid, *Dermatologica*, 142, 133, 1971.
36. **Fry, L., Macdonald, A., and McMinn, R. M. H.**, Effect of retinoic acid in psoriasis, *Br. J. Dermatol.*, 83, 301, 1970.
37. **Macdonald, A. and Fry, L.**, Retinoic acid in the treatment of psoriasis, *Br. J. Dermatol.*, 86, 524, 1972.
38. **Günther, S.**, The therapeutic value of retinoic acid in chronic discoid, acute guttate and erythrodermic psoriasis, clinical observations on twenty-five patients, *Br. J. Dermatol.*, 89, 515, 1973.
39. **Runne, U., Orfanos, C. E., and Gartmann, H.**, Perorale applikation zweir derivate der Vitamin A Saure zur internen psoriasis therapie, *Arch. Dermatol. Forschung.*, 247, 171, 1973.
40. **Fritsch, P., Henigsmann, H., Jasokko, E., and Wolff, K.**, Puva in psoriasis. Pharmacological augmentation of aromatic retinoid. Data presented at *European Society for Dermatological Research*, 1978.
41. **Prutkin, L.**, The effect of Vitamin A acid on hyperkeratinization and the keratoa canthoma, *J. Invest. Dermatol.*, 49, 165, 1967.
42. **Prutkin, L.**, Modification of the effect of Vitamin A acid on the skin tumour, kerato anthoma by applications of Actinomycin D, *Cancer Res.*, 31, 1080, 1971.
43. **Bollag, W.**, Effects of Vitamin A acid on transplantable and chemically induced tumours, *Cancer Chemother. Rep.*, 55, 53, 1971.
44. **Bollag, W. and Ott, F.**, Vitamin A acid benign and malignant epithelial tumours of the skin, *Acta Derm. Venerol.*, Suppl., 74, 163, 1975.
45. **Schumacher, A. and Stuttgen, G.**, Vitamin A Saüre bei hyperkeratosen epithelialen tumoren und akne dtsch, *Med. Wochenschr.*, 40, 1547, 1971.
46. **Belisario, J. C.**, Recent advances in topical cytotoxic therapy of skin cancer and precancer, in Melanoma and Skin Cancer, *Proc. Int. Cancer Conf., Sydney*, 1972, 349.
47. **Prutkin, L.**, The effects of Vitamin A acid on hyperkerinization and the keratoacanthoma, *J. Invest. Dermatol.*, 49, 165, 1967.
48. **Rook, A.**, in *Textbook of Dermatology*, Rook, A., Wilkinson, D. S., and Ebling, F. J. E., Eds., Blackwell Scientific Publications, Oxford, 1972.
49. **Mirrer, E. and McGuire, J.**, Lamellar ichthyosis. Response to retinoic acid (tretinoin). A case report. *Arch. Dermatol.*, 102, 548, 1975.
50. **Gunther, S.**, Topical treatment with retinoic acid in various dyskeratoses of childhood keratosis, plantaris verrucous naevus and autosomal dominant ichthyosis, *Dermatol. Monatsschr.*, 159, 721, 1973.
51. **Stüttgen G.**, Zur lokalbehandlung von keratosen mit Vitamin A Säurt, *Dermatologica (Basel)*, 124, 65, 1962.
52. **Beer, P.**, Untersuchungen uber die Wirkung der Vitamin A Säure, *Dermatologica (Basel)*, 124, 192, 1962.
53. **Grice, K., Sattar, H., and Baker, H.**, Urea and retinoic acid in ichthyosis and their effect on transepidermal water loss and water holding capacity of stratum corneum, *Acta Derm. Venerol.*, 53, 114, 1973.
54. **Muller, S. A., Belcher, R. W., and Esterly, N. B.**, Keratinizing Dermatoses, *Arch. Dermatol.*, 113, 1052, 1977.
55. **Stüttgen, G.**, Oral Vitamina A acid therapy, *Acta Dermatolovener (Stockh.)*, Suppl. 74, 174, 1975.
56. **Peck, G. L. and Yoder, F. W.**, Treatment of lamellar ichthyosis and other keratinising dermatoses with an oral synthetic retinoid, *Lancet*, 2, 1172, 1976.
57. **Fulton, J. E., Gross, P. R., Cornelius, C. E., and Kligman, A. M.**, Darier's disease. Treatment with topical Vitamina A acid, *Arch. Dermatol.*, 98, 396, 1968.
58. **Goette, D. K.**, Zosteriform keratosis follicularis cleared with topically applied Vitmin A acid, *Arch. Dermatol.*, 107, 113, 1973.
59. **Orfanos, C. E., Kurka, M., and Strunk, V.**, Oral treatment of keratosis follicularis with a new aromatic retinoid, *Arch. Dermatol.*, 114, 1211, 1978.
60. **Heiss, H. B. and Gross, P. R.**, Keratosis pulmaris et plantaris treatment with topically applied Vitamin A acid, *Arch. Dermatol.*, 101, 100, 1970.

61. **Günther, S. H.**, Vitamin A acid in the treatment of palmoplantar keratoderma, *Arch. Dermatol.*, 106, 854, 1972.
62. **Touraine, R. and Revuz, J.**, Topical treatment with keratosis palmaris et plantaris with tretinoin, *Acta Dermatovener (Stockh.)*, Suppl. 74, 152, 1975.
63. **Günther, S. H.**, Retinoic acid versus placebo in linear verrucose naevi, scaly lichenified eczema and verrucae plantaris, *Br. J. Dermatol.*, 89, 317, 1973.
64. **Mills, O. H. and Kligman, A. M.**, Topically applied tretinoin in the treatment of trichostasis spinulosa, *Arch. Dermatol.*, 108, 378, 1973.
65. **Günther, S. H.**, Treatment of skin and mucosa lichen planus with retinoic acid. Results of clinical and histological examinations on 40 patients, *Bull. Soc. Fr. Dermatol. Syphiligr.*, 79, 477, 1972.
66. **Günther, S. H.**, Vitamin A acid in treatment of oral lichen planus, *Arch. Dermatol.*, 107, 277, 1973.
67. **Stuttgen, G., Ippen, H., and Mahrle, G.**, Oral Vitamin A acid in treatment of dermatoses with pathologic keratinization, *Int. J. Dermatol.*, 16, 500, 1977.
68. **Folkes, E. and Tafelkruyer, J.**, Subcorneal postulas dermatosis (Sneddon-Wilkinson disease) therapeutic problems, *Br. J. Dermatol.*, 98, 681, 1978.
69. **Papa, C. M. and Berger, R. S.**, Veneral Herpes-like molluscum contagiosum: Treatment with tretinoin, *Cutis*, 18, 537, 1976.
70. **Steigleder, G. K.**, Behandlung von-mollusca contagiosa mit Vitamin A Säure — Salbe, *Hautarzt*, 22, 165, 1971.
71. **Fanta, D. and Kokoschka, E. M.**, Mollusca contagiosa ubersicht und kritsche betrachtung einer neven behandlungsmethode, *Wien. Klin. Wochenschr.*, 87, 154, 1975.
72. **Fell, H. B., Lucy, J. A., and Dingle, J. T.**, Studies on the mode of action of Vitamin A. I. Metabolism composition degradation of chick-limb cartilage in vitro, *Biochem. J.*, 78, 11, 1961.
73. **Dingle, J. T., Gauert, A. M., Daniel, M., and Lucy, J. A.**, Vitamin A and membrane systems. I. The action of the vitamin on the membranes of cells and intracellular particles, *Biochem. J.*, 84, 76, 1962.
74. **Dingle, J. T. and Lucy, J. A.**, Studies on the mode of action of excess of Vitamin A. V. The effect of Vitamin A on the stability of the erythrocyte membrane, *Biochem. J.*, 84, 611, 1962.
75. **Blough, H. A.**, A molecular approach to teratogenesis effect of Vitamin A on influenza virus in vivo, *Nature (London)*, 199, 33, 1963.
76. **Günther, S. H.**, Topical administration of Vitamin A acid (retinoic acid) in palmar keratoses callosities, hyperkeratotic eczema, hypertrophic lichen planus, pityriasis rubra pilaris, *Dermatologica*, 145, 344, 1972.

Chapter 9

CLINICAL TOXICOLOGY OF VITAMIN SUPPLEMENTS

Fiona Cumming, Michael Briggs, and Maxine Briggs

TABLE OF CONTENTS

I. Introduction 188

II. Vitamin A 188
- A. Introduction 188
- B. Animal Toxicity 190
- C. Clinical Toxicology 190

III. Vitamin C 195
- A. Animal Studies 195
- B. Human Toxicity 196

IV. Vitamin D 198
- A. Animal Studies 198
- B. Clinical Toxicology 201

V. Vitamin E 204
- A. Animal Toxicology 204
- B. Clinical Studies 205

VI. Vitamin K 207
- A. Animal Studies 207
- B. Human Toxicity 208

VII. B Vitamins and Other Water-Soluble Factors 209
- A. Thiamin 209
- B. Vitamin B_6 212
- C. Vitamin B_{12} 212
- D. Folic Acid 213
- E. Riboflavin 214
- F. Biotin 214
- G. Pantothenic Acid 214
- H. Niacin (Nicotinic Acid) and Niacinamide (Nicotinamide) 214
- I. Choline 216
- J. Inositol 216

References 217

I. INTRODUCTION

Vitamins are one of the world's most widely used pharmaceutical products. By definition they are essential for human health and for this reason are often considered to be nontoxic. This is certainly untrue and adverse reactions in humans, sometimes life-threatening, have been reliably reported following administration of high doses of many individual vitamins.[155]

While oral vitamin supplements are available in many countries without medical prescription, upper limits on the vitamin content of such products have been introduced into the legislation of some nations.[37] This is to minimize the chances of people taking daily doses of vitamins many times the recommended allowance.[551] It is important to note that many vitamin supplements are now formulated with synthetic derivatives, rather than natural vitamins. Often these derivatives are simple salts or esters, though occasionally major structural changes have been made to the vitamin molecule (Table 1).[481]

In medical practice, vitamins are frequently given by both oral and parenteral routes, often with little objective therapeutic effect, in a surprisingly wide range of conditions, many totally unrelated to vitamin deficiency. The literature abounds with reports of mega-vitamin therapy in patients who may sometimes have diseases affecting the mechanisms for vitamin absorption, transport, detoxification, and extretion.

Individual idiosyncratic reactions to particular vitamins are also well documented, and may occasionally lead to anaphylactic shock in susceptible persons.[3,6,47,57,69,116,143,181,227,271,282,358,418,444,453,484,494,515,627,635,688,702,743,757,764,831] It is our purpose in this review to present an overview of clinically adverse reactions to vitamin treatment and to relate this, where possible, to toxicity signs reported in animal toxicity studies with each particular vitamin. An up-to-date comprehensive bibliography has been included. The interested reader is also referred to several previously published reviews[46,64,343,346,350,365,406,749,792] on this topic, as well as to such traditional sources as *Martindale's Extra Pharmacopoedia*,[836] *Mehyler's Side Effects of Drugs*,[837] the new *Side Effects of Drugs Annual*,[838] and the multivolume *The Vitamins*.[289]

II. VITAMIN A

A. Introduction

A large number of naturally occurring and synthetic substances display vitamin A activity, while carotenoids, particularly β-carotene, are commonly considered provitamins as they are converted in vivo to retinol. Most systematic investigations of vitamin A toxicity have been conducted with vitamin A esters, particularly retinyl palmitate, though there are a number of investigations with carotenes. Hypervitaminosis A has occurred in animals and man[152,153] due to the ingestion of large amounts of vitamin A-active compounds contained within the livers of certain animals consumed as food (liver from the polar bear, seal, halibut, whale, tuna, cod, shark, and fox).

Previous reviews specifically on vitamin A toxicity have been by Jenkins,[401] Korner and Vollm,[428] Muenter et al.,[531] Bartolozzi et al.,[62] Clarke,[151] Jaffe and Filler,[394] Jeghers and Marraro,[399] Naha,[544] Nieman and Klein,[557] Moore,[521] Knudsen and Rothman,[423] Stimson,[744] and Rodriguez and Irwin.[652]

It is important to realize that vitamin A toxicity can appear as either an acute or chronic form. Generally, acute toxicity follows the administration of one, or a few massive doses of the vitamin; in contrast, hypervitaminosis results from the long-term intake of smaller, though excessive doses.[69,126]

Table 1
NATURALLY OCCURRING VITAMINS AND SYNTHETIC DERIVATIVES

Vitamin	Principal naturally occurring forms	Synthetic derivatives in clinical use
A[521]	Retinol (3,7-dimethyl-9-(2,6,6-trimethyl-cyclohex-1-enyl)-nona-2,4,6,8-tetraen-1-ol) Retinal (retinaldehyde) Retinol-2 (dehydro-retinol) Alpha-carotene (pro-vitamin)[294] Beta-carotene (pro-vitamin) Cryptoxanthin (pro-vitamin)	Retinol acetate Retinol palmitate Retinoic acid[711] 13-cis-Retinoic acid Retinoic ethylamide Aromatic retinoid ethylamide
C	L-Ascorbic acid (enolic 3-oxo-L-gulofuranolactone) Dehydro-L-ascorbic acid	Sodium ascorbate
D[191]	Ergocalciferol (vitamin D_2: 9,10-secoergosta-5,7,10(19),22-tetraen-3 beta-ol) Cholecalciferol (vitamin D_3)	Calcifediol (25-hydroxy cholecalciferol 1-Alpha,25-dihydroxy-cholecalciferol 1-Alpha-hydroxycholecalciferol Tachysterol
E[681]	(+)-Alpha-tocopherol (d-2,5,7,8-tetramethyl-2-(4,8,12-trimethyltridecyl)chroman-6-ol) Many other naturally occurring tocopherols show weak vitamin E activity	(±)-Alpha-tocopherol Alpha-tocopheryl acetate Alpha-tocopheryl acid succinate Alpha-tocopherol calcium succinate
K[46]	Phytomenadione (vitamin K_1: 2-methyl-3-phytyl-1,4-naphthoquinone) Menaquinone (vitamin K_2)	Menaphthone (vitamin K_1) Acetomenaphthone Menaphthone sodium bisulfite Menadiol sodium diphosphate
B-vitamins	Thiamin (vitamin B_1: 3-(4-amino-2-methylpyrimidin-5ylmethyl)-5-(2-hydroxyethyl)-4-methylthiazolium chloride)	Thiamin hydrochloride Thiamin mononitrate Thiamin propyl disulfide
	Riboflavine (vitamin B_2: 7,8-dimethyl-10-(D-ribit-1-yl)isoalloxazine	Riboflavine sodium phosphate
	Pyridoxine (vitamin B_6: 3-hydroxy-4,5-di(hydroxymethyl)-2-methylpyridine) Pyridoxal Pyridoxamine	Pyridoxine hydrochloride
	Hydroxycobalamin (vitamin B_{12}: alpha-(5,6-dimethylbenzimidazol-1-yl)hydroxocobamide	Cyanocobalamin
	Niacinamide (nicotinamide: pyridine-3-carboxamide) Niacin (nicotinic acid) Tryptophan (pro-vitamin)	Aluminium hydroxydinicotinate
	Folic acid (N-p-(2-amino-4-hydroxypterid-6-ylmethylamino) benzoyl-L-(+)glutamic acid)	Calcium folinate
	Pantothenic acid ((+)-beta-alpha, gamma-dihydroxy-beta, beta-dimethylbutyramido) propionic acid)	Calcium pantothenate d-Pantothenyl alcohol

Table 1 (continued)
NATURALLY OCCURRING VITAMINS AND SYNTHETIC DERIVATIVES

Vitamin	Principal naturally occurring forms	Synthetic derivatives in clinical use
	d-Biotin (cis-delta-(hexahydro-2-oxo-1H-thieno(3,4-d)imidazol-4-yl)valeric acid)	?
	Myo-inositol (meso-inositol: cyclo-hexane-1,2,3,4,5,6-hexol)	?

B. Animal Toxicity

From the data reviewed by Hayes and Hegsted[343] there seems no doubt that there is a wide range of tolerance to excess vitamin A among mammalian species. From the high concentration found in their livers,[650] it seems likely that animals such as the polar bear and seal tolerate very large daily doses without clinical effects. Experimental studies indicate that the dog[145] and cat[249] can tolerate daily doses (20,000 to 60,000 μg/kg) without harmful effects, while much smaller doses (up to 3000 μg/kg daily) produces adverse clinical side effects in the human,[22,33,34,91,130,353] the calf,[296,315,344] and the pig.[797]

While the clinical signs of vitamin A excess in experimental animals vary somewhat with the experimental conditions,[649] species, routes of administration,[446] etc., they commonly include anemia, loss of hair, skin lesions, hemorrhages related to a secondarily induced vitamin K deficiency, raised intercranial pressure,[469] together with degenerative atrophy of various organs. There is depressed growth with hypervitaminosis A in growing animals,[629] while congenital malformations[275,692,760] are seen if vitamin A is administered in excess dose[527] during pregnancy to the mother (Table 2). Effects of excess oral vitamin A in animal experiments are summarized in Table 3.

Effects of vitamin A on neoplasia are variable.[521] Under some circumstances, large amounts of vitamin A may be protective, though under other other circumstances there may be an increase in the number and growth of tumors. A specific retinoic acid-binding protein has been identified in a number of human neoplasia.[99]

C. Clinical Toxicology

Humans have been exposed to large doses of vitamin A on either an acute or chronic basis, following treatment of various diseases with large doses of vitamin A,[664] or by consuming meals based on the liver of animals particularly rich in this vitamin.[466] Megadoses of vitamin A have been used by medical practitioners, particularly for the treatment of certain skin disorders, and in some cases this has led to a hypervitaminosis. Oral, self-administered doses of vitamin A are also used by food faddists who may sometimes continue the dosage for long periods of time.[451] It is generally possible to distinguish between acute vitamin A intoxication following the administration of one or a few megadoses.[248,302,568,731,771,814] Chronic toxicity generally follows the administration of lower, though still high, doses of the vitamin or its precursors.[79,200,225,278,305,407,483,609,689,715,723,738,745,755,780,823,827]

Korner and Vollm[428] have reviewed 517 cases of hypervitaminosis A occurring in the published literature, which suggests that 75% of cases reported over the past four decades have been acute. Despite this, there is probably more current concern with chronic toxicity. It is pointed out that young children are the principal victims,[624,821] though in recent years there has been a decline in the incidence of vitamin A intoxication as the result of changes in medical practice. Acute adverse reactions[449] to vitamin A are usually transitory,[573] while the chronic hypervitaminosis follows almost equally

Table 2
TERATOGENICITY OF VITAMIN A IN ANIMALS

Species	Dose	Effect	Ref.
Rat	100,000—200,000 IU on 12—16 day	Cleft palate, growth disorganization of mandibular and maxillary system, macroglossia	828 160 284,286
	10,000—40,000 IU/kg day on 8—10 day	Prenatal and postnatal mortality, growth retardation, abnormal gait, behavioral hypoactivity, decreased learning ability	791
	50,000 IU 7—10 day	Neural tube abnormalities and malformations of skull	561 440
Mouse	80 mg retinoic acid/kg 10—16 day or up to 100,000 i.u. vitamin A days 8—16	Focomelia and micromelia, cleft palate	148,325,424 534
Guinea pig	75,000—400,000 IU/kg	Retarded tail and limb development, hydronephrosis, anephrosis, imperferate ani	647
Rabbit	80,000—150,000 IU/kg	'50% abortions with malformations at lower dose; total abortions at higher dose	240
Hamster	20,000 IU on 8th day	Mesodermal alterations and somite necrosis	478
	Up to 400,000 IU/kg	Retarded limb and tail development with abnormal kidneys	647
Pig	?	Anophthalmos	310
Chick	20 mg/100 g body weight/day	Inhibition of osteoblasts with deformed bone and increased epiphyseal cartilage; inflamed skin	88
Duck	Up to 25,000 IU/day for 19 days	Decreased growth with abnormal epiphyseal cartilage and accelerated endocondral ossification	817

from medical overprescription and self-overdose. In the survey the duration of administration of vitamin A ranged from 41 to 3600 days, while average daily doses ranged from 2000 to 60,000 IU/kg body weight. Chronic manifestations of vitamin A overdosage particularly included desquamation of the skin[578] and mucus membranes, though skeletal disorders,[408] thyroid suppression,[461] elevation of cerebrospinal fluid pressure,[63,220,232,235,301,498,523,781] and enlarged liver,[289,370,658] together with a wide range of subjective symptoms, have also been described. The effects of excess vitamin A on skin are poorly understood;[412] however, changes in skeletal tissue, liver,[290] and lipid metabolism are more open to investigation.[427,428]

Young children suffering from chronic hypervitaminosis A show abnormalities in bone which include premature epiphyseal closing, retarded growth of long bones, and thickening of the cortical regions (Table 4). Fortunately, permanent bone malformations seem relatively rare according to the investigation of Ruby and Mital.[659] Despite the latter finding, it would seem that the bone changes induced by excess vitamin A are among the most long lasting side effects (Hayes and Hegsted).[343] Overdosing with vitamin A for relatively short periods during infancy can lead to subsequent manifestations of bone disorder many years later. An increased tendency for fractures has been noted, while deformities may range from flexion contractures and short stature through to discrepancies in the length of the lower extremities. Vertebral irregularities are also noted. It has been shown that vitamin A enhances the activation and release

Table 3
ANIMAL TOXICITY OF VITAMIN A

Species	Oral intake	Summary of major effects	Ref.
Rat	7.5—30 mg/day	Acute doses had no effect in young or adult animals; treatment for 6—8 days induced a large increase in cerebrous spinal fluid	469
	500 IU/gm body weight 3—4 weeks	Adverse effects on heart	782
	250,000 IU	Significantly reduced hemoglobin, hematocrit, erythrocytes, platelets and concentrations of coagulation factors II, V, VII, X	718,725
	30,000 IU retinol 2 days	Liver concentrations of glycogen, cholesterol and triglycerides increased, together with plasma free fatty acids; adrenal increased while cholesterol content decreased	709
	50,000—75,000 IU/day 17 days	Increased mortality, calcification of lungs, bone lesions and damage to heart, kidney and liver	683,173
	10,000 IU/100 g body weight 30 days	Bone changes involved decreased collagen and mucopolysaccharides	150
Guinea pig	2.5—3 million IU	Loss of weight and appetite with alopecia and spastic walk; death within 17 days	120
	330 mg/kg diet for 40 days	Calcinosis of kidney and liver with severe hepatic atrophy; reduced growth	281
Pig	20,000 ug/kg body weight/day 5 weeks	Cutaneous erythroma, hemorrhage and anorexia and posterior lameness; bone deformities, including fragility	795 819
Cat	Commercial food containing high content of vitamin A derived mainly from fish liver	Hemorrhages and death	279
	Commercial food with high vitamin A content derived mainly from bovine liver estimated at up to 35 ug/g body weight daily	Lipid infiltration of liver, lung, spleen, with deforming cervical spondylosis	680
	Commercial diets with high bovine liver or synthetic retinol supplements	Effects include marked irritability, exopophalamos, changes in skin and hair; spinal rigidity, hyperesthesia, lameness and abnormalities of gait; loss of fertility in males	679
Goat	Circa 18,000 μg/kg/day for 16 weeks	Bone and epidermal lesions	257
Calf	Up to 32 μg/kg/day for 12 weeks	Increase in heart rate, decreased spinal fluid pressure, hyperplasia of kidneys, heart and liver; change in bone composition	352 344
Chick	20 mg/100 g body weight/day	Inhibition of osteoblasts with deformed bone and increased epiphyseal cartilage; inflamed skin	88
Duck	Up to 25,000 IU/day for 19 days	Decreased growth with abnormal epiphyseal cartilage and accelerated endocondral ossification	817

Table 4
EFFECTS OF HYPERVITAMINOSIS A IN YOUNG CHILDREN

	Dose Age	Effects	Ref.
Infants	17,000 IU/day 3 months	Fontanales bulged, widened metaphysis of wrist and ankle	605
	350 IU	Intense bulging of the fontanale within 12 hr of ingestion, vomiting and agitation	476
	Up to 500,000 IU/day up to 15 months	Hyperirritability, swelling and pain of the extremities, pruritis	432
	Up to 600,000 IU/day for approx. 33 months	Long bone changes, increased intracranial pressure	523,572
	15,000 IU/day for 7 months	Severe alopecia, bone tenderness, enlarged head, shiny skin, with permanent deformity of bones and scoleosis	659
Children	Up to 460,000 IU/day for 6 weeks	Increased intracranial pressure with central nervous symptoms	572
	57,000 IU/day for 1 year	Enlarged liver and spleen, large head, alopecia, painful walking, stiff neck	706
	625,000 IU/week for more than 4 years	Enlarged liver and spleen with hepatic dysfunction; anemia, alopecia, anorexia with dry skin	658
Adult	Up to 300,000 IU/day year	Psuedotumour cerebri, with severe headache, palsy, diplopia	441
	Prolonged intake of variable dose from 5000 IU—200,000 IU/day	Hypercalcemia with skeletal pain; emotional signs, weight loss, periostal calcification and demineralization of sella turcia	249
	100,000—300,000 IU/day for approx 4 months	Headache, muscular stiffness, anorexia, generalized pruritis, severe hepatic fibrosis and fatty changes	531
	Up to 150,000 IU/day for almost 2 years	Severe toxic psychotic reaction, headache, blurred vision, insomnia, irregular menses, tinitis, weight loss	637
	300,000 IU/day 2 years	Dry and squamous skin, holo-cranial murmur, eye defects, sleep disturbances, and easy fatigue	462
	50,000 IU—400,000 IU/day for 2 years	Severe headaches, dermatitis and scaling, hepatomegaly, anorexia	714
	120,000 IU/day	Periferal nervous lesions	796
	1,300,000 IU within 3 days	Blurring of vision, exfoliation of superficial skin, headache, vomiting, nausea	266
	30,000—125,000 IU/day for several months	Dry and scaley skin, pigmentation around mouth, alopecia, depression, leg pain	305
	100,000 IU/day 6 months	Bilaterial papilledema, suffusions, and nystagmus when looking aside	462
	Up to 1,500,000 IU/day 3 months together with large doses of carrot juice for approx. 6 years	Death accompanied by severe jaundice with liver enlargement	451
	Up to 1,250,000 IU/day 5 years	Enlarged liver and spleen with ascites, dry skin, fatigue, alopecia	293,663

of proteolytic enzymes by cell organelles, particularly lysosomes. This increase in proteolytic activity may damage the bone matrix in subjects with hypervitaminosis A.[67,180] It is possible, for example, to explain the reduction in longitudinal growth of bones by degenerative changes in the cartilaginous epiphyseal plates by this mechanism.

Related to the changes in bone physiology are alterations of calcium metabolism. Human subjects with hypervitaminosis A show persistent negative calcium balance. Katz and Tzagournis[417] have suggested that the excess vitamin A increases mobilization of endogenous calcium, and increased urinary calcium excretion has been reported in human subjects taking large doses of vitamin A (Weiland et al.).[805] Hypercalcemia has been repeatedly encountered in patients with hypervitaminosis A by Katz and Tzagournis[417] and Weiland et al.[805] In many cases the change in serum calcium concentration is similar to that encountered in hypervitaminosis D, and may be accompanied by generalized skeletal pains. Clearly, in subjects who are voluntarily ingesting overdoses of vitamin A the possibility of concurrent high intake of vitamin D has to be investigated, but has generally been ruled out (Frame et al.[249]). Muenter[532] has particularly commented on the availability of multivitamin preparations containing both vitamins A and D that may make differential diagnosis of hypercalcemia particularly difficult.

As in hypervitaminosis D, intoxication with vitamin A accompanied by hypercalcemia will lead to calcium deposits in appropriate tissues, such as kidneys, lungs, liver, myocardial muscle, and arterial walls (Fisher and Skillern[239]). There may also be renal failure.[241,242,829]

It has been pointed out that the soft tissue calcification occurring with excess intake of either vitamin A or D results from different primary lesions. Vitamin A excess affects particularly the organic matrix of bone, while overdosing with vitamin D results in dissolution of the mineral matrix of bone. Nevertheless, both overdoses result in skeletal tissue release of calcium into the general circulation.

Increases in the size of the liver and spleen are a common finding in clinical hypervitaminosis A,[428] though the nature of the hyperplasia of the two organs is still uncertain,[83] particularly in the case of the spleen. A detailed examination of liver changes occurring in hypervitaminosis A has been published by Russell et al.[663] Examination of liver biopsies has revealed hepatic fibrosis, lipid deposition, and obstruction of portal blood flow with portal hypertension; there was also sclerosis of the central vein.[439]

There is no doubt that hypervitaminosis A in experimental animals is teratogenic (see previous section). Almost all major body systems appear to be affected, depending on the stage of gestation at which the vitamin A excess is administered. It is not known from animal studies whether these teratogenic effects are due to direct effects of vitamin A on the embryo, or are secondary due to maternal hypervitaminosis A. It is known that malformations can be induced in cultured embryos by the addition of excess vitamin A to the culture medium, but it is not known whether excess vitamin A is transmitted via the placenta. As with other aspects of acute or chronic vitamin A intoxication, it is possible that the changes are due to alterations in lysosomal membrane permeability.[807]

Several[267,268,464,506] have expressed concern that indiscriminitory use of vitamin A during human pregnancy may result in malformations, but there is little evidence to suggest any real effect in the human species. Accounts by Pilotti and Scorta[612] and Bernhardt and Dorsey[29] of children with renal anomalies born to mothers who had received excessive amounts of vitamin A during pregnancy may well be isolated incidents.

Perhaps the most detailed investigation of vitamin A in relationship to human congenital malformations is that of Gal et al.[267] After reviewing previous investigations of vitamin A in teratology, the authors report their own investigation into vitamin A concentrations in postpartum maternal blood and fetal liver in relationship to congenital malformations. An investigation is presented in a pilot study involving 14 patients with CNS malformations which was followed up by 56 investigations on mothers with live born spina-bifida babies and then on 36 further cases with mothers with all types of neural tube malformations born alive or dead. The authors conclude that vitamin A reaches the human embryo during early stages of organogenesis and that under physiological circumstances the amount of vitamin A in fetal liver may be important in the development of malformations. Very young and also elderly mothers show relatively high concentrations of vitamin A in the liver of their fetus, while malformed fetuses apparently show a higher concentration than those in normal fetuses. It is also shown, however, that the standardization of sampling techniques is particularly difficult and there are variations, for example, in the liver lobe concentration of vitamin A. Marked differences in vitamin A also occur in maternal plasma even within women in the same age group. For these reasons no definite conclusions could be drawn from this particular investigation, though a plea for further investigation is made.

An important aspect of vitaminosis A is the physiological transport of retinol within the organism. It is now clear that blood plasma contains a retinol-binding protein, a protein of low molecular weight secreted by the liver.[111,808] However, this protein on entry into the circulation immediately complexes with free albumin, so that vitamin A in blood is largely in combination with this protein complex. Vitamin A esters and carotene are not carried by this complex, but appear to be in weak association with the various lipoporteins and albumin.

During hypervitaminosis A there are disturbances to this retinol transport system.[41] The ability of liver to take up retinol may be exceeded and an inhibition of retinol-binding protein secretion may be suppressed by as much as 50% (Mallia et al).[472] Due to the excess intake of retinol, together with the decrease in the specific binding protein in the plasma, there is an increase in the free retinol in the circulation. This appears to be the fraction capable of direct interaction with the various tissues and induction of cellular damage.

III. VITAMIN C

A. Animal Studies

Animal tissues contain L-ascorbic acid, together with its metabolite dehydro-L-ascorbic acid, which two forms are interconvertible. Ascorbic acid is a compound of very low toxicity, while the salts of ascorbic acid (sodium or calcium derivative), which are used in many therapeutic preparations, are even better tolerated than the free acid.

Table 5 summarizes the results of animal investigations into the toxicity of ascorbic acid and dehydro-ascorbic acid. A major difficulty facing such studies is the fact that very few animal species lack the ability to synthesize ascorbic acid from glucose. Aside from the human species and primates, the only convenient laboratory animal is the guinea pig. It is uncertain whether the effects of exogenous ascorbic acid are the same in a species with a high endogenous synthesis rate, as in a species with an absolute dietary requirement for this vitamin. The concentration of the noradrenaline in the brain and CNS of either species is unaffected by the vitamin. Hypercholesterolemia due to ascorbic acid has been described by Klevay.[421]

Evidence for effects on bone metabolism are available for the chick and the pig.[119] High doses of ascorbic acid in either species stimulates release of calcium and phosphorous from bone while collagen turnover is enhanced and there is increased urinary output of hydroxyproline. Large amounts of ascorbic acid shorten the gestation period in the pregnant guinea pig[342] and increase stillbirths.[666] However, no adverse effects of high dose ascorbic acid treatment were noted in either pregnant rats or mice in the investigation of Von Frohberg et al.[788]

Structural similarity between dehydro-ascorbic acid and alloxan has been noted by

Table 5
ANIMAL TOXICITY OF VITAMIN C

Compound	Species	Effects of high doses	Ref.
Ascorbic acid	Rat	Damaged liver lysosomes	458
		Thyroid inhibition	186
		Decreased β-carotene utilization	485
	Guinea pig	Thyroid inhibition	186,474
		Heart noradrenaline decreased	183
		Scorbutic effect of deficient diet enhanced	295,563
		Conceptions decreased, dead fetus	555
		Gestation shortened, stillbirths	529
		Abortions and stillbirths	666
	Pig	Bone collagen turnover increased	119
	Chicken	Bone resorption increased	762
Dehydro-ascorbic acid	Rat	Diabetes of the Alloxan-type	591
		Diabetic-type cateracts	592

Patterson,[591] who has reported dehydro-ascrobic acid is diabetogenic in the rat with the induction of pancreatic damage similar to that seen following alloxan treatment. Prolonged ingestion of dehydro-ascorbic acid in the rat led to the development of diabetic-type cateracts.[592]

Schrauzer et al.[672] have reported reduced high altitude resistance in laboratory animals overdosed with ascorbic acid.

B. Human Toxicity

Clinical toxicity of high-dose ascorbic acid treatment is rarely encountered, though there have been some adverse reports, a few of them serious.[218] According to Hoffer[356] the most common side effect of high-dose vitamin C for long periods is diarrhea. In the carefully designed double-blind study of high-dose ascrobic acid versus a placebo conducted by Anderson et al.[30,31] a low incidence of diarrhea was noted, together with other vague abdominal symptoms, but the incidence was the same in both treated and placebo groups. Two cases of skin rash in the ascorbic acid-treated group were seen and could be due to individual allergic response. There are three reports of skin rash associated with vitamin C treatment in an Australian report[49,163] of suspected drug reactions in 1964 to 1976.

Vickery[784] has described a woman with a personal and family history of urinary stone formation, who while taking high doses of vitamin C developed blockage of the ileocecal valve which required surgical correction. The intestinal stones contained large amounts of ascorbic acid and it is possible that the particular formulation ingested by this patient was not absorbed.

There is no doubt that ascorbic acid is partially converted to oxalic acid in humans, and that some of the oxalic acid is excreted in the urine (Lamden and Chrystowski).[438] The latter investigation demonstrated a dose-related increase in urinary oxalate excretion, though the effect was small. A reinvestigation of this matter by Briggs et al.[113,117] revealed the presence of a few individuals within a population who responded to large doses of ascorbic acid with a massive rise in 24-hr urinary oxalate excretion (up to 8 mmol). Despite this subpopulation who appear to have a metabolic abnormality in ascorbate metabolism, reports of urinary stones following megadose ascorbic acid are extremely rare. There is a secondary report by Briggs[110] of a university teacher who developed a urinary stone following a short course in vitamin C at 2 g daily.

Relevant to this discussion is the work of McMichael[542] who has investigated hospitalizations for kidney stones in Australia over the period 1966 to 1976 and has tried to relate these to changes in vitamin C consumption. As vitamin C is not manufactured in Australia and is therefore imported, it is possible to obtain reliable figures on different levels of intake over this time period.

It is concluded that the incidence of kidney stones has risen during the period approximately similar to the mean individual consumption of vitamin C. McMichael[542] suggests that individuals at increased risk of urinary stone formation related to high intakes of vitamin C are those consuming several grams daily or with innate characteristics, such as metabolic disorders, which enhance susceptibility to oxalate formations.

An investigation by Kallner[413] on the effects of 5 g ascorbic acid daily in 14 persons during 1 month has found very large increases in 24-hr urinary oxalate output. The mean pretreatment values was 49 ± 16 μmol/24 hr, but this rose to 1540 ± 601 μmol/24 hr by the end of the treatment period. This study also presents investigations on changes on in individual serum bile acids. A significant increase in chenodeoxycholate concentration was found on interruption of vitamin C supplementation, but no other changes in bile acid concentrations were significant.

While most investigations of possible relationships between ascorbic acid ingestion and urinary stones have related to oxalate, there is also a possibility of an interaction between high ascorbate ingestion and uric acid stones. A paper by Stein et al.[736] points out that high doses of ascorbic acid can induce uricosuria. The subjects were maintained on a purine-free diet, and drugs known to affect uric acid metabolism or excretion were not used. During the daily ingestion of 4 g ascorbic acid there was an approximate doubling of uric acid clearance. Smaller single doses of 0.5 g or 2.0 g did not significantly alter the clearance.

To investigate the mechanism of this interaction studies were made on urate-binding to serum proteins, but no change was noted with ascorbic acid. Similarly, there was no alteration of creatinine clearance and the glomeralur filtration rate was not changed. The authors suggest that altered tubular function is the most likely explanation of their findings. During the receipt of the ascorbic acid doses, serum uric acid concentrations declined in the test subjects. It is concluded that high doses of ascorbic acid may be relevant to the formation of urate stones in particular individuals.

Studies in guinea pigs[295] indicate that sudden withdrawal of high-dose vitamin C supplementation results in a more rapid development of scurvy when the animals are switched to a scorbutic diet. Similar effects may occur in humans, and a higher incidence of scurvy among people suddenly deprived of an adequate diet who had previously received vitamin C supplements is described by Jakovlieu,[395] while Rhead and Schrauzer[638,673] have described two cases of frank scurvy in otherwise healthy men who abandoned the use of vitamin C supplements and returned to a normal dietary intake. The results of Rhead and Schrauzer suggests that a conditioning to high-dose vitamin C supplementation may occur in otherwise healthy adults who gradually develop lower blood and red cell ascorbic acid concentrations, together with increased urinary excretion, during continued treatment with the supplement.

There is also some evidence for adverse hematological effect of ascorbic acid. Horrobin[364] has described his own experience of taking a single large dose of ascorbic acid at night (3 g) and awakening the next morning with a suspected deep vein thrombosis.

An investigation in 538 geriatric patients by Andrews and Wilson[32] in which 200 mg daily vitamin C was compared with placebo in a double-blind investigation revealed a high incidence of thrombotic episodes associated with ascorbic acid than with the placebo.

Cowen et al.[172] have described adverse in vitro effects of high concentrations of ascorbic acid on human platelet structure and function.

An adverse effect of ascorbic acid on red blood cells is discussed by Mengel and Green.[489] Their healthy volunteers received 5 g ascorbic acid daily in divided doses. Blood was collected and the lytic sensitivity of the red cells to hydrogen peroxide was measured. While none of the subjects showed any evidence of in vivo hemolysis, the in vitro lysis rose from 3% prior to ascorbic acid supplementation to 9% while the supplement was being received. This report is probably relevant to a fatal case of disseminated intravascular coagulation associated with high doses of i.v. ascorbic acid discussed by Campbell et al.[128] The patient was a 60-year-old black man admitted for treatment of acute renal railure. The patient had also recently suffered second degree burns to the hand and 80 g of ascorbic acid was administered on each of two consecutive days. The man proved to have approximately 50% deficiency of glucose-6-phosphate-dehydrogenase, and electrophoresis revealed the isoenzyme GdA which is typical of individuals of African descent.

Hanck[315] has investigated the effect of 1 g daily ascorbic acid in healthy young men on urinary excretion of calcium, iron, and manganese. These three elements were all increased, but there are reduced urinary excretion of copper and zinc. Unfortunately, wide individual variations were noticed and it is not clear whether the changes are due to any real physiological action of the vitamin supplement.

Of greater significance is the investigation by Cook and Monsen[166] who have investigated iron absorption in 63 men treated with ascorbic acid supplements in the range 25 mg:1 g daily. Significant increase in radio-iron absorption was noted with the vitamin supplement in a dose-related manner. The authors point out that while the effect may be of value in normal individuals, patients with idiopathic hemochromotosis, thalassaemia major, sideroblastic anemia, or other conditions in which iron absorption regulation is abnormal, there may be adverse effects from high-dose vitamin C supplements.

That large amounts of ascorbic acid may disrupt fertility in young women has been raised by Briggs.[109] It was suggested that mucus of the uterine cervix may have its disulfide bonding of glycoprotein fibrils disturbed with a resulting impedence of sperm penetration following exposure to large amounts of ascorbic acid. Details of women who failed to conceive while receiving vitamin C supplements are given.

Samborskaia[666] mentions abortion in a small series of pregnant women who received 6 g daily ascorbic acid for 3 days. Full clinical details are lacking.

Drug interactions with excess ascorbic acid, together with interference of vitamin C with clinical chemistry measurements, are reviewed by Briggs[114] and are summarized in Table 6.

Finally, mention must be made of the report by Stick et al.[741] who found mutagenicity following treatment of human fibroblasts with ascorbic acid in vitro. It is not clear whether the range of supplements used in clinical practice is sufficient to produce the type of effect reported in these studies.

IV. VITAMIN D

A. Animal Studies

There have been rapid and important developments in the understanding of the mode of action and regulation of vitamin D synthesis in mammalian systems in recent years.[51,154,191,328,330] Vitamin D (cholecalciferol) is formed in mammalian skin by the effect of short-wave ultraviolet light acting upon 7-dehydro-cholesterol, which results in rupture of the steroid B ring. The cholecalciferol so formed within the skin enters

Table 6
INTERACTIONS OF ASCORBIC ACID WITH DRUGS AND WITH CLINICAL CHEMISTRY MEASUREMENTS

		Effects of excess ascorbate
Drug interactions	Aminosalicylic acid	Enhanced risk of crystalluria
	Amphetamines	Decreased renal tubular reabsorption[318]
	Warfarin	Impaired response[657,710,734]
	Tricyclic antidepressants	Decreased renal tubular reabsorption[318]
	Salicylates	Increased renal tubular reabsorption[318]
	Vitamin B_{12}	Substantial losses[352]
Clinical chemistry interactions	Serum bilirubin	Result low[729]
	Serum aminotransferases (transaminases)	Result low[708,729]
	Serum lactate dehydrogenase	Result low[729]
	Serum glucose (blood sugar)	Result low[651]
	Serum uric acid	Result high[138]
	Urine glucose (sugar)	Result high or low depending on method[486]
	Urine 17-hydroxycortico-steroids	Result low
	Urine urobilinogen	Result low
	Stool occult blood	Result low[393]

the circulation and on passage through the liver is converted to 25-hydroxy -cholecalciferol,[732] which again enters the circulation,[307] probably on a specific carrier protein.[102] As this monohydroxylated metabolite passes through the kidneys it is acted upon by a specific enzyme with the formation of 1α,25-dihydroxy cholecalciferol.[107,190] This latter metabolite is thought to be the compound which produces most or all of the effects normally ascribed to vitamin D. A number of other di- and tri-hydroxylated derivatives have been identified in a variety of systems and at this stage are of unknown physiological significance. The therapeutic use and side effects of the various natural and synthetic analogues of vitamin D are discussed elsewhere in this volume.

Formation of cholecalciferol from 7-dehydrocholesterol within the skin appears to be self-limiting and vitamin D toxicity from endogenous over-synthesis due to ultraviolet light does not appear to have been reported. The action of ultraviolet on the basal cell layers of the skin is to increase melanin pigmentation and distribution, and this probably limits the penetration of ultraviolet light and consequently the rate of formation of cholecalciferol. It has been suggested by Lunaas et al.[468] that melanin pigmentation of the skin as a result of exposure to ultraviolet light may be a physiological mechanism for limiting vitamin D formation.

Large amounts of preformed cholecalciferol are present in a variety of animal foods, particularly fish, liver, and related oils. Plant tissues contain the structurally related ergocalciferol. Excess vitamin D is known to uncouple oxidative phosphorylation,[678] though other toxic actions also occur.[730]

Most cases of hypervitaminosis D reported in the literature are due to overdosing with cholecalciferol, ergocalciferol, or dihydro-tachysterol.

Clinical and physiological manifestations of vitamin D toxicity in laboratory animals, especially the rat, dog, or rabbit, are similar to those in humans overdosed with this vitamin. The major effect is hypercalcemia, with extensive calcification of extraskeletal tissues, particularly blood vessel walls[84,86] and the kidney.[60] Haussler et al.[331]

have reported the occurrence of 1,25-dihydroxy-cholecalciferol-glycoside in the Argentinian plant, *Solanum malacoxylon*. Cattle feeding on this plant are at particular risk of vitamin D toxicity. Comprehensive reviews on the literature on hypervitaminosis D have been published by Hayes and Hegsted,[343] 1973, Volume 3 of Sebrell and Harris, 1971, and the National Academy of Sciences National Research Council,[549] 1975. A recent summary of the evidence is published by Harrison,[325] 1978.

An interesting question that is still unresolved is the nature of the toxic metabolite in hypervitaminosis D. This could be unmetabolized vitamin D itself, 25-hydroxy vitamin D, 1α,25-dihydoxy vitamin D, or some other metabolite. The evidence of Haussler and McCain[330] suggests that vitamin D itself in normal doses lacks biological activity, while the synthesis of 25-hydroxy and 1,25-dihydroxy-vitamin D is carefully regulated and is under negative feedback control by high concentrations of calcium and phosphate ions, together with the metabolites themselves. A possible explanation between the lack of physiological activity of vitamin D at normal concentrations and its potential activity in hypervitaminosis D may relate to alterations in binding to plasma proteins, such as occurs during hypervitaminosis A (Smith and Goodman[714]). Under ordinary physiological conditions vitamin D is transported on a specific plasma binding globulin, while up to 40% is transported by lipoproteins. During hypervitaminosis D these binding sites may become saturated with an increase in the free pool of vitamin D allowing it to gain access to many cell membranes.

A study by DeLangen and Donath[188] has shown that vitamin D significantly alters the formation of atheromata in the arterial wall of rabbits. A comparison was made of diets containing large amounts of cholesterol and/or large amounts of vitamin D. A combination of vitamin D and cholesterol was more effective in producing atheromatosis than cholesterol alone. It is pointed out that high doses of vitamin D significantly alter concentrations of both cholesterol and total lipids in blood serum, as well as inducing hyperphosphatemia. These are all factors favoring the formation of atheroma.

A further investigation of arterial disease in rabbits associated with hypervitaminosis D has been published by Hass et al.[327] The lesions induced were characterized by medial calcific degeneration, with secondary mesenchymal reactions.

An important investigation by Friedman and Roberts[254] has demonstrated transplacental passage of excess vitamin D in pregnant rabbits. At birth the offspring had significantly higher serum calcium levels than untreated offspring, while aortic abnormalities similar to supravalvular aortic stenosis were noted.

Investigations in the rat by Eisenstein and Groff[224] indicate similar effects to those in the rabbit. Massive doses of vitamin D produce a marked increase in serum calcium concentration, together with increases in seromucoid levels. There is increased calcareous deposition in the heart, kidneys, and blood vessels, together with accumulation of polysaccarides in those locations. The calcium deposits appear to occur in a matrix containing acid mucopolysaccarides. Despite deposition of calcium in the heart, the associated myocarditis did not correlate well with the degree of calcium deposition as assessed histologically. Of possible relevance to the mechanism of hypervitaminosis D in the rat are the observations of Schacter et al.[668] When D_3 was added directly into the lumen of loops of rat duodenum obtained from vitamin D deficient animals, there was a significant increase in calcium transport by slices of the loop in vitro. The same dose of vitamin given intravenously or placed in a duodenal loop had little or no effect on the duodenal tissue and the authors suggest that vitamin D in the rat may act directly on the small intestine without prior activation in another organ.[323]

The interaction between vitamin A and D during hypervitaminosis has been recently discussed by Arnrich.[46] She points out that the major pathological impact of excess vitamin D is related primarily to the stimulation of calcium absorption and the mobilization of calcium from bone. This leads to abnormal calcification of organs and vessels following hypercalcemia, together with depletion of calcium in skeletal tissues. Hypervitaminosis A also has an important effect on bone structure and metabolism, though in this case the primary target appears to be the organic matrix of the skeleton, so that reduced mineralization occurs secondarily to matrix dissolution. Studies in rats by Clark and Bassett[149] and Bélanger and Clark[71] and in chicks by Taylor et al.[752] have shown that a large excess of vitamin A will protect young rats from vitamin D toxicity. This antagonism between excess of the two nutrients has been further investigated by Taylor et al.[752] who were able to distinguish between three categories of biochemical response in the chick. Antagonism was particularly marked on plasma calcium, phosphate, and acid phosphatases, whereas plasma lysosomal enzymes and packed cell volume were affected by the excess vitamin A only.

B. Clinical Toxicology

Animal investigations of hypervitaminosis D in the common laboratory species are predictive of vitamin D toxicity in humans. There can be little doubt that vitamin D is the most toxic of all the vitamins in humans. Indeed in some developed countries it is probably fair to say that cases of vitamin D toxicity now exceed cases of rickets.[215,543,545,830] In a recent review of knowledge of vitamin D, the U.S. National Research Council[550] confirms the hazardous effects of excess vitamin D and suggests that only individuals with disease affecting vitamin D absorption or metabolism require more than 400 IU daily. Such individuals need to be identified by clinical evaluation and treatment should be specifically recommended and supervised by physicians.

An anonymous comment[37] published by the Journal of the American Medical Association in 1973 points out that the U.S. Food and Drug Administration has tightened control on the sale and use of all vitamins from October 1973, and that with regard to vitamin D a prescription is now required for single doses larger than 400 IU daily. It is almost impossible to consume a toxic dose of vitamin D from natural sources. The liver oil of fishes, particularly those of the *Percomorph* order, contain the highest concentration and at one time were used as dietary supplements. Pharmaceutical preparations containing vitamin D are now almost entirely based on ergocalciferol or cholecalciferol, especially the former. Under normal circumstances the concentration of vitamin D (including its metabolites) in human plasma is between 1 to 2 IU/mℓ.[794] This is almost entirely present as 25-hydroxy-vitamin D. During hypervitaminosis D the concentration of vitamin D activity in plasma increases by 10 to 30 times.[793] The investigations of Rosenstreich et al.[656] indicate that vitamin D_3 is stored in high concentrations in adipose tissue, though the liver is not a major storage site in the mammal, unlike the fish. Using radio ligand receptor assay techniques Hughes et al.[37] have investigated human patients with hypervitaminosis D for plasma concentrations of 25-hydroxy-vitamin D and 1α,25-dihydroxy-vitamin D. No significant increase in the dihydroxylated metabolite was noted, though significant increases in the monohydroxylated form were observed.

Hypervitaminosis D results in toxicity primarily due to the hypercalcemia caused by the increased intestinal absorption of calcium, together with increased bone mineral mobilization. The rapid consequence is deposition of calcium phosphate in extraskeletal soft tissue[802] and renal tubule lumen.[691] If the ingestion of excess vitamin D continues there is a progressive reduction in renal function as a result of scouring and eventually renal insufficiency results. Other tissues influenced by calcification include blood vessels,[628] gastric mucosa, pancreas, lungs,[770] and heart. Severe hypertension with complicating encephalopathy may also result. Obvious and nonspecific symptoms of the hypercalcemia include nausea, vomiting, anorexia, weakness, and constipation. Inhibition of the action of vasopressin on distal tubules may result in a vasopressin resistant polyuria.[27] Where hypervitaminosis D is fatal, the usual cause of death is renal insufficiency.

Hypervitaminosis D in adults can occur either from the therapeutic use of large doses of vitamin D for conditions such as pulmonary or cutaneous tuberculosis, in various allergic states, or especially in rheumatoid arthritis. Self-administration of excess D in the form of vitamin supplements has also been described. It has been pointed out by Stanbury[733,734] that the plasma half-life of vitamin D is very long and that following cessation of vitamin D therapy the concentration of plasma vitamin D activity decays extremely slowly. It is suggested that large doses of vitamin D may contribute to high plasma concentrations for periods up to several years following withdrawal of the medication.

A protracted case of vitamin D intoxication in a 56-year-old man is described by Shetty et al.[693] High concentrations of vitamin D in the plasma were measured on many occasions, and normocalcemia was maintained only by continuous corticosteroid therapy.

There has been considerable interest in correlation between serum lipids and vitamin D administration.[14] There seems little doubt from studies reviewed by Linden and Seeleg[457] that vitamin D is hyperlipidemic in healthy adults. Schroll et al.[674] have described an inverse correlation between serum triglyceride and calcium concentration in subjects aged over 60 years. Seelig and Heggtveit[683] have pointed out that magnesium deficiency is also associated with hypolipidemia and that vitamin D excess induces magnesium loss.

It is pointed out by Counts et al.[171] that patients with chronic renal insufficiency develop resistance to vitamin D and that because of this resistance such patients are frequently treated with extremely large doses. Despite bilateral nephrectomy, hypercalcemia may develop following vitamin D, which suggests that this effect is due not to the dihydroxylated metabolite thought to be formed only by renal tissue.

Individual hypersensitive reactions to vitamin D have been noted (the anonymous comment in Medical Letter[38]) and such individuals may experience toxic effects even from relatively low doses. Gegick and Danowski[276] have described a 47-year-old woman who post-thyroidectomy was resistant to vitamin D. She received a course of 25-hydroxy-cholecalciferol and experience headaches, nausea, vomiting, diarrhea, and fever. On substituting 1α-hydroxy cholecalciferol there were no clinical problems, but when she received 1α,25-dihydroxy-choleclaciferol the symptoms recurred within 48 hr. The authors suggest that this may be an idiosyncratic reaction to the presence of the hydroxyl group of carbon-25 of vitamin D.

Possible correlation between vitamin D intake and increased mortality from ischemic heart disease are provided by Dalderup .[179] It is pointed out that various investigations have revealed increases in serum cholesterol concentrations in men aged 35 to 54 years receiving vitamin D preparations.

Leeson and Fourman[447] describe a 40-year-old woman with parathyroid deficiency who was accidentally overdosed with vitamin D and who developed acute pancreatitis and epilepsy. The authors attribute the pancreatitis to the hypercalcemia induced by the vitamin D overdose and the epilepsy to the corresponding fall in serum magnesium concentration. It is pointed out that patients with parathyroid deficiency usually tolerate very high doses of vitamin D that would be particularly toxic in normal persons. They make the interesting observation, however, that two of their adult patients who had severe tetany from parathyroid deficiency, who were given vitamin D at progressively increasing doses, developed adverse side effects. Following the appearance of these adverse toxic effects, both patients then responded to much smaller doses of vitamin D than had previously been possible. It is proposed that exposure to a massive dose of vitamin D in some way increases sensitivity of bone to a much smaller dose.

There has been much interest in possible relationships between vitamin D overdose[81,369,748] and infantile hypercalcemia[176,250,712] and its complications (see particularly the comprehensive review by the American Academy of Pediatrics Committee on Nutrition[20,21] and reviews by Taussig,[751] Friedman,[253] and Seelig[682]). Hypercalcemia in infants appears as different clinical forms depending upon the degree of abnormality of the serum calcium concentration. Children with mild hypercalcemia fail to thrive and the condition usually has a sudden onset at the third to seventh month after birth.[546] The long-term prognosis is good and the condition is reversible, often without specific medication, though occasionally children may die in acute hypercalcemic phases. When treatment is necessary vitamin D and sunlight must be eliminated at least temporarily while calcium intake is restricted. Occasional cases may require corticosteroids.[477,811] Severe cases of infantile hypercalcemia have been described by a number of authors, particularly Black and Bonham-Carter.[92] It is possible that the condition may arise in utero.[236,245,246] The infants are of low birth weight, have peculiar faces that are occasionally noted at birth, and often appear with systolic murmers. The prognosis for such infants is poor and mental defects are likely to be permanent. Cardiovascular and renal problems are rarely reversible,[618] and a significant number of infants are likely to die in the neonatal period.[100] There is sufficient evidence to suggest that there are at least two different syndromes involved, both of which include supravalvular aortic stenosis.[43,85,87,216,411,576,809] Denie and Verhaught[194] suggest that there is a familial form of this condition in which aortic stenosis is the only abnormality.[490] When aortic stenosis is combined with severe mental defects, peculiar facial appearance and abnormalities of dentition[272] are likely to be a condition of different etiology.[93,526] The role of vitamin D toxicity in the intrauterine development of idiopathic hypercalcemia is still uncertain, though placental transfer of the vitamin from the mother may well occur and may be damaging to a particularly sensitive fetus.

A study designed to experimentally explore any relationships between exposure to excess vitamin D during pregnancy and the development of a cranio-facial complex and abnormal dentition in offspring has been published by Friedman and Mills.[255,256] They point out that infants with the supravalvular aortic stenosis syndrome have a characteristic cranial facial appearance as well as many dental abnormalities. Particularly common is mandibular hyperplasia, together with congenital absence of particular teeth, especially the lateral incisors and second premolars of the maxilla. A comparison is made between infants with this syndrome and offspring of rabbits receiving high doses of vitamin D during pregnancy. The authors conclude that the human syndrome may be related to deranged vitamin D metabolism during pregnancy.

Complications of high-dose vitamin D in patients with renal disease have been published by Feest et al.,[233] Bouillon et al.,[102] Balsan et al.,[55] Pierides et al.,[610] Mallick and Berlyne,[473] Paunier et al.,[598] Brickman and Norman,[108] Dent and Friedman,[196] Henderson et al.,[348] and Russell et al.[661,662] Complications of long-term use of vitamin D or its analogues in hypoparathyroidism or pseudohypoparathyroidism are published by Leeson,[448] Parfitt,[579] Spaulding and Yendt,[728] Bell and Stern,[70] Wagner and Fulkers,[792] Avioli,[50] Hossain,[366] Ireland et al.,[379] Kooh et al.,[426] Winterhorn,[813] Haussler et al.,[329] and Russell et al.[661] An investigation of 43 patients with familial hypophosphatemic vitamin D resistant rickets is published by Stickler et al.[742] Hypercalcemia induced by large doses of vitamin D occurred at least once in 23 of these subjects, but it was noticed that the dose of vitamin D per unit of body weight not causing hypercalcemia was higher in the smaller subjects. One patient suffered irreversible renal damage, while two had severe hypercalcemia due to vitamin D intoxication.

In summary, it may be said that vitamin D intoxication occurs readily in both adults[182,195,247,416] and infants[127,189,243,629,655,707] receiving excessive amounts of cholecalciferol, ergocalciferol, or one of their metabolites, or a synthetic analogue.[140,219] The major effects are hypercalcemia[414] and soft tissue calcification,[380] and these may lead to long-term effects,[261,607] including renal insufficiency,[101,622] aortic stenosis,[194,652] mental deficiency,[204,221] and even death.

Individual susceptibility to vitamin D excess appears to vary markedly, and there is evidence to suggest that patients with particular conditions may develop tolerance to high doses. Transplacental transfer of vitamin D, or its active metabolites, has been demonstrated an animal experiments and almost certainly occurs in the human. For this reason vitamin D use during pregnancy should be considered with considerable caution and any offspring exposed to vitamin D in utero must be followed up with particular care.

The hypercalcemia of sarcoidosis may be due to enhanced sensitivity to normal circulating levels of vitamin D, or to the production of an abnormally active metabolite.[350]

V. VITAMIN E

A. Animal Toxicology

Several naturally occurring tocopherols possess varying degrees of vitamin E activity.[681] The most active natural compound is α-tocopherol, which is also the most widely used synthetic form. This compound has three centers of asymmetry and naturally occurring α-tocopherol has the 2R, 4R, 8R configuration. It is popularly known as D-α-tocopherol. Commercial preparations frequently contain esters rather than the free alcohol.

Vitamin E activity is also possessed by a series of synthetic tocopherol derivatives, particularly tocopherol amines.[88,677] The animal toxicology of the various natural and synthetic tocopherols is probably similar, and all display only very low toxicity.[115] A summary of the results of investigations into the toxicology of α-tocopherol in laboratory animals is provided in Table 7. Very large doses, representing 100 to 200 times the estimated nutritional requirement, appear to produce no adverse effects.[193,214] Examinations of tocopherols for inherent carcinogenicity have similarly failed to produce any positive findings,[195,229] though there is some evidence of a potential procarcinogenicity.

Experiments by Telford[754] in which mice received dibenzanthracene revealed a higher incidence of lung cancer in animals simultaneously receiving 2 mg α-tocopherol every 2 days than in a matched group fed a vitamin E deficient diet. Similarly, the incidence of s.c. tumors was increased in the α-tocopherol-treated group.

An examination by Hook et al.[362] for potential teratogenic effects of α-tocopherol found 1 malformed animal out of a total of 91 offspring from 7 litters of ICR strain mice who received 591 IU D-α-tocopherol by stomach tube on days 7 to 11 of pregnancy. The malformed animals displayed exencephaly, open eye, and micrognathia. The authors comment that these malformations are not seen in untreated control mice of this strain, though they are known to be inducable in the strain by various teratogens.

Fatty infiltration of the liver, with deposition of cholesterol, has been reported by Marxs et al.[482] in rats receiving 50 mg of a vitamin E concentrate orally every 7 days. Aortic lesions were also seen in these animals, with overdevelopment of collagenous tissue at the base of the aortic valve and in the medial coat of the aorta, together with intimal sclerosis of the aortic vessels.

Table 7
ANIMAL TOXICOLOGY OF α-TOCOPHEROL

Species	Effects of high doses	Ref.
Mouse	No adverse effects (up to 50 g/kg)	193
	Lung and subcutaneous tumors increased in dibenzanthracene treated animals receiving supplementary tocopherol	754
	One malformed offspring in pregnant mice receiving tocopherol day 7—11 of pregnancy	361
Rat	No adverse effects	214
	Lack of carcinogenicity	201
	Liver fatty infiltration with aortic intimal sclerosis	229,482
	Increased liver cholesterol	11
	Development of fatty liver by ethanol enhanced	454
	Coagulopathy and increased mortality	488
Chick	Increased clotting time, depressed thyroid function, decreased growth	475

A more recent investigation by Alfin-Slater et al.[11] also found high levels of liver cholesterol in rats fed high doses of α-tocopherol. Possible toxic synergistic effects between tocopherols and other substances have been reported. An investigation by Levander et al.,[454] in which rats were fed a diet, together with 20% ethanol in water as a drinking fluid, demonstrated greatly enhanced deposition of triglycerides in the liver if the animals also received a tocopherol supplement.

Increased mortality and coagulopathy has been demonstrated by Mellette and Leone[488] in rats fed irradiated beef supplemented with high doses of α-tocopherol. Depression of plasma prothrombin was demonstrated in the supplemented animals. Increased clotting times in chicks receiving high doses of α-tocopherol has also been demonstrated by March et al.[475] The latter group also found depression of thyroid function in vitamin E supplemented chicks, as well as increased requirements of both vitamin D and K. An interaction between vitamin A and vitamin E has been documented by several authors.[65,89,402,434,436,449]

B. Clinical Studies

Prophylactic dose of vitamin E is 5 to 30 IU/day. While there are no clearly defined symptoms of vitamin E deficiency, and there is a lack of convincing evidence for any therapeutic value of large doses[90], vitamin E has nevertheless been widely used in the treatment of muscular dystrophy, habitual abortion, cardiovascular disease, and numerous other conditions.[365] Probably the best indication for the use of vitamin E is in the management of certain anemias of newborn children, and of anemias associated with malabsorption syndromes.[116]

While vitamin E preparations are generally of very low systemic toxicity, they are not entirely free of undesirable side effects in certain circumstances.[39,42,378,538] A summary of adverse side effects associated with high dose tocopherols is given in Table 8.

In 1973 H. M. Cohen[156,157] described severe weakness and fatigue as side effects induced in a series of healthy adults treated in his practice with 800 IU daily α-tocopherol. It was claimed that vitamin E-associated fatigue was more common than fatigue associated with either anaemia or hypothyroidism in this practice.

Toone[765] has stated that elderly men with ischemic heart disease receiving 1600 IU

Table 8
SUMMARY OF ADVERSE SIDE EFFECTS ASSOCIATED WITH HIGH DOSE TOCOPHEROLS

Adverse effects	Ref.
Fatigue and weakness	112,116,151
Increased urinary creatine excretion	112,354
Allergic contact dermatitis	6,118,506
Ecchymoses and/or altered coagulation factors	169,522
Elevated plasma lipids	177,231
Reduced response to iron therapy	487
Increased urinary androgen excretion	616

α-tocopherol daily did not experience fatigue, but a double-blind study of 800 IU α-tocopherol daily versus an inert placebo conducted by Briggs[112] had to be terminated early due to the development of severe fatigue and muscular weakness into otherwise healthy young men receiving the vitamin supplement. The symptoms were associated with increased serum creatine kinase activity and with a large increase in 24-hr urinary excretion of creatine. Interestingly, Hillman[354] has also described a marked creatinuria in a young man taking high doses of vitamin E, though in this case there were apparently no clinical signs or symptoms.

It would seem likely that the induction of clinically notable fatigue and weakness by high-dose vitamin E is a rare event occurring in only a minority of particularly susceptible individuals. It appears not to be a psychological reaction, but to have a biochemical and clinical basis suggestive of muscular damage.

Vitamin E has been used in a large number of cosmetic sprays and creams. Brodkin and Bleiberg[118] were among the first to report sensitivity in particular individuals to topically applied vitamin E, with the appearance of a moderate to severe allergic contact dermatitis.

Similar reports have now been published by Minkin et al.[506] and Aeling et al.[6] One of the early investigations of wheat-germ oil therapy by Shute[705] reported occasional skin rashes associated with gastrointestinal tract irritation, but the role of components other than tocopherols could not be ruled out.

Other side effects of vitamin E therapy have been reported in patients with diagnosed diseases in whom vitamin E was being used therapeutically. A study by Baer et al.[52] of 22 patients with cardiac conditions receiving up to 400 mg vitamin E daily for several weeks reported that a few became definitely worse during treatment, though this may be due to natural progression of their condition rather than to side effects of tocopherol excess.

Levy and Boas,[456] together with Vogelsang et al.,[787] have all reported nonspecific symptoms including headache, giddiness, and intestinal pain in cardiac patients receiving high dose vitamin E, though Anderson[29] has pointed out that similar reports occur in patients receiving only an inert placebo.

Effects of vitamin E excess on hematology and coagulation parameters are somewhat uncertain and the literature is conflicting.[40,217] Farrell et al.[231] investigated a group of men habitually consuming up to 800 IU vitamin E per day and measured a variety of laboratory parameters, including coagulation. No significant changes were noticed.

In contrast, Corrigan and Marcus[169] observed prolonged prothrombin time and ecchymoses in an elderly male patient taking up to 1200 IU vitamin E per day. The patient was also receiving warfarin and clofibrate. Discontinuation of the warfarin and vitamin E resulted in clearing of the ecchymoses and a reduction in prothrombin time.

These changes were maintained even when the warfarin treatment was reinstituted. After the patient had been stabilized for 2 months he was again challenged with 800 IU vitamin E daily for 42 days. Prothrombin time showed a progressive increase during this test period, while coagulation factors dependent upon vitamin K declined progressively and reached their lowest point on the final day of vitamin E ingestions. At this time multiple ecchymoses and a hematoma appeared and the vitamin E supplement was discontinued. Prothrombin time and coagulation factors gradually returned to normal after the withdrawal of vitamin E. The authors suggest there may be a synergism between clofibrate and vitamin E.

A Swedish investigation[536] of nine male patients who had myocardial infarctions receiving 300 mg α-tocopherol per day and observed at intervals for up to 64 weeks revealed a tendency towards prolonged plasma clotting time and decreased fibrinolysis. None of the patients were receiving drugs that could influence serum lipids or hemostasis.

Melhorn and Gross[487] have reported that the hematological response of children with iron deficiency anemia to treatment with iron dextran is impaired if vitamin E supplements are given simultaneously.

A group of elderly subjects receiving 300 mg daily α-tocopherol is briefly discussed by Dahl[177] who observed a highly significant increase in fasting serum cholesterol concentration in these patients. No such change was noted, however, in healthy young men receiving larger daily doses.

Farrell and Bieri,[231] however, did note an increase in serum lipids in some healthy adults receiving megavitamin E supplements, and also observed significant increase in plasma carotenoid concentration.

Treatment by Pinelli et al.[616] of patients with porphyria cutanea tarda with 1 g α-tocopherol daily for 3 months led to marked increases in urinary androgen excretion, though 24-hr urinary pregnandiol excretion dropped dramatically. Such alterations may be of importance in patients with dermatological conditions sensitive to androgens (i.e., acne) or in persons with hormone-sensitive neoplasms. A summary of these adverse clinical associations is given in Table 8.

VI. VITAMIN K

A. Animal Studies

The toxicity of vitamin K has been recently reviewed by Barash[58] and also by Arnrich.[46]

The first systematic examination of the oral and parenteral toxicity of various vitamin K compounds was published in 1940 by Molitor and Robinson,[518] while a detailed investigation of the various synthetic forms of vitamin K, i.e., menaphthone and related compounds, was published later.[26,641,694,716] It is immediately seen that the various synthetic derivatives have a significantly greater toxicity than the naturally occurring forms of vitamin K, though high doses may be administered orally. The oral LD_{50} in mice of menaphthone is approximately 500 mg/kg. In contrast, phytomenadione may be administered to this and other species at doses up to 25 g/kg without fatality.

In dogs, cats, and monkeys, hematological abnormalities result at relatively low doses following menadione treatment. While no anatomical pathological abnormalities are noted postmortem in animals treated with doses up to 5 mg/kg, many showed a mild anemia. Oral doses of menadione greater than 25 mg/kg produce a marked anemia accompanied with hemoglobinuria, urobilinuria, and urogilinoguria in the majority of animals. The studies of Richards and Shapiro[641] indicate that anemia induced by high doses of vitamin K_3 are reversible on withdrawal of the compound. The same authors also administered large but sublethal amounts of menadione to dogs with the induction of methemoglobinuria and cyanosis. The animals showed high prothrombin concentrations, while liver and kidney damage was noted at postmortem. An early response to vitamin K_3 was a decrease in prothrombin time, but prolonged treatment with high doses resulted in an elevation in this parameter.[776]

A review by Finkel[237] on the relative toxicity of natural and synthetic vitamin K analogues reveals that high doses of vitamin K_3 in the rat produce fatty infiltration with enlargement of the liver, together with hemosiderosis. There was also extensive renal tubular damage.

The cause of death in mice receiving lethal levels of vitamin K_3 appeared to be mainly related to respiratory paralysis, though effects on the eye were particularly marked, with many animals showing exophthalmos and excessive lacrimation.

An investigation by Wynn[824] into the etiology of the anemia produced by excess menadione and its water-soluble analogues in newborn and adult rats suggests that the compounds act as oxidizing hemolysins. In the newborn animal the excess bilirubin formed by the hemolysis produces a special stress factor on the liver due to defective glucuronide conjugation. In addition, there is competition between the vitamin K_3 and bilirubin for the available conjugation system.[801]

An interaction between vitamin K and vitamin E has been described by Allison et al.[13] When rats were deprived of vitamin E and then treated with excess menadione there was a particularly severe hemolysis hemoglobinuria and decrease in serum hemoglobin. In contrast, phytonadione had no particular effect in the vitamin E deficient animals. It has suggested, therefore, that vitamin E may protect against oxidative effects of menadione analogues. This latter finding is of importance in aspects of vitamin K_3 toxicity in the treatment of hematological defects in pregnant women with vitamin K_3 or the uses of large doses of menadione, in the hemorrhagic disease of the newborn.

B. Human Toxicity

A 1966 review of Deutsch[198] on the use of vitamin K in medical practice in the treatment of adults concludes that even high doses appear to be innocuous. This is a generalization derived mainly from the use of the naturally occurring fat-soluble form, but it is apparent that the water-soluble forms of vitamin K_3 have produced serious side effects in high doses,[746] especially in newborn infants.[12,197] It is probably fair to claim that the major effects of excess vitamin K in adults are circulatory, whereas in newborn infants they are hematologic.[35,36]

In many forms of hemolytic anemia there is a formation of a characteristic cytoplasmic inclusion known as a Heinz body, which is indicative of impending cellular disintegration.[61] The treatment of premature infants with vitamin K was reported by Gasser[273] to result in Heinz body formation and eventually hemolytic anemia. Uses of large doses of menadione in the treatment of hemorrhagic disease of the newborn was shown by Allison[12] to result in kernicterus, combined with hemolytic anemia. These findings, confirmed by Laurance[442] and others,[392,403,491] showed that the effect was dose-related to menadione administration. It is quite clear from these and other studies, particularly those summarized by the Committee on Nutrition of the American Academy of Pediatrics,[162] that the outcome of excessive vitamin K in newborns is occasionally fatal.

It is important to note that these effects on the newborn can be achieved not only by the use of large doses directly on the infants, but also by the treatment of their mothers with large doses of menadione during labor.[465] The studies of Asteriadou-Samartzis and Leikin[48] indicate that naturally occurring vitamin K even in very high doses is unable to induce the effects seen with menadione therapy.

The investigations of Murphy[538] indicate that prothrombin synthesis cannot be stimulated by vitamin K in the newborn due to the limited functional capacity of the immature liver. For this reason many authors, including Snyderman and Holt[722] and others,[207,213,228,251,297,404,520,620,785] recommend only low doses of vitamin K after birth as a specific for the treatment of hypoprothrombinemia in the newborn. It is fairly clear that premature infants have greater intolerance to excess vitamin K than full-term infants.[175]

Aside from the possible interaction with vitamin E mentioned above under animals studies, it has also been suggested by Zinkham[835] that reductions in glutathione may favor susceptibility to vitamin K induced hemolysis. Vest[783] has also suggested that vitamin K excess may interfere with redox systems of the erythrocytes and produce a decrease in intracellular osmotic pressure favoring hemolysis.

A variety of cardiac and pulmonary signs and symptoms have been reported in adults treated with large doses of vitamin K. Intravenous administration of vitamin K_1 was reported by Beamish and Storrie[68] to produce hypotension, jaundice, anemia, rigor, and hemoglobinuria in some patients. Complications such as dyspnoea, flushing, pain in the chest, or even cardiovascular collapse, has been reported by others (Cohn[161] and Douglas and Brown[205,206]) when vitamin K has been given intravenously by injection. It is only fair to comment, however, that in almost all these cases there is a lack of evidence to show causal effect of the vitamin K administration on the symptoms and signs reported. Perhaps the best documented case is that published by Barash et al.[57] who reported on acute cardiovascular collapse following i.v. phytonadione in a patient with severe liver disease who was being extensively monitored during surgery for cancer of the vocal chords. In this case the vitamin K was given intravenously at a slow rate (1 mg/min) and a drastic decrease in blood pressure, associated with periferal vascular collapse, was noted 5 min following the injection. As the patient had two subsequent vitamin K injections without side effects, it is unlikely that the response reported was due to an anaphylactic reaction. The authors of the report suggest that either the colloid emulsion or the vitamin had a profound periferal vasodilating effect on this particular individual.

Smith and Custer[711] have suggested that abnormalities in clotting time following the administration of vitamin K_3 may be related to hepatic dysfunctions. An abnormal prothrombin time is not fully corrected following vitamin K administration in patients with severe liver disease.[621, 703, 704,772,773] Paradoxically, vitamin K administration in patients with preexisting subclinical hepatic dysfunction, and a normal prothrombin time, often results in a prolongation.

There are occasional reports of skin reactions to vitamin K.[66,139,212,510,758,759] A summary of signs of vitamin K toxicity is provided in Table 9.

VII. B VITAMINS AND OTHER WATER-SOLUBLE FACTORS

A. Thiamin

A recent comprehensive review on the toxicology of thiamin in both animals and humans has been published by Itokawa.[381] He concludes that there is a very wide range between the therapeutic dose and the toxic dose of thiamin in most laboratory animals,[264,460,470,517] with pharmocodynamic effects being noted only after the administration of > 1000 times the usual therapeutic dose.[345,505,532,533,717-721] The oral LD_{50} of thiamine for common laboratory rodents ranges from 2000 to 3000 mg/kg/p.o., 80 to 120 mg/kg/i.v., and 300 to 500 mg/kg/i.p.[99,311,312,504,516] A summary of the major signs of thiamin toxicity in common laboratory species receiving massive doses of the vitamin is given in Table 10. Thiamin is an essential component of cocarboxylase.[685,686,766]

Table 9
VITAMIN K TOXICITY

Subject	Signs of excess vitamin K
Newborn infants	Heinz body formation[271]
	Hemolytic anemia[321]
	Kernicterus[299]
	Bilirubinemia[103]
Adults	Hypotension
	Anemia
	Rigor
	Hyperbilirubinemia
	Hemoglobinurea
	Dyspnoea
	Cardiovascular collapse
	Prolongation of prothrombin time

By virtue of its thiozolium ring structure, thiamin is a quaternary ammonium compound and related to a variety of neuro-muscular and ganglionic blocking agents.[1,2,7,45,119,199,202,222,248,288,290,333-337,339,340,342,376,382-392,415,507-509,547,556,569,606,624,667,818] An i.v. dose of 20 mg/kg in the cat or 80 mg/kg in the dog is sufficient to cause a complete neuromuscular paralysis with up to 95% ganglionic block.[202,789,790] Thiamin may also induce liver microsomal enzymes.[300]

In humans, thiamin is of extremely low toxicity, especially when administered by the oral route.[167,168,208,293,405,492,493,535-537,540,613-615,622,806,832] Serious clinical consequences have been reported occasionally, however, following large parenteral doses.[182,397,443,444,450,489,502,503,513,623,633,669,670,687,688,702,726,737,757,806] There are at least seven fatal reactions to high doses of parenteral thiamin reported in the literature, the cause of death being anaphylactic shock. It has been noted that the patient affected in this manner has frequently been accidentally sensitized to thiamin by administration of the vitamin over a long period of time. A subsequent injection has exceeded the latent period and precipitated an anaphylactoid reaction. Desensitization to thiamin has been demonstrated by a number of workers, especially Mitrani[513] and Kawasaki et al.[418]

To avoid the problem of possible sensitization, it has been recommended that where massive parenteral doses of thiamin are indicated, for example in the treatment of certain physiatric emergencies, the risk of adverse reactions can be considerably reduced if a single large dose of thiamin is given in combination with other members of the vitamin B complex.

A review of Gould[298] describes the signs and symptoms of overdosage with parenteral thiamin (more than 400 mg) as including acute mental alertness, though without pressure of ideas, which lasts for a few hours. Subacute symptoms include lethargy, solemness, mild ataxia, heaviness in the limbs, and a diminution of gut tone. Similar signs and symptoms are seen with chronic overdose of parenteral thiamin, though are generally less severe and the patient is likely to complain of nausea with loss of appetite. Patients receiving long-term treatment with high-dose thiamin develop the characteristic smell of the compound (yeast-like) in newly washed skin and also in their excreta. All these signs and symptoms are readily reversible on withholding vitamin B supplements and returning to a normal dietary intake.

Other clinical manifestations of thiamin overdose in humans include acute hypotension and the development of hyperthyroid-like symptoms. Intraspinal injections of thiamin[338,341] have been related to the development of weakness, conjunctivitis, anorexia, buzzing in the ear, muscle stiffness, fever, headache, vomiting, meningeal irritation, back stiffness, and leg pains. It is uncertain whether these are related to thiamin

Table 10
TOXICITY OF THIAMIN, VITAMIN B_{12} AND BIOTIN IN MAN AND ANIMALS

Vitamin	Animal results	Human results
Thiamin (B_1)	Respiratory inhibition,[332] vasodilation, muscular spasms, convulsions, cyanosis and anoxia disturbed coenzyme A metabolism[23-25]	Anaphylactic shock, lethargy and ataxia, nausea, hypotension, riboflavin depletion
B_{12}	Lung congestion	Allergic reactions, hyperuricemia(?)
Biotin	Delayed growth, infertility	—

itself, to some other component of the injection material, or to trauma during administration. Increases in protein concentration in cerebrospinal fluid, associated with increased cell counts, have also been noted following intraspinal injections of thiamin.[648,740] Despite this multitude of effect, no serious or persisting damage has yet been reported following this method of administration.[570] Alterations in EEG may occur in some patients receiving thiamin.[314]

An interesting possible interaction between thiamin and riboflavin has been reported by Klopp et al.[422] Shortly following administration of thiamin in high dose it is reported that there is a transitory increase in urinary riboflavin excretion, however, no clinical signs or chemical evidence of riboflavin deficiency were observed in patients who had received large doses of thiamin for up to 73 days. A confirmation of these findings is reported by Fujiwara[262] who found that doses of 5 to 100 mg of thiamin increased riboflavin excretion and reduced blood concentration of riboflavin, especially in patients with liver disease, tuberculosis, or anemias. Again, clinical signs of riboflavin deficiency were not noted.

An investigation by Inoue et al.[377] into the effects of massive parenteral doses of thiamin propyl disulfide given for 1 to 4 weeks in patients with neurological disorders has suggested the induction of clinical signs of riboflavin deficiency. The treated subjects developed cheilosis and a beefy red tongue, together with reduced concentrations of blood riboflavin. These changes were all reversed following administration of riboflavin.[553,554]

There is some evidence to suggest a protective effect of vitamin B_6 against the effects of excessive thiamin in the rat.[263,524,525,640,676,695-701,778] Thiamin is able to inhibit the vasoconstrictor action of the nicotine.[309,391,608,750,826,827]

B. Vitamin B_6

The toxicology of vitamin B_6 has been recently reviewed by Haskell[326] who summarizes the acute LD_{50} of the various forms of vitamin B_6, together with its metabolites in six species. By any route of administration the B_6 compounds are of low systemic toxicity.[10,44,56,163,187,322,367,372,630] By the oral route the compound of highest toxicity (approximately 2000 mg/kg body weight) is pyridoxal hydrochloride.[359,431,776,777,824]

No teratogenic effect of high dose vitamin B_6 has been detected in any species.[419] High doses have been administered during human pregnancies as an antinauseant without malformations being reported.[258]

There is experimental evidence that under certain conditions very large amounts of vitamin B_6 may inhibit prolactin secretion[131] and consequently reduce milk production in a lactating woman. Such a decrease in milk supply could have detrimental effects on an infant.

C. Vitamin B_{12}

All compounds in the vitamin B_{12} series are of extremely low systemic toxicity in both laboratory animals and humans.[123,774] For many years the most widely used therapeutic substance was cyanocobalamin, though recently this has tended to be replaced by hydroxycobalamin, largely on the theoretical grounds of reducing cyanide exposure.

Adverse reactions to vitamin B_{12} administrations are infrequent and are almost entirely confined to allergic reactions in sensitive individuals.[209,238,282,396,452,560,631]

Attempts to induce anaphylactic shock in guinea pigs by multiple injections of small doses of crystallized vitamin B_{12} was attempted by Traina,[767] but the experiments were unsuccessful. Similar experiments with albino mice resulted in fatalities from congestion of the lungs following injections of 3 mg/kg/body weight. A detailed study by

Lipton and Steigman[459] has demonstrated immunological sensitization of guinea pigs to vitamin B_{12} and evidence was produced to show that this was due to the vitamin itself and not to contaminants in the preparation injected.

Clinically, the most frequent adverse reaction to vitamin B_{12} is a mild skin reaction, though more severe cases, including anaphylactic shock,[590] have been described, usually in association with urticaria. A possible relationship to thrombosis has been suggested.[420]

An investigation of skin reactions to intradermal injection of vitamin B_{12} preparation has been reported by Bedford.[69] He found that vitamin B_{12} derived from *Streptomyces griseus* produced positive skin reactions and that this side effect was twice as common in people who had previously received antibiotics. He suggests that some of the reactions to vitamin B_{12} preparations may be due to impurities and that antibiotic therapy may be a factor in inducing idiosyncratic responses. Aside from possible contaminant allergins in vitamin B_{12} manufactured from mold cultures, the preservative phenylcarbinol, which is commonly included in vitamin B_{12} injections, has been shown to be a potential sensitizing agent in a minority of subjects.

A theoretical complication of vitamin B_{12} therapy is a consequence of the rapid maturation of cells following vitamin B_{12} administration in subjects with pernicious anemia. It is known that this maturation increases nucleic acid degradation and may produce a sudden rise in serum uric acid concentration and in urinary excretion. Patients susceptible to gout, or with pre-existing hyperuricemia, may be at potential risk of exacerbation of their symptoms on the receipt of vitamin B_{12} for this indication.

An interaction between folic acid and vitamin B_{12} may occur.[373,786]

D. Folic Acid

While generally considered a nontoxic vitamin, folic acid supplements are not entirely free of hazard[94,280,690] (Preuss[626]). In animal studies, it has been reported that doses of 45 to 120 mg given intravenously to young rats will induce epilepsy (Hommes et al.,[361] Obbens and Hommes[567]). Epilepsy has been seen in rats following doses of folic acid as low as 1 mg, but only in animals with intracortical electrodes or brain lesions which may alter the blood-brain barrier.[361] Hommes and Obbens[361] have shown that the epileptic-inducing properties of folic acid are eliminated, or reduced, in animals with hepatic portal vein ligature or a partial hepatectomy, suggesting that the active compound is a metabolite from the liver.

Other effects of folic acid in animal studies are marked changes in blood pressure and cardiac contracture that can also be demonstrated on in vitro preparations (Jenkins and Spector[400]). Studies by Haddow[308] and by Threlfall et al.[763] suggest that folic acid supplements significantly increase the weight of kidneys following administration to various rodent species. The weight increase is partially due to nucleic acids, and it has been shown (Tilson[764]) that folic acid stimulates DNA synthesis in kidney, though not in gut. The folic acid treatment results in a mild degree of kidney tubule damage[627] which is nonfatal.[104,671,675]

There are no reports of comparable adverse side effects following the use of folic acid in human medicine.[174] According to Hunter et al.[374] the most common human side effects of regular oral treatment with 15 mg folic acid were mild gastrointestinal disturbances, irritability, malaise, hyperexcitability and disturbed sleep with vivid dreams. Two studies which attempted to confirm these findings were unable to do so.[347,634] Folic acid supplements mask the anemia but not the neurological manifestations of pernicious anemia.[164]

Hypersensitivity reactions have been described in occasional patients.[512]

E. Riboflavin

The general low toxicity of riboflavin is confirmed in a recent review by Rivlin,[644] who points out that the relatively low solubility of this vitamin at body temperature (less than 190 mg/ℓ[398]) in aqueous media led to its being administered either as a suspension, or dissolved in a variety of organic solvents. Kuhn and Boulanger[435] have suggested that early toxic findings with riboflavin in laboratory animals were due to the effects of the solvent, rather than the vitamin. Studies by Demole[192] on a wide range of species revealed no harmful effects, while Unna and Greslin[779] reported a lack of toxicity in rats receiving 10 g/kg by mouth, or 5 g/kg s.c., nor in dogs receiving 2 g/kg p.o.

Probably the first reliable report of animal toxicity for riboflavin was an investigation of rats receiving 0.6 g/kg i.p. (Unna and Greslin[779]), where the animals became anuric and vitamin crystals were found in the renal tubules.

The lack of toxicity following oral administration can probably be explained by the limited capacity of intestinal absorption mechanisms.[146,410] Attempts to increase tissue levels of riboflavin-containing cofactors generally are unsuccessful, even when the vitamin is given parenterally.[226] This suggests the existence of some regulatory mechanisms,[121] but which is at present unidentified.

Suspected riboflavin deficiency in humans is routinely treated with oral doses of up to 20 mg daily (Goldsmith[292]), and no adverse reactions have been reported. Larger doses of riboflavin have been given to humans by the parenteral route, with only a very small number of reactions reported.[726]

Farhangi and Osserman[230] have described a patient with multiple myeloma whose plasma contained a specific monoclonal immunoglobulin which showed high affinity and specificity for riboflavin. The woman has a previous history of multivitamin tablet ingestion and showed yellow pigmentation of the skin and hair. Her turnover and excretion of dietary riboflavin were impaired.[617]

Ribovlavin is known to interact strongly with boric acid,[653,654] with the formation and excretion in urine of a complex of the two substances. Riboflavin depletion is readily induced by boric acid,[645] while massive riboflavin adminstration may be of use in boric acid poisonings.

While flavoproteins are intimately involved in drug-metabolizing enzymes,[129,646] it is not known whether riboflavin excess causes any significant alteration in drug clearance rates.

F. Biotin

An extensive recent review on biotin toxicity in animals and humans has been published by Paul.[593] Like many others of the B vitamin complex, biotin is a substance of low toxicity in most animals.[106,511,756] Extremely large overdoses of biotin, however, may alter metabolic processes concerned with growth,[368] development, and reproduction.[4,72,158,159,455,514,594-597,611,684] The effect may be at least partially due to inhibition of the pentose phosphate pathway of metabolism.[320]

There is an absence of evidence of adverse effects of high doses of biotin in humans.

G. Pantothenic Acid

This member of the vitamin B complex is similarly of exceedingly low toxicity and adverse reactions have not been reported following high doses in animals or humans.

H. Niacin (Nicotinic Acid) and Niacinamide (Nicotinamide)

Aside from their use in the treatment of pellegra, niacin, and niacinamide have been widely used in the treatment of sprue, psychiatric conditions such as schizophrenia,[357,816] and in a wide range of circulatory disorders including hypertension, cerebral thrombosis, and intermittent claudication. The doses used are often greatly in excess of recommended dietary allowances. Niacin and related synthet-icanalogues,[304,467,727] but not niacinamide, are used widely in the treatment of hyperlipidemia.[4,5,59,76,137,178,269,291,355,375,580,585,587-589,643,833] This topic is reviewed elsewhere in this volume as well as in a number of other publications.[15,16,18,19,170,223,234,252,259,260,500,501,575,632,815]

Niacin is well absorbed following oral administration and is transported in the blood mainly in association with red cells.[265] There is a little body storage of niacin at pharmacological doses,[283,445] and most can be recovered from urine in the form of niacinamide, N-methyl-niacinamide, 2- and 4-pyridones, 6-hydroxy-niacinamide, and 6-hydroxy-niacin.[132] There are marked species variations in the nature of urinary metabolites of niacin.[142,244,603,639,798-800] A recent review on the pharmacology and toxicology of niacin and related derivatives has been published by Waterman.[797]

Generally, naicinamide is 2 to 3 times more toxic than niacin, though both compounds have low systemic toxicity. The acute LD_{50} in mice or rats is 4.5 to 7 g/kg by the oral route, 2.8 to 5 g/kg subcutaneously and 2.5 g/kg intravenously.[105,260,775] In long-term experiments, rats given niacin at 2% of the diet developed fatty livers,[316] though this was not confirmed in another study.[548] Young dogs receiving 1 g/kg daily showed normal growth, while no adverse reactions were seen in adult dogs given 2 g/kg daily p.o. In other studies,[562] dogs received up to 4 g daily for 3 months without toxic reactions and liver biopsies were normal. However, one study reported loss of condition and occasional deaths in dogs given only 2 g niacin daily.[143] No abnormalities were seen in offspring of rabbits given 0.3 g niacin daily from before mating, through pregnancy, and lactation.[16]

The use of niacin in the treatment of hyperlipidemia arises from observations of Altschul that blood cholesterol levels are lowered following inhalation of oxygen,[17] or exposure to ultraviolet light.[14] It was hypothesized that increased levels of the important respiratory coenzymes (NAD and NADP) might similarly lower cholesterol by stimulating its oxidation to various metabolities. Niacin was given to enhance tissue levels of these coenzymes. In fact, the mechanism for the hypocholesterolemic actions of niacin are more complicated and involve alteration of a wide range of metabolic processes,[98,122,133,134,136,240,270,319,494-496,499,558,559,564,584,600,663] some of which appear species specific.[135,144,211,433,565,566,820]

Summaries of adverse side effects of niacin following human administration have been published by several authors.[75,77] The major reports are listed in Table. 11.

Niacin has a marked peripheral vasodilatory effect in humans, though not in some other species. The flushing commonly reported following niacin therapy is accompanied by increased skin temperature and can be shown to be positively related to the concentration of niacin in the blood.[747] There is often a simultaneous increase in cardiac output. The flushing may persist in a high percentage of treated patients, some of whom also experience pruritis and itching.

Disturbance of gastrointestinal function is relatively common following oral niacin. There seems little doubt that the compound has significant irritant properties that occasionally lead to ulceration, particularly in patients with a previous history of such conditions.

The reported changes in liver function[53] are usually mild and readily reversible on withdrawing niacin. There are occasional reports of jaundice[642] or even of severe hepatic dysfunction, though this is rare.[425]

A high percentage of nondiabetic subjects show altered glucose tolerance during niacin treatment,[78,306,810] though again this is reversible on stopping. Diabetic patients

Table 11
CLINICAL TOXICOLOGY OF NIACIN
(NICOTINIC ACID)

Adverse association	Ref.
Initial flush	73,74,147,271,582,583,586
Prolonged flushing	74
Pruritis	73,74,147
Anorexia	74
Nausea/vomiting	74
Nervousness/panic	73
Abnormal glucose metabolism	74,147,271,497
Abnormal plasma uric acid	74,147
Abnormal liver function tests	74,147,271,586,577,812
Gastric ulceration	73,581
Anaphylaxis	601
Circulatory collapse	625
Skin changes	660,768

often show increased insulin requirements following niacin.[519] Serum uric acid level tends to rise during niacin treatment, and this may precipitate attacks of gout in susceptible individuals. Adverse mental changes have sometimes been reported by individual patients.[54,73,349]

I. Choline

In experimental animals the prevention or curing the fatty infiltration of the liver can be demonstrated with high doses of choline. Deficiency of choline, however, results only when both the intake of protein, particularly that containing methylamine, is limited. An absolute requirement for choline in all circumstances has not been demonstrated in either animals or humans. There is evidence for an interaction between choline and vitamin B_6.[203]

The toxicity of choline has recently been reviewed by Byington.[125] As with many other water-soluble vitamins, the acute toxicity of choline is relatively low, being about 5 g/kg for the rat, which is similar to that of the esters of choline and other related quaternary ammonium compounds.

No adverse side effects were noted in sows fed 3.4 g choline/day, though broiler chicks receiving approximately 0.5 mg/lb or more for 7 weeks showed signs similar to those of vitamin B_6 deficiency. The clinical signs were reversed by the addition of additional vitamin B_6 or by withdrawal of the choline supplement.

Single oral doses as large as 10 g produced no obvious defects in healthy humans.[735] However, a larger dose of 16 g choline administered daily for 8 days in a patient with tardive dyskinesia[185] induced symptoms of cholinergic stimulation, such as sweating, salivation, etc.

In cats receiving 40 mg/kg choline to relieve arterial spasm, there was depolarization and blockage of skeletal muscle with respiratory paralysis. There was transient hypertension with copious secretion of mucus and saliva following injection of this quantity.

J. Inositol

Inositol is a compound of relatively low toxicity and its effects on animals and humans has been recently reviewed by Tomita. Acute administration of inositol to laboratory animals is likely to produce diarrhea (Martin et al.,[480] Bly et al.,[97] Anderson,[28] Maeda[470]). This appears to be due to a stimulation of the motility of the gastrointestinal tract. Chronic ingestion of large doses of inositol may lead to growth inhibition in young animals (Agranoff and Fox,[8] and Natume[552]). The concentration of inositol in a standard diet is approximately 10 mg/kg/day. At 1 g/kg/day young rats grow significantly less, but all the animals aged 3 to 5 months do not exhibit any growth inhibition, even when fed up to 5 g/kg/day or myoinositol.

An interaction between inositol and choline has been described by a number of authors (Agranoff and Fox[8], Handler,[317] Jukes,[409] and Gavin and McHenry[274]). Generally, excess of inositol will aggravate symptoms such as renal necrosis or perosis due to choline deficiency.[8] With respect to the development of fatty liver, However, inositol and choline act synergistically (Gavin and McHenry,[274] Best et al.,[82] and Kotaki et al.[430]), but have opposite actions on the growth of yeast (Taylor and McKibbin[753]).

Inositol has been little used in human pharmacology and its toxicity in humans has not been the subject of systematic study.

REFERENCES

1. **Abderhalden, E.**, Beitrag zum Problem der fur die B (Aneurin bzw. Thiamin) Avitaminose charakterischer Erscheinungen, *Pfluegers Arch. Gesamte Physiol. Menschen. Tiere.*, 240, 647, 1938.
2. **Abderhalden, E. and Abderhalden, R,.** Vitamin B_1 and acetylcholine, *Klin. Wochenschr.*, 17, 1480, 1938.
3. **Acharya, V., Store, S.D., and Golwalla, A. F.**, Anaphylaxis following ingestion of aneurine hydrochloride, *J. Indian Med. Assoc.*, 52, 84, 1969.
4. **Achor, R. W. P. and Berge, K. G.**, Treatment of hypocholesterolemia with large oral doses of nicotinic acid, *Med. Clin. North Am.*, 871, 1958.
5. **Achor, R. W. P., Berge, K. G., Barker, N. W., and McKenzie, B. F.**, Treatment of hypercholesterolemia with nicotinic acid, *Circulation*, 17, 497, 1958.
6. **Aeling, J. L., Panagotacos, P. J., and Andreozzi, R. J.**, Allergic contact dermatitis to vitamin E aerosol deodorant, *Arch. Dermatol.*, 108, 579, 1973.
7. **Agid, R. and Balkanyi, J.**, Action de la vitamin B_1 sur le coeur isolé de grenouille, *C. R. Soc. Biol.*, 127, 680, 1938.
8. **Agranoff, B. W. and Fox, M. R. S.**, Antagonism of choline and inositol, *Nature (London)*, 183, 1259; 1959.
9. **Ahrends, L. G., Kienholz, E. W., Shutze, J. V., and Taylor, D. D.**, Effect of supplemental biotin on reproductive performance of turkey breeder hens and its effect on the subsequent progeny's performance, *Poult. Sci.*, 50, 208, 1971.
10. **Alder, S. and Zbinden, G.**, Use of pharmocological screening tests in subacute neurotoxicity studies of isoniazid, pyridoxine HCL and hexachlorophene, *Agents Actions*, 3, 233, 1973.
11. **Alfin-Slater, R. B., Aftergood, L., and Kishineff, S.**, Investigations on hypervitaminosis E in rats, *Abst. Int. Congr. Nutr.*, IX, 191, 1972.
12. **Allison, A. C.**, Danger of vitamin K to the newborn, *Lancet*, 1, 669, 1955.
13. **Allison, A. C., Moore, T., and Sharman, I. M.**, Haemoglobinuria in vitamin E deficient rats after injections of vitamin K substitutes, *Br. J. Haematol.*, 2, 197, 1956.
14. **Altschul, R.**, Lowering of serum cholesterol by ultraviolet irradiation, *Geriatrics*, 10, 208, 1955.
15. **Altschul, R.**, Influence of nicotinic acid on cholesterol metabolism, *Circulation*, 14, 494, 1956.
16. **Altschul, R.**, Influence of nicotinic acid (niacin) on hypercholesterolemia and hyperlipemia and on the course of atherosclerosis, in *Niacin in Vascular Disorders and Hyperlipemia*, Altschul, R., Ed., Charles C Thomas, Springfield, Ill., 1964, 3.
17. **Altschul, R. and Herman, I. H.**, Influence of oxygen inhalation on cholesterol metabolism, *Arch. Biochem. Biophys.*, 51, 308, 1954.
18. **Altschul, R. and Hoffer, A.**, The effect of nicotinic acid on hypercholesterolemia, *Can. Med. Assoc., J.*, 82, 783, 1960.
19. **Altschul, R., Hoffer, A., and Stephen, S. O.**, Influence of nicotinic acid on serum cholesterol in man, *Arch. Biochem.*, 54, 558, 1955.

20. American Academy of Pediatrics, Committee on Nutrition, The prophylactic requirement and the toxicity of vitamin D, *Pediatrics,* 31, 512, 1963.
21. American Academy of Pediatrics, Committee on Nutrition, The relation between infantile hypercalcemia and vitamin D — public health implications in North America, *Pediatrics,* 40, 1050, 1967.
22. **Ammann, P., Herwig, K., and Baumann, T.,** Vitamin A excess, *Helv. Paediatr. Acta,* 23, 137, 1968.
23. **Ando, H.,** Effects of excessive administration of thiamine derivatives on the coenzyme A metabolism in rat. I. Effect of excessive administration of thiamine derivatives on hepatic coenzyme A concentration in young rat, *Vitamins,* 38, 158, 1968.
24. **Ando, H.,** Effect of excessive administration of thiamine derivatives on the coenzyme A metabolism in rat. II. Effect of excessive administration of thiamine derivatives on the hepatic coA, pantothenic acid and vitamin B_6 concentration in young rat, *Vitamins,* 38 163, 1968.
25. **Ando, H.,** Effect of excessive administration of thiamine derivatives on the coenzyme A metabolism in rat. III. Effect of thiamine on the phosphorylation of pantothenic acid and effect of excessive administration of thiamine derivatives on the pantothenate-4-phosphotransferase activity in rat, *Vitamins,* 38, 169, 1968.
26. **Anasbacher, S., Corwin, W. C., and Thomas, B. G. H.,** Toxicity of menadione, menadiol and esters, *J. Pharmacol. Exp. Ther.,* 75, 111, 1942.
27. **Anderson, D. C., Cooper, A. F., and Naylor, G. J.,** Vitamin D intoxication, with hypernatraemia, potassium and water depletion, and mental depression, *Br. Med. J.,* 4, 744, 1968.
28. **Anderson, R. J.,** The utilization of inosite in the dog, *J. Biol. Chem.,* 25, 391, 1916.
29. **Anderson, T. W.,** Vitamin E in angina pectoris, *Can. Med. Assoc. J.,* 110, 401, 1974.
30. **Anderson, T. W., Reid, D. B. W., and Beaton, G. H.,** Vitamin C and the common cold: a double-blind trial, *Can. Med. Assoc., J.,* 107, 503, 1972.
31. **Anderson, T. W., Surany, G., and Beaton, G. H.,** The effect on winter illness of large doses of vitamin C, *Can. Med. Assoc., J.,* 111, 31, 1974.
32. **Andrews, C. T. and Wilson, T. S.,** Vitamin C and thrombotic episodes, *Lancet,* 2, 39, 1973.
33. **Anon.,** Hypervitaminosis A, *Nutr. Rev.,* 9, 183, 1951.
34. **Anon.,** Hypervitaminosis A, *Med. J. Aust.,* 2, 505, 1953.
35. **Anon.,** Uses and hazards of vitamin K drugs, *Med. Lett.,* 5, 98, 1963.
36. **Anon.,** Today's drugs. Vitamin K, *Br. Med. J.,* 2, 40, 1969.
37. **Anon.,** New rules on vitamins, *JAMA,* 225, 1174, 1973.
38. **Anon.,** New developments in pharmacology of vitamin D, *Med. Lett., Drugs Therapeu.,* 16, 15, 1974.
39. **Anon.,** Vitamin E, *Med. Lett. Drugs Therapeu.,* 69, 1975.
40. **Anon.,** Hypervitaminosis E and coagulation, *Nutr. Rev.,* 33, 269, 1975.
41. **Anon.,** The transport of vitamin A in hypervitaminosis A, *Nutr. Rev.,* 34, 119, 1976.
42. **Anon.,** Scientific status summary by the Institute of Food Technologists expert panel on Food Safety and Nutrition and the Committee on Public Information. "Vitamin E", *Nutr. Rev.,* 35, 57, 1977.
43. **Antia, A. V., Wiltse, H. E., Rowe, R. D., Pitt, E. L., Levin, S., Ottesen, O. E., and Cooke, R. E.,** Pathogenesis of the supravalvular aortic stenosis syndrome in children, *J. Pediatr.,* 71, 431, 1967.
44. **Antopol, W. and Tarlov, I. M.,** Experimental study of the effects produced by large doses of vitamin B_6, *J. Neuropathol. Exp., Neurol.,* 1, 330, 1942.
45. **Arnett, R. W. and Cooper, J. R.,** The role of thiamine in nervous tissue: effect of antimetabolites of the vitamin on conduction in mammalian nonmyelinated nerve fibers, *J. Pharmacol. Exp. Ther.,* 148, 137, 1965.
46. **Arnrich, L.,** Toxic effects of megadoses of fat-soluble vitamins, in *Nutrition and Drug Interrelations,* Hathcock, J. N., and Coon, J., Eds., Academic Press, New York, 1978, 751.
47. **Assem, E. S. K.,** Anaphylactic Reaction to thiamine, *Practitioner,* 211, 565, 1973.
48. **Asteriadou-Samartzis, E. and Leikin, S.,** The relation of vitamin K to hyperbilirubinaemia, *Pediatrics,* 21, 397, 1958.
49. Australian Drug Evaluation Committee. Report of suspected adverse drug reactions. Aust. Govern. Pub. Serv. No. 4, 1964—1976.
50. **Avioli, L. V.,** The therapeutic approach to hypoparathyroidism, *Am. J. Med.,* 57, 34, 1974.
51. **Avioli, L. V. and Haddad, J. G.,** Vitamin D: current concepts, *Metab. Clin. Exp.,* 22, 507, 1973.
52. **Baer, S., Heine, W. I., and Gelfond, D. B.,** The use of vitamin E in heart disease, *Am. J. Med. Sci.,* 215, 542, 1948.
53. **Bagenstoss, A. H., Christensen, N. A., Berge, K. B., Balders, W. P., Spiekerman, R. E., and Ellefson, R. D.,** Fine structural changes in liver in hypercholesterolemic patients receiving long term nicotinic acid therapy, *Mayo Clin. Proc.,* 42, 385, 1967.
54. **Baker, J. R., Howell, J., and Thompson, J. N.,** Hypervitaminosis A in the chick, *Br. J. Exp. Pathol.,* 48, 507, 1967.

55. **Balsan, S., Garabedian, M., Sergniard, R., Holic, M. F., and DeLuca, H. F.**, 1,25-Dihydroxy vitamin D_3 and 1α-hydroxyvitamin D_3 in children: biologic and therapeutic effects in nutritional rickets and different types of vitamin D resistance, *Paediatr. Res.*, 9, 586, 1975.
56. **Barber, G. W. and Spaeth, G. I.**, The successful treatment of homocystinuria with pyridoxine, *J. Pediatr.*, 75, 463, 1969.
57. **Barash, P., Kitahata, L., and Mandel, S.**, Acute cardiovascular collapse after intravenous phytonadione, *Anesth. Analg.*, 55, 304, 1976.
58. **Barash, P.G.**, Nutrient toxicities of vitamin K, in *Handbook in Nutrition and Food,* Recheigl, M., Jr., Ed., CRC Press, Boca Raton, Fla., 1978, 97.
59. **Barker, N. W.**, The effect of niacin on the blood cholesterol, *Ill. Med. J.*, 116, 138, 1959.
60. **Barrucand, D., Mayer, G., and Mollet, E.**, Néphropathie hypercalcémique après traitement d'un hypoparathyroidisme primitif, *Ann. Méd. Nancy,* 7, 1311, 1968.
61. **Barnes, C. A.**, Erythrocytic inclusions, their diagnostic significance and pathology, *JAMA*, 198, 171, 1966.
62. **Bartolozzi, G. G., Bernini, G., Marianelli, L., and Corvaglia, E.**, Chronic vitamin A excess in infants and children. Description of two cases and a critical review of the literature, *Rev. Clin. Pediatr.*, 80, 231, 1967.
63. **Bass, M. H.**, The relation of vitamin A intake to cerebrospinal fluid pressure: a review, *J. Mt. Sinai Hosp. N.Y.*, 24, 713, 1957.
64. **Basu, T. K. and Dickerson, J. W. T.**, Interrelations of nutrition and the metabolism of drugs, *Chem. Biol. Interact,* 8, 193, 1974.
65. **Bauernfeind, J. C., Newmark, H., and Brin, M.**, Vitamins A and E nutrition via intramuscular or oral route, *Am. J. Clin. Nutr.*, 27, 234, 1974.
66. **Bazex, A., Dupré, A., Christol, B., and Serres, D.**, Réactions sclérodermiformes lombo-fessierès après injection de vitamine K_1: presentation de 2 observations. Constations histologiques, *Bull. Soc. Franc. Derm. Syph.*, 79, 578, 1972.
67. **Bazin, S. and Delavnay, A.**, Influence of vitamin A excess on the constitution of the connective tissue and on the collagenolytic activity of normal or inflamed tissues, *Ann. Inst. Pasteur,* 110, 487, 1966.
68. **Beamish, R. E. and Storrie, V. M.**, Severe haemolytic reaction following the intravenous administration of emulsified vitamin K (mephyton), *Can. Med. Assoc., J.*, 74, 149, 1956.
69. **Bedford, P. D.**, Side effects of a preparation of vitamin B_{12}, *Br. Med. J.*, 1, 690, 1952.
70. **Bell, N. H. and Stern, P. H.**, Hypercalcemia and increases in serum hormone value during prolonged administration of 1α, 25-dihydroxyvitamin D, *N. Engl. J. Med.*, 298, 1241, 1978.
71. **Bélanger, L. F. and Clark, I.**, Radiographic and histological observations on the skeletal effects of hypervitaminoses A and D, *Anat. Rec.*, 158, 443, 1967.
72. **Benschoter, C. A. and Paniagua, R. G.**, Reproduction and longevity of Mexican fruit flies, *Anastrepha ludens* (Diptera Tephritidae) fed biotin in the diet, *Ann. Entomol. Soc. Am.*, 59, 289, 1966.
73. **Belle, M. and Halpern, M. M.**, Oral nicotinic acid for hyperlipemia, *Am. J. Cardiol.*, 2, 449, 1958.
74. **Berge, K. G., Achor, R. W.P., Christensen, N. A., Mason, H. L., and Barker, N. W.**, Hypercholesterolemia and nicotinic acid: a long-term study, *Am. J. Med.*, 31, 24, 1961.
75. **Berge, K. G.**, Side effects of nicotinic acid in the treatment of hypercholesterolemia, *Geriatrics,* 16, 416, 1961.
76. **Berge, K. G., Achor, R. W. P., Barker, N. W., and Paver, M. H.**, Comparison of the treatment of hypercholesterolemia with nicotinic acid, sitosterol and safflower oil, *Am. Heart J.*, 58, 849, 1959.
77. **Berge, K. G., Achor, R. W. P., Christensen, N. A., Power, M. H., and Barker, N. W.**, Hypercholesterolemia and nicotinic acid: a long-term study, *Circulation,* 20, 671, 1959.
78. **Berge, K. C. and Molnar, G. D.**, Effects of nicotinic acid on clinical aspects of carbohydrate metabolism, in *Niacin in Vascular Disorders and Hyperlipemia,* Charles C Thomas, Springfield, Ill., 136.
79. **Bergen, S. S. and Roels, O.**, Hypervitaminosis A: a report of a case, *Am. J. Clin. Nutr.*, 16, 265, 1965.
80. **Bernhardt, I. B. and Dorsey, P. J.**, Hypervitaminosis A and congenital renal anomalies in a human infant, *Obstet. Gynecol.*, 43, 750, 1974.
81. **Berio, A. and Moscatelli, P.**, Hypervitaminosis D, *Minerva Pediatr.*, 19, 972, 1967.
82. **Best, C. H., Lucas, C. C., Patterson, J. M., and Ridout, J. H.**, The rate of lipotropic action of choline and inositol under special dietary conditions, *Biochem. J.*, 48, 452, 1951.
83. **Best, C. H., Lucas, C. C., and Ridout, J. H.**, Vitamins and the protection of the liver, *Br. Med. Bull.*, 12, 9, 1956.
84. **Beuren, A. J., Apitz, J., Stoermer, J., Schlange, H., Kaiser, B., Von Berg, W., and Jörgensen, G.**, Vitamin D-hypercalcamische Herz- und Gefässerkrankung, *Dtsch. Med. Wochenschr.*, 91, 881, 1966.
85. **Beuren, A. J., Apitz, J., and Harmjanz, D.**, Supravalvular aortic stenosis in association with mental retardation and certain facial appearance, *Circulation,* 26, 1235, 1962.

86. **Beuren, A. J., Apitz, J., Stoermer, J., Kaiser, B., Schlange, H., von Berg, W., and Jorgensen, G.,** Vitamin D-hypercalcamische Herz-und Gefäszerkrankung, *Monatsschr. kinderheilkd.*, 114, 457, 1966.
87. **Beuren, A. J., Schulze, C., Eberle, P., Harmjanz, D., and Apitz, J.,** Syndrome of supravalvular aortic stenosis, peripheral pulmonary stenosis, mental retardation and similar facial appearance, *Am. J. Cardiol.*, 13, 471, 1964.
88. **Bieri, J. G.,** Biological activity and metabolism of N-substituted tocopheramines: implications on vitamin E function, in *The Fat-Soluble Vitamins,* DeLuca, H. F. and Suttie, J. W., Eds., University of Wisconsin Press, Madison, 1969, 307.
89. **Bieri, J. G.,** Effect of excessive vitamins C and E on vitamin A status, *Am. J. Clin. Nutr.*, 26, 382, 1973.
90. **Bieri, J. G.,** Vitamin E, *Nutr. Rev.*, 33, 161, 1975.
91. **Bifulco, E.,** Vitamin A intoxication: report of a case in an adult, *N. Engl. J. Med.*, 248, 690, 1953.
92. **Black, J. A. and Bonham-Carter, R. E.,** Association between aortic stenosis and facies of severe infantile hypercalcaemia, *Lancet,* 2, 745, 1963.
93. **Black, J. A., Butler, N. R., and Schlesinger, B. E.,** Aortic stenosis and hypercalcemia, *Lancet,* 2, 546, 1965.
94. **Blair, J. A.,** Toxicity of folic acid, *Lancet,* 1, 360, 1970.
95. **Block, H. S.,** Chronic hypervitaminosis A: report of a probable case in man, *Minn. Med.*, 38, 627, 1955.
96. **Blumberg, R., Forbes, G., Fraser, D., Hansen, A., May, C., Smith, C., Smith, N., Sweeney, M., and Fomon, S.,** Vitamin K compounds and the water soluble analogues, *Pediatrics,* 28, 501, 1961.
97. **Bly, C. G., Heggeness, E. W., and Nasset, E. S.,** The effects of pantothenic acid and inositol added to whole wheat bread on evacuation time, digestion, and absorption in the upper gastrointestinal tract of dogs, *J. Nutr.*, 26, 161, 1943.
98. **Boberg, J., Carlson, L. A., Forberg, S., Olsson, A., Oro, L., and Rossner, S.,** Effects of chronic treatment with nicotinic acid on intravenous fat tolerance and post-heparin lipoprotein lipase activity in man, in *Metabolic Effects of Nicotinic Acid and its Derivatives,* Gey, K. F. and Carlson, L. A., Eds., Hans Huber, Bern, Switzerland, 1971, 465.
99. **Boissier, J. R., Tillement, J. P., Viars, P., Merlin, L., and Simon, P.,** Quelques aspects de la toxicite et du metabolism de la thiamine adminstreé à fortes doses, *Anesth. Analg. Reanim.*, 24, 515, 1967.
100. **Bongiovanni, A. M., Eberlein, W. R., and Jones, I. T.,** Idiopathic hypercalcemia of infancy with failure to thrive; report of three cases with a consideration of the possible etiology, *N. Engl. J. Med.*, 257, 951, 1957.
101. **Bonnet, H.,** L'atteinte tubulaire au cours des maladies métaboliques et des intoxications, *Méd. Infant.*, 76, 337, 1969.
102. **Bouillon, R., Van Baelen, H., and De Moore, P.,** The measurement of the vitamin D-binding protein in human serum, *J. Clin. Endocrinol. Metab.*, 45, 225, 1977.
103. **Bound, J. P. and Tefler, T. P.,** Effect of vitamin K dosage on plasma bilirubin levels in premature infants, *Lancet,* 1, 720, 1956.
104. **Brade, W., Herken, H., and Merker, H. J.,** Disturbance of function and stimulation of growth of the kidneys by large doses of folic acid, *Klin. Wochenschr.*, 46, 1232, 1968.
105. **Brazda, F. G. and Coulson, R. A.,** Toxicity of nicotinic acid, *Proc. Soc. Exp. Biol. Med.*, 62, 19, 1946.
106. **Brewer, L. E. and Edwards, H. M.,** Jr., Studies on the biotin requirement of broiler breeders, *Poult. Sci.*, 51, 619, 1972.
107. **Brickman, A. S., Coburn, J. W., and Norman, A. W.,** Action of 1,25-dihydroxycholecalciferol: a potent, kidney-produced metabolite of vitamin D_3 in uremic man, *N. Engl. J. Med.*, 287, 891, 1972.
108. **Brickman, A. S. and Norman, A. W.,** Treatment of renal osteodystrophy with calciferol (vitamin D) and related steroids, *Kidney Int.*, 4, 161, 1973.
109. **Briggs, M. H.,** Fertility and high-dose vitamin C, *Lancet,* 2, 1083, 1973.
110. **Briggs, M. H.,** Side effects of vitamin C, *Lancet,* 2, 1439, 1973.
111. **Briggs, M. H.,** Vitamin A and the teratogenic risks of oral contraceptives, *Br. Med. J.*, 2, 170, 1974.
112. **Briggs, M. H.,** Vitamin E supplements and fatigue, *N. Engl. J. Med.*, 290, 579, 1974.
113. **Briggs, M. H.,** Vitamin C induced hyperoxaluria, *Lancet,* 1, 154, 1976.
114. **Briggs, M. H.,** Effect of specific nutrient toxicities in animals and man: vitamin C, in *Handbook of Nutrition and Food,* Rechcigl, M., Ed., CRC Press, Boca Raton, Fla., 1978, 16.
115. **Briggs, M. H.,** Effect of specific nutrient toxicities in animals and man: tocopherols, in *Handbook of Nutrition and Food,* Rechcigl, M., Ed., CRC Press, Boca Raton, Fla., 1978, 91.
116. **Briggs, M. H. and Briggs, M.,** Are vitamin E supplements beneficial?, *Med. J. Aust.*, 1, 434, 1974.
117. **Briggs, M. H., Garcia-Webb, P., and Johnson, J.,** Dangers of excess vitamin C, *Med. J. Aust.*, 2, 48, 1973.

118. **Brodkin, R. H. and Bleiberg, J.**, Sensitivity to topically applied vitamin E, *Arch. Dermatol.*, 92, 76, 1965.
119. **Brown, R. G.**, Possible problems of large intakes of ascorbic acid, *JAMA*, 224, 1529, 1973.
120. **Brusa, A. and Testa, F.**, Lesions of the central nervous system of guinea pigs with hypervitaminosis A, *Int. Z. Vitaminforsch.*, 25, 55, 1953.
121. **Burch, H. B., Lowry, O. H., Padilla, A. M., and Combs, A. M.**, Effects of riboflavin deficiency and realimentation on flavin enzymes of tissues, *J. Biol. Chem.*, 223, 29, 1956.
122. **Butcher, R. W.**, Effects of nicotinic acid on cyclic AMP Levels in rat adipose tissue, in *Metabolic Effects of Nicotinic Acid and Its Derivatives*, Gey, K. F. and Carlson, L. A., Eds., Hans Huber, Bern, Switzerland, 1971, 347.
123. **Butterworth, C. E., Scott, C. W., Magnus, E., Santini, R., and Dempsey, H.**, Metabolic changes associated with recovery from vitamin B_{12} deficiency, *Med. Clin. North Am.*, 50, 1627, 1966.
124. **Byer, J. and Harpuder, K.**, The sensitizing effect of thiamine for acetylcholine, *J. Pharmacol. Exp. Ther.*, 70, 328, 1940.
125. **Byington, M.**, Effect of nutrient toxicities in animals and man: choline, in *Handbook in Nutrition and Food*, Rechcigl, M., Jr., Ed., CRC Press, Boca Raton, Fla., 1978, 59.
126. **Caffey, J.**, Chronic poisoning due to excess of vitamin A, *Am. J. Roentgenol. Radium Ther. Nucl. Med.*, 65, 12, 1951.
127. **Calandi, C., Calzolari, C., DiMaria, M., and Pierro, U.**, Vitamin D poisoning in infancy (12 clinical cases), *Riv. Clin. Pediatr.*, 77, 3, 1966.
128. **Campbell, G. D., Jr., Steinberg, M. H., and Bower, J. D.**, Ascorbic acid-induced hemolysis in G6PD deficiency, *Ann. Intern. Med.*, 82, 810, 1975.
129. **Campbell, T. C. and Hayes, J. R.**, Role of nutrition in the drug-metabolizing enzyme system, *Pharmacol. Rev.*, 26, 171, 1974.
130. Canadian Paediatric Society, Report of the Nutrition Committee, The use and abuse of vitamin A, *Can. Med. Assoc. J.*, 104, 521, 1971.
131. **Canales, E. S., Soria, J., Zarate, A., Mason, M., and Molina, M.**, The influenue of pyridoxine on prolactin secretion and milk production in women, *Br. J. Obstet. Gynaecol.*, 83, 387, 1976.
132. **Carlson, L. A.**, Nicotinic acid: its metabolism and its effects on plasma free fatty acids, in *Metabolic Effects of Nicotinic Acid and Its Derivatives*, Gey, K. F. and Carlson, L. A., Eds., Hans Huber, Bern, Switzerland, 1971, 157.
133. **Carlson, L. A., Freyschuss, U., Kjellberg, J., and Ostman, J.**, Suppression of splanchinic ketone body production in man by nicotinic acid, *Diabetologia*, 3, 494, 1967.
134. **Carlson, L. A. and Micheli, H.** Stimulatory effect of prostaglandin E_1 on fat mobilizing lipolysis in adipose tissue of rats treated with nicotinic acid, in *Metabolic Effects of Nicotinic Acid and Its Derivatives*, Gey, K. F. and Carlson, L. A., Eds., Hans Huber, Bern, Switzerland, 1971, 995.
135. **Carlson, L. A., and Nye, E. R.**, Acute effects of nicotinic acid in the rat. I. Plasma and liver lipids and blood glucose, *Acta Med. Scand.*, 179, 453, 1966.
136. **Carlson, L. A. and Oro, L.**, The effect of nicotinic acid on the plasma free fatty acids. Demonstration of a metabolic type of sympathicolysis, *Acta Med. Scand.*, 172, 641, 1962.
137. **Carlson, L. A. and Walldius, G.**, Serum and tissue lipid metabolism and effect of nicotinic acid in different types of hyperlipidemia, *Adv. Exp. Med. Biol.*, 26, 165, 1972.
138. **Carroll, J.**, A simplified alkaline phosphotungstate assay for uric acid in serum, *Clin. Chem.*, 17, 158, 1971.
139. **Carton, M. F. X.**, Reaction allergique au cours d'un traitement: Vitamin K_1 + extrait de foie, *Bull.Soc. Franc. Derm. Syph.*, 72, 228, 1965.
140. **Catto, G. R. D., MacLeod, M., Pelc, B., and Kodicek, E.**, 1α-Hydroxycholecalciferol: a treatment for renal bone disease, *Br. Med. J.*, 1, 12, 1975.
141. **Chanarin, I., Fenton, J. C. B., and Mollin, D. L.**, Sensitivity to folic acid, *Br. Med. J.*, 1, 1162, 1957.
142. **Chang, W. M. L., and Johnson, B. C.**, Metabolism of C^{14}-nicotinic acid in pigs and sheep, *J. Nutr.*, 76, 512, 1962.
143. **Chen, K. K., Rose, C. L., and Robbins, F. B.**, Toxicity of nicotinic acid, *Proc. Soc. Exp. Biol. Med.*, 38, 241, 1938.
144. **Chmelar, M. and Chmelarova, M.**, Interactions of nicotinic acid with hormone-sensitive lipases of different mammalian tissue, in *Metabolic Effects of Nicotinic Acid on Its Derivatives*, Gey, K. F. and Carlson, L. A., Eds., Hans Huber, Bern, Switzerland, 1971, 236.
145. **Cho, D. Y., Frey, R. A., Guffy, M. M., and Leipold, H. W.**, Hypervitaminosis A in the dog, *Am. J. Vet. Res.*, 36, 1597, 1975.
146. **Christensen, S.**, The biological fate of riboflavin in mammals, *Acta Pharmacol. Toxicol.*, 32 (Suppl. 11) 1, 1973.

147. **Christensen, N. A., Achor, R. W. P., Berge, K. G., and Mason, H. L.,** Nicotinic acid treatment of hypercholesterolemia. Comparison of plain and sustained-action: preparations and report of two cases of jaundice, *JAMA*, 177, 546. 1961.
148. **Chytil, F. and Ong, D. E.,** Mediation of retinoic acid-induced growth and antitumour activity, *Nature (London)*, 260, 49, 1965.
149. **Clark, I. and Bassett, C. A. L.,** Amelioration of hypervitaminosis D in rats with vitamin A, *J. Exp. Med.*, 115, 147, 1962.
150. **Clark, I. and Smith, M. R.,** Effects of hypervitaminosis A and D on skeletal metabolism, *J. Biol. Chem.*, 239, 1266, 1964.
151. **Clark, L.,** Hypervitaminosis A: a review, *Aust. Vet. J.*, 47, 568, 1971.
152. **Cleland, J. B. and Southcott, R. V.,** Illnesses following the eating of seal liver in Australian waters, *Med. J. Aust.*, 1, 760, 1969.
153. **Cleland, J. B. and Southcott, R. V.,** Hypervitaminosis A in the Australian Antartic expedition of 1911—1914: a possible explanation of the illness of Mertz and Mawson, *Med. J. Aust.*, 1, 1337, 1969.
154. **Coburn, J. W., Hartenbower, D. I. and Norman, A. W.,** Metabolism and action of the hormone vitamin D., *West. J. Med.*, 121, 22, 1974.
155. **Cochrane, W. A.,** Over-nutrition in prenatal and neonatal life, *Can. Med. Assoc., J.*, 93, 893, 1965.
156. **Cohen, B. M.,** Fatigue caused by vitamin E?, *Calif. Med.*, 199, 72, 1973.
157. **Cohen, H. M.,** Effects of vitamin E: good and bad, *N. Engl. J. Med.*, 289, 980, 1973.
158. **Cohen, E. and Levinson, H. Z.,** Disrupted fertility of the hide beetle, *Dermester maculatus* (Deg) due to dietary overdose of biotin, *Experientia*, 24, 367, 1968.
159. **Cohen, E. and Levinson, H. Z.,** Studies on the chemosterilizing effects of biotin on the hide beetle, *Dermestes maculatus* (Dermestidae; Coleoptera), *Comp. Biochim. Physiol. B*, 43, 143, 1972.
160. **Cohlan, S. Q.,** Excessive intake of vitamin A as a cause of congenital anomalies in the rat, *Science*, 117, 535, 1953.
161. **Cohn, V. A.** Fats soluble vitamins III, vitamin K and vitamin E, in *The Pharmacologic Basis of Therapeutics*, Goodman, L. S. and Gilman, A., Eds., Macmillan, New York, 1970, 1690.
162. Committee on Nutrition, Vitamin K compounds and the water soluble analogues, *Paediatrics*, 28, 501, 1961.
163. Committee on Nutrition, Vitamin B_6 requirements in man, *Pediatrics*, 38, 1068, 1966.
164. **Conley, C. L. and Krevans, J. R.,** Development of neurologic manifestations of pernicious anaemia during multivitamin therapy, *N. Engl. J. Med.*, 245, 529, 1951.
165. **Conley, C. L. and Krevans, J. R.,** New developments in the diagnosis and treatment of pernicious anaemia, *Ann. Int. Med.*, 43, 758, 1955.
166. **Cook, J. D. and Monsen, E. R.,** Vitamin C, the common cold, and iron absorption, *Am. J. Clin. Nutr.*, 30, 235, 1977.
167. **Cooper, J. R., Itokawa, Y., and Pincus, J. H.,** Thiamine triphosphate deficiency in subacute necrotizing encephalomyelopathy, *Science*, 164, 368, 1969.
168. **Cooper, J. R. and Pincus, J. H.,** Treatment of Leigh's disease (subacute necrotizing encephalomyelopathy) with thiamine derivatives, in *Abstr. 4th Int. Meet. Int. Soc. Neurochemistry*, The Local Organizing Committee of I.S.N., Tokyo, 1973, 382.
169. **Corrigan, J. J., and Marcus, F. I.,** Coagulopathy associated with vitamin E ingestion, *JAMA*, 230, 1300, 1974.
170. Council on Drugs, New and unofficial drugs. Use of nicotinic acid in hypercholesterolaemia, *JAMA*, 168, 1773, 1958.
171. **Counts, S. J., Baglink, D. J., Shen, F., Sherrard, D. J., and Hickman, R. O.,** Vitamin D intoxication in an anephric child, *Ann. Int. Med.*, 82, 196, 1975.
172. **Cowan, D. H., Graham, R. C., Shook, P., and Griffiths, R.,** Influence of ascorbic acid on platelet structure and function, *Thromb. Diath.Haemorrh.*, 34, 50, 1975.
173. **Cox, R. P., Deuel, H. J., Jr., and Ershoff, B. H.,** Potentiating effects of DPPD, bile salts and sulfasuxidine on hypervitaminosis A in the rat, *Exp. Med. Surg.*, 15, 328, 1957.
174. **Crosby, W. H.,** The daily dose of folic acid, *J. Chron. Dis.*, 12, 583, 1960.
175. **Crosse, V. M., Meyer, T. C. and Gerrard, J. W.,** Icernicterus and prematurity, *Arch. Dis. Child.*, 30, 501, 1955.
176. **Cuthbertson, W. F. J.,** The vitamin D activity of plasma of children with idiopathic hypercalcaemia, *Br. J. Nutr.*, 17, 627, 1963.
177. **Dahl, S.,** Vitamin E in clinical medicine, *Lancet*, 1, 465, 1974.
178. **Dalton, C.,** Antilipaemic effect of nicothiamide, *Nature (London)*, 216, 825, 1967.
179. **Dalderup, L. M.,** Ischaemic heart disease and vitamin D, *Lancet*, 2, 92, 1973.
180. **Danes, B. S. and Bearn, A. G.,** The effect of retinol (vitamin A alcohol) on urinary excretion of mucopolysaccharides in the Hurler Syndrome, *Lancet*, 1, 1029, 1967.

181. **Danilov, L. N.**, Anaphylactic shock during vitamin B_6 treatment, *Klin. Med.*, 51, 139, 1973.
182. **Danilovic, V. and Ljaljevic, M.**, Unsere beobachtungen uber die arzneimittelallergien mit besonderer Berucksichtigung der penicillin und vitamin B_1 allergien, *Allergie Asthma*, 11, 185, 1965.
183. **Dashman, T., Horst, D., Bautz, G., and Kamm, J. J.**, Ascorbic acid: effect of high doses on brain and heart catecholamine levels, *Experientia*, 29, 832, 1973.
184. **Davies, P.**, Vitamin D poisoning: a report of two cases, *Ann. Intern. Med.*, 53, 1250, 1960.
185. **Davis, K. L., Berger, P. A., and Hollister, L. E.**, Choline for tardive dyskinesia, *N. Engl. J. Med.*, 293, 152, 1975.
186. **Deb, C. and Mallick, N.**, Effect of chronic graded doses of ascorbic acid on thyroid activity, protein bound iodine of blood and deiodinase enzyme of peripheral tissues of guinea pigs, *Endocrinologie*, 63, 231, 1974.
187. **Degkwitz, R., Frowein, R., Kulenkampff, C., and Mohs, U.**, Uber die wirtungen des L-dopa beim menschen und deren beeinflussung durch reserpin, chlorpromazin, iproniazid und vitamin B_6, *Klin. Wochenschr.*, 38, 120, 1960.
188. **DeLangen, C. D. and Donath, W. F.**, Vitamin D sclerosis of the arteries and the danger of feeding extra vitamin D to older people, with a view on the development of different forms of arterosclerosis, *Acta Med. Scand.*, 156, 317, 1956.
189. **DeLuca, G. and Cozzi, M.**, Syndrome of vitamin D excess, *Minerva Pediatr.*, 16, 210, 1964.
190. **De Luca, H. G.**, The kidney as a endocrine organ for the production of 1,25-dihydroxyvitamin D_3, a calcium mobilising hormone, *N. Engl. J. Med.*, 289, 359, 1973.
191. **De Luca, H. G.**, Recent advances in the metabolism and function of vitamin D, *Fed. Proc. Fed. Am. Soc. Exp. Biol.*, 28, 1678, 1969.
192. **Demole, V. V.**, Vertagltchkeit des Lactoflavins, *Z. Vitam. Horm. Fermentforsch.*, 7, 138, 1938.
193. **Demole, V. V.**, Pharmacology of vitamin E, *Int. Z. Vitaminforsh.*, 8, 338, 1939.
194. **Denie, J. J. and Verhaught, A. P.**, Supravalvular Aortic Stenosis, *Circulation*, 18, 902, 1958.
195. **Dent, C. E.**, Dangers of vitamin D intoxication, *Br. Med. J.*, 1, 834, 1964.
196. **Dent, C. E. and Friedman, M.**, Hypophosphateric osteomalacia with complete recovery, *Br. Med. J.*, 1, 1676, 1964.
197. **Denton, R. L.**, Vitamin K for the newborn?, *Ped. Clin. North Am.*, 8, 455, 1961.
198. **Deutsch, E.**, Vitamin K in medical practice: adult, *Vitam. Horm. N.Y.*, 24, 665, 1966.
199. **Dias, M. V.**, Action of thaimine applied directly to the cerebral cortex, *Science*, 105, 211, 1947.
200. **Di Benedetto, R. J.**, Chronic hypervitaminosis A in an adult, *JAMA*, 201, 700, 1967.
201. **Dingemanse, E. and Van Eck, W. S.**, Wheat-germ oil and tumor formation, *Proc. Soc. Exp. Biol. Med.*, 41, 622, 1939.
202. **Di Palma, J. R. and Hitchcook, P.** Neuromuscular and ganglionic blocking action of thiamine and its derivatives, *Anesthesiology*, 19, 762, 1958.
203. **Saville, D. G., Solvyns, A., and Humphries, C.**, Choline induced pyridoxine deficiency in broiler chickens, *Aust. Vet. J.*, 43, 346, 1967.
204. **Dolmierski, R.**, Zaburzenia psychiezne wywolane hiperwitaminoza D, *Neurol. Neurochir. Psychiatr. Pol.*, 15, 859, 1965.
205. **Douglas, A. S.**, *Anticoagulant Therapy*, Blackwell, Oxford, 1962, 337.
206. **Douglas, A. S. and Brown, A.**, Effect of vitamin K preparations on hypoprothrombinaemia induced by Dicoumarol and Tromexan, *Br. Med. J.*, 1, 412, 1952.
207. **Douglas, A. S. and Davies, P.**, Hypothrombinaemia in the newborn, *Arch. Dis. Child.*, 30, 509, 1955.
208. **Dreyfus, P. M.**, Thiamine deficiency encephalopathy: thoughts on its pathogenesis, in *Thiamine*, Gubler, C. J., Fujiwara, M., and Drefus, P. M., Eds., John Wiley & Sons, New York, 1976, 229.
209. **Dugois, P., Amblard, P., Imbert, A. R., and De Bignicourt, B.**, Acnés à la vitamine B_{12} *Bull. Soc. Franc. Derm. Syph.*, 76, 382, 1969.
210. **Duke, P. S.**, Ocular side effects of drug therapy, *Med. J. Aust.*, 1, 927, 1967.
211. **Duncan, C. H. and Best, M. M.**, Lack of nicotinic acid on pool size and turnover of taurocholic acid in normal and hypothyroid dogs, *J. Lipid Res.*, 1, 159, 1960.
212. **Duperrat, B. and Noury, J. Y.**, Granulome annulaire au decours de la vitaminotherapie, *Bull. Soc. Franc. Derm. Syph.*, 79, 672, 1972.
213. **Dyggvr, H.**, Bilirubin studies in premature infants who received menadione derivatives or vitamin K_1 at birth, *Acta Paediatr.*, 49, 230, 1960.
214. **Dysmsze, H. A. and Park, J.**, Excess dietary vitamin E in rats, *Fed. Proc. Fed. Am. Soc. Exp. Biol.*, 34, 912, 1975.
215. Editorial, Vitamin D as a public health problem, *Br. Med. J.*, 1, 1654, 1964.
216. Editorial, Hypercalcaemia and supravalvar aortic stenosis, *Lancet*, 1, 606, 1967.
217. Editorial, Hypervitaminosis E and coagulation, *Nutr. Rev.*, 33, 269, 1975.

218. Editorial, Vitamin C toxicity, *Nutr. Rev.*, 34, 236, 1976.

219. Editorial, Clinical use of 1α-hydroxyvitamin D_3, *Br. Med. J.*, 1571, 1978.

220. **Edmunds, C., Behrens, M., Lewis, L., and Lennon, R.**, Pseudotumor cerebri and low vitamin A intake, *JAMA*, 226, 674, 1973.

221. **Ehrhart, A. A. and Money, J.**, Hypercalcemia. A family study of psychologic functioning, *Johns Hopkins Med. J.*, 121, 14, 1967.

222. **Eichenbaum, J. W. and Cooper, J. R.**, Restoration by thiamine of the action potential in ultraviolet irradiated nerves, *Brain Res.*, 32, 258, 1971.

223. **Eisaman J. L.**, Roniacol timespan in the treatment of peripheral vascular and allied disorders, *Curr. Ther. Res.*, 2, 527, 1960.

224. **Eisenstein, R. and Groff, W. A.**, Experimental hypervitaminosis D: hypercalcemia hypermucoproteinemia and metastatic calcification, *Proc. Soc. Exp. Biol. Med.*, 94, 441, 1957.

225. **Elliot, R. A. and Dryer, R. L.**, Hypervitaminosis A: a report of a case in an adult, *JAMA*, 161, 1157, 1956.

226. **Ellis, L. N., Zmachinsky, A., and Sherman, H. C.**, Experiments upon the significance of liberal levels of intake of riboflavin, *J. Nutr.*, 25, 153, 1943.

227. **Engelhardt, H. T. and Baird, V. C.**, Sensitivity to thiamine hydrochloride. A potential hazard in a common office procedure, *Ann. Allergy*, 4, 291, 1946.

228. **Erlandson, M. E. and Hilgartner, M.**, Hemolytic disease in the neonatal period and early infancy, *J. Pediatr.*, 54, 566, 1959.

229. **Evans, H. M. and Emerson, G. A.**, Failure to produce abdominal neoplasms in rats receiving wheat-germ oil extracted in various ways, *Proc. Soc. Exp. Biol. Med.*, 41, 318, 1939.

230. **Farhangi, M. and Osserman, E. F.**, Myeloma with xanthoderma due to an IgG monoclonal antiflavin antibody, *N. Engl. J. Med.*, 294, 177, 1976.

231. **Farrell, P. M. and Bieri, J. G.**, Megavitamin E supplementation in man, *Am. J. Clin. Nutr.*, 28, 1381, 1975.

232. **Fedotin, M. S.**, Hypervitaminosis causing pseudotumour cerebri, *JAMA*, 212, 628, 1970.

233. **Feest, T. G., Ward, M. K., Kerr, D. N. S., Postlethwaite, R. J., and Houston, I. B.**, Impairment of renal function in patients on 1α-hydroxycholecalciferol, *Lancet*, 2, 427, 1978.

234. **Feldman, E. B.**, Nicotinic acid in the management of frank disorders of lipid metabolism, in *Niacin in Vascular Disorders and Hyperlipemia*, Altschul, R., Ed., Charles C Thomas, Springfield, Ill., 1964, 208.

235. **Feldman M. H. and Schlezinger, N. S.**, Benign intracranial hypertension associated with hypervitaminosis A, *Arch. Neurol.*, 22, 1, 1970.

236. **Fellers, F. X. and Schwartz, R.**, Etiology of the severe form of idiopathic hypercalcemia of infancy: a defect in vitamin D metabolism, *N. Engl. J. Med.*, 259, 1050, 1958.

237. **Finkel, M. J.**, Vitamin K, and the Vitamin K analogues, *Clin. Pharmacol. Ther.*, 2, 794, 1961.

238. **Fisher, A. A.**, B_{12} shots — still another side of the coin, *JAMA*, 233, 21, 1975.

239. **Fisher, G. and Skillern, P. G.**, Hypercalcemia due to hypervitaminosis A, *JAMA*, 227, 1413, 1974.

240. **Fitzgerald, O., Heffernan, A., Brennan, P., Mulcahy, R., Fennelly, J. J., and McFarlane, R.**, Effect of nicotinic acid on abnormal serum lipids, *Br. Med. J.*, 1, 157, 1964.

241. **Foldi, E.**, Renal tubular failure after treatment with high doses of vitamin A, *JAMA*, 235, 1631, 1976.

242. **Foldi, E., Ehlers, B., and Moeller, J.**, Tubulare insuffizienz nach hochdosierter vitamin-A behandlung, *Dtsch. Med. Wochenschr.*, 101, 205, 1976.

243. **Foman S. J., Younoszai, M. K., and Thomas, L. N.**, Influence of vitamin D on linear growth of normal fullterm infants, *J. Nutr.*, 88, 345, 1966.

244. **Fontenot, R., Redetski, H., and Deupree, R.**, Effects of nicotinic acid and nicotinamide on serum cholesterol and erythrocyte nicotinamide adenine dinucleotide levels of rabbits, *Proc. Soc. Exp. Biol. Med.*, 119, 1053, 1965.

245. **Forfar, J. O., Balf, C. L., Maxwell, G. M., and Tompsett, S. L.**, Idiopathic hypercalcemia of infancy: clinical and metabolic studies with special reference to aetiological role of vitamin D, *Lancet*, 1, 981, 1956.

246. **Forfar, J. O., Tompsett, S. L., and Forshall, W.** Biochemical studies in idiopathic hypercalaemia of infancy, *Arch. Dis. Child.*, 34, 525, 1959.

247. **Fournier, A., Pauli, A., Cousin, J., Barry, M., Paoli, M., and Lefebvre, F.**, Hypercalcemie iatrogene, *J. Sci. Med. Lille*, 85, 319, 1967.

248. **Fox, J. M. and Duppel, W.**, The action of thiamine and its di- and triphosphates on the slow exponential decline of the ionic currents in the node of ranvier, *Brain Res.*, 89, 287, 1975.

249. **Frame, B., Jackson, C. E., Reynolds, W. A., and Umphrey, J. E.**, Hypercalcemia and skeletal effects in chronic hypervitaminosis A, *Ann. Int. Med.*, 80, 44, 1974.

250. **Fraser, D., Langford Kidd, B. S., Sang Whay Kooh, and Paunier, L.,** A new look at infantile hypercalcemia, *Pediatr. Clin. North Am.*, 503, 1966.
251. **Fresh, J. W., Adams, H., and Morgan, F. M.,** Vitamin K blood clotting studies during pregnancy and prothrombin and proconvertin levels in the newborn, *Obstet. Gynecol.*, 13, 37, 1959.
252. **Friedman, M. and Byers, S. O.,** Evaluation of nicotinic acid as an hypocholesteramic and antiatherogenic substance, *J. Clin. Invest.*, 38, 1328, 1959.
253. **Friedman, W. F.,** Vitamin D as a cause of the supravalvular aortic stenosis syndrome, *Am. Heart J.*, 73, 718, 1967.
254. **Friedman, W. F. and Roberts, W. C.,** Vitamin D and the supravalvular aortic stenosis syndrome: the transplacental effects of vitamin D on the aorta of the rabbit, *Circulation*, 34, 77. 1966.
255. **Friedman, W. F. and Mills, L. S.,** The production of "elfin facies" and abnormal dentition by vitamin D during pregnancy. Relationship to the supravalvular aortic stenosis syndrome, *Proc. Soc. Pediatr. Res.*, 37, 80, 1967.
256. **Friedman, W. F. and Mills, L. S.,** The relationship between vitamin D and the craniofacial and dental anomalies of the supravalvular aortic stenosis syndrome, *Paediatr.*, 43, 12, 1969.
257. **Frier, H. I., Gorgacz, E. J., Hall, R. C., Jr., Gallina, A. M., Rousseau, J. E., Eaton, H. D., and Nielsen, S. W.,** Formation and absorption of cerebrospinal fluid in adult goats with hypo- and hypervitaminosis A, *Am. J. Vet. Res.*, 35, 45, 1974.
258. **Frimpter, G. W., Andelman, R. J., and George, W. F.,** Vitamin B_6 dependency syndrome, *Am. J. Clin. Nutr.*, 22, 794, 1969.
259. **Froberg, S. O., Boberg, J., Carlson, L. A., and Eriksson, M.,** Effect of nicotinic acid on the diurnal variation of plasma levels of glucose, free fatty acids, triglycerides, and cholesterol and of urinary excretion of catecholamines, in *Metabolic Effects of Nicotinic Acid and its Derivatives*, Gey, K. I., and Carlson, L. A., Eds., Hans Huber, Bern, Switzerland, 1971, 167.
260. **Fromherz, K. and Spielgelberg, H.,** Pharmakologische wirkungen des β-pyridylcarbinols und verwandter β-pyridylverbindungen, *Helv. Physiol., Pharmacol, Acta.*, 6, 42, 1948.
261. **Frykholm, A.,** Emaljhypoplasier efter intoxikation med D-vitamin, *Sven. Tandlaek. Tidskr*, 60, 415, 1967.
262. **Fujiwara, M.,** Influence of thiamine on riboflavin metabolism, *Vitamins*, 7, 206, 1954.
263. **Fukutomi, H.,** Clinical and experimental studies on the excessive administration of thiamine, V. Excessive administration of thiamine for pyridoxine deficient rats, *Vitamins*, 33, 156, 1966.
264. **Fukumoto, H.,** Clinical and experimental studies on the excessive administration of thiamine. IV. On the absorption, excretion and accumulation of thiamine after the excessive administration, *Vitamins*, 33, 151, 1966.
265. **Fumagalli, R.,** Pharmacokinetics of nicotinic acid and some of its derivatives, in *Metabolic Effects of Nicotinic Acid and its Derivatives*, Gey, K. F. and Carlson, L. A., Eds., Hans Huber, Bern, Switzerland, 1971, 33.
266. **Furman, K. I.,** Acute hypervitaminosis A in an adult, *Am. J. Clin. Nutr.*, 26, 575, 1973.
267. **Gal, I., Sharman, I. M., and Pryse-Davies, J.,** Vitamin A in relation to human congenital malformations, in *Advances in Teratolory*, Vol. 5, Woottam, D. H. M., Ed., 143.
268. **Gal, I.,** Vitamin A, pregnancy, and oral contraceptives, *Br. Med. J.*, 2, 560, 1974.
269. **Galbraith, P. A., Perny, W. F., and Bearush, R. E.,** Effect of nicotinic acid on serum lipids in normal and atherosclerotic subjects, *Lancet*, 1, 222. 1959.
270. **Gamble, W. and Wright, L. D.,** Effect of nicotinic acid and related compounds on incorporation of mevalonic acid into cholesterol, *Proc. Soc. Exp. Biol. Med.*, 107, 160, 1961.
271. **Gant, Z. N., Solomon, H. M., and Miller, O. N.,** The influence of antilipemic doses of nicotinic acid on carbohydrate tolerance and plasma insulin levels in man, in *Metabolic Effects of Nicotinic Acid and Its Derivatives*, Gey, K. I. and Carlson, L. A., Eds., Hans Huber, Bern, Switzerland, 1971, 923.
272. **Garcia, R. E., Friedman, W. F., Kaback, M. M., and Rowe, R. D.,** Idiopathic hypercalcemia and supravalvular aortic stenosis: documentation of a new syndrome, *N. Engl. J. Med.*, 271, 117, 1964.
273. **Gasser, C.,** Heinz body anaemia and related phenomena, *J. Pediatr.*, 54, 673, 1959.
274. **Gavin, G. and McHenry, E. W.,** Inositol: a lipotropic factor, *J. Biol. Chem.*, 139, 485, 1941.
275. **Geelen, J. A. G.,** Skullbase malformations in rat fetuses with hypervitaminosis A-induced exencephaly, *Teratology*, 7, 49, 1973.
276. **Gegick, C. G., and Danowski, T. S.,** Idiosyncratic reaction to 25-hydroxylated vitamin D_3; *Ann. Int. Med.*, 80, 416, 1974.
277. **Gelpke, P. M.,** Vitamin A intoxication, *Can. Med. Assoc. J.*, 104, 533, 1971.
278. **Gerber, A., Raab, A. P., and Sobel, A. E.,** Vitamin A poisoning in adults: with description of a case, *Am. J. Med.*, 16, 729, 1954.
279. **Gershoff, S. N.,** Nutritional problems of household cats, *J. Am. Vet. Med. Assoc.*, 166, 455, 1975.

280. **Gibberd, F. B., Nicholls, A., Dunne, J. F., and Chaput De Sainton, D. M.,** Toxicity of folic acid, *Lancet,* 1, 360, 1970.
281. **Gil, A., Briggs, G. M., Typpo, J., and MacKinney, G.,** Vitamin A requirement of the guinea pig, *J. Nutr.,* 96, 359, 1968.
282. **Gillhespy, R. O.,** Reaction to vitamin B_{12}, *Lancet,* 1, 1076, 1955.
283. **Ginoulhaic, E., Tenconi, L. T., and Chiancone, F. M.,** 3-Pyridine acetic acid and nicotinic acid: blood levels, urinary elimination and excretion of nicotinic acid derivatives in man, *Nature (London),* 193, 948, 1962.
284. **Giroud, A., and Martinet, M.,** Fentes du palais chez l'embryon de rat par hypervitaminose, A, *C. R. Soc. Biol.,* 148, 1742, 1954.
285. **Giroud, A. and Martinet, M.,** Diverse malformations of the rat fetus in relation to the time of administration of vitamin A in excess, *C. R. Soc. Biol.,* 149, 1088, 1955.
286. **Giroud, A. and Martinet, M.,** Vitamin A hypervitaminose A und anomalien beim foetus der ratte, *Int. Z. Vitaminforsch.,* 26, 10, 1955.
287. **Giroud, A. and Martinet, M.,** Vitamin A effects of hypervitaminosis A on rabbit embryos, *C. R. Soc. Biol.,* 152, 931, 1958.
288. **Glick, D. and Antopol, W.,** The inhibition of choline esterase by thiamine (vitamin B_1), *J. Pharmacol. Exp. Ther.,* 65, 389, 1939.
289. **Harris, R. S. and Sebrell, W. H., Eds.,** *The Vitamins,* Vol. 1, 1st ed., Academic Press, New York, 1954.
290. **Goldberg, D. J., Bergenisich, T. B., and Cooper, J. R.,** Effects of thiamine antagonists on nerve conduction. II. Voltage clamp experiments with antimetabolites, *J. Neurobiol.,* 6, 453, 1975.
291. **Goldner, M. G. and Vallan, L. E.,** Marked and sustained blood cholesterol lowering effect by medication with niacin and pyridoxine, *Am. J. Med. Sci.,* 236, 341, 1958.
292. **Goldsmith, G.,** Riboflavin deficiency, in *Riboflavin,* Rivlin, R. S., Ed., Plenum Press, New York, 1975, 221.
293. **Goodhart, R. and Jolliffe, N.,** Effect of vitamin B (B_1) therapy on the polyneuritis of alcohol addicts, *JAMA,* 110, 414, 1938.
294. **Goodman, D. S. and Huang, H. S.,** Biosynthesis of vitamin A with rat intestinal enzymes, *Science,* 149, 879, 1965.
295. **Gordonoff, T.,** Can water-soluble vitamins be overdosed? Research on vitamin C, *Schweiz. Med. Wochenschr.,* 90, 726, 1960.
296. **Gorgacz, E. J., Nielsen, S. W., Frier, H. I., Eaton, H. D., and Rousseau, J. E., Jr.,** Morphologic alterations associated with decreased cerebrospinal fluid pressure in chronic bovine hypervitaminosis A, *Am. J. Vet. Res.,* 36, 171, 1975.
297. **Gottsegen, C.,** Use of vitamin K in the newborn, *Lancet,* 1, 1010, 1956.
298. **Gould, J.,** The use of vitamins in psychiatric practice, *Proc. R. Soc. Med.,* 47, 215, 1954.
299. **Govan, A. D. T. and Scott, J. M.,** Kernicterus and prematurity, *Lancet,* 1, 611, 1953.
300. **Grasse, W., III and Wade, A. E.,** The effect of thiamine consumption on liver microsomal drug-metabolizing pathways, *J. Pharmacol. Exp. Ther.,* 176, 758, 1971.
301. **Greer, M.,** Management of benign intracranial hypertension (pseudotumor cerebri), *Clin. Neurosurg.,* 15, 161, 1968.
302. **Gribetz, D., Silverman, S. H., and Sobel, A. E.,** Vitamin A poisoning, *Pediatrics,* 7, 372, 1951.
303. **Grossman, L. A.,** Increased intracranial pressure: consequence of hypervitaminosis A, *South Med. J.,* 65, 916, 1972.
304. **Guarneri, B.,** Esperienze terapeutiche e risultati di laboratorio un psoriasci trattati con prodotti ad azione normolipemica ed eudermica, *Minerva Med.,* 62, 691, 1971.
305. **Gupta, M. C. and Kumar, S,.** Chronic hypervitaminosis A in an adult, *J. Assoc. Physicians India,* 22, 865, 1974.
306. **Gurian, H. and Aldersberg, D.,** The effect of large doses of nicotinic acid on evaluating lipids and carbohydrate tolerance, *Am. J. Med. Sci.,* 237, 12, 1959.
307. **Haddad, J. G. and Stamp, T. C. B.,** Circulating 25-hydroxy-vitamin D in man, *Am. J. Med.,* 57, 57, 1974.
308. **Haddow, A.,** *Chemistry and Biology of Pteridines,* Churchill, Livingstone, London, 1954.
309. **Haimovici, H. and Pick, E. P.,** Inhibitory effect of thiamine on vasoconstrictor action of nicotine tested in the Laewen-Trendelenburg preparation, *Proc. Soc. Exp. Biol. Med.,* 62, 234, 1946.
310. **Hale, F.,** The relation of vitamin A to anophthalmus in pigs, *Am. J. Ophthalmol.,* 18, 1087, 1935.
311. **Haley, T. J., A.,** comparison of the acute toxicity of two forms of thiamine, *Proc. Soc. Exp. Biol. Med.,* 68, 153, 1948.
312. **Haley, T. J. and Flesher, A. M.,** A toxicity study of thiamine hydrochloride, *Science,* 104, 567, 1946.
313. **Hall, R. C., Jr., Rousseau, J. E., Jr., Gorgacz, E. J., and Eaton, H. D.,** Sodium and potassium in supraspinatus muscle from hypervitaminotic A Holstein calves, *J. Dairy Sci.,* 56, 252, 1973.

314. **Hamamoto, E., Inaba, M., Oka, E., Ohno, M., Bon, T., and Ohtahara, S.,** Studies on the effects of large amounts of thiamine and thiamine propyldisulfide on the EEG of mentally retarded children, *Vitamins,* 39, 145, 1969.
315. **Hanck, A. B.,** Der Einfluss von 1000 mg Vitamin C pro tag auf das renale Ausscheidungsverhalten einiger 1-elektrolyte im Harn des gesunden Meuschen, *Int. Z. Ern. Forsch.,* 43, 34, 1972.
316. **Handler, P. and Dunn, W.,** Inhibition of rat growth by nicotinamide, *J. Biol. Chem.,* 146, 357, 1942.
317. **Handler, P.,** Factors affecting the occurrence of hemorrhagic kidneys due to choline deficiency, *J. Nutr.,* 31, 621, 1946.
318. **Hansten, P. D.,** *Drug Interactions,* 2nd ed., Lea & Febiger, Philadelphia, 1972.
319. **Hardy, R. W. F., Gaylor, J. L., and Bauman, C. A.,** Biosynthesis of sterols and fatty acids as affected by nicotinic acid and related compounds, *J. Nutr.,* 71, 159, 1960.
320. **Harley, J. D., Robin, H.,** Role of the pentose phosphate pathway in the regulation of methaemoglobinaemia, *Nature (London),* 198, 397, 1963.
321. **Harley, J. D., Robin, H., Sass-Kortsach, A., Thalme, P., and Ernster, L.,** Haemolytic activity of vitamin K_3: evidence for a direct effect on cellular enzymes, *Nature (London),* 193, 478, 1962.
322. **Harris, J. W. and Horrigan, D. L.,** Pyridoxine-responsive anemia: prototype and variations on the theme, *Vitam. Horm. (N.Y.),* 22, 721, 1964.
323. **Harrison, H. E. and Harrison, H. C.,** Transfer of Ca^{++} across the intestinal wall *in vitro* in relation to action of vitamin D and cortisol, *Am. J. Physiol.,* 199, 265, 1960.
324. **Harrison, H. E.,** Effect of nutrient toxicities in animals and man: vitamin D, in *Handbook in Nutrition and Food,* Rechcigl, M., Jr., Ed., CRC Press, Boca Raton, Fla., 1978, 87.
325. **Harrison, S. D., Hixon, E. J., Burdeshaw, J. A., and Denine, E. P.,** Effect of aspirin administration on retinoic acid in mice, *Nature (London),* 269, 511, 1977.
326. **Haskell, B. E.,** Toxicity of vitamin B_6, in *Handbook in Nutrition and Food,* Rechcigl, M., Jr., Ed., CRC Press, Boca Raton, Fla., 1978, 43.
327. **Hass, G. M., Trueheart, R. E., and Hemmens, A.,** Experimental athero-arteriosclerosis due to calcific medial degeneration and hypercholesteremia, *Am. J. Pathol.,* 38, 289, 1961.
328. **Haussler, M. R.,** Vitamin D: mode of action and biomedical applications, *Nutr. Rev.,* 32, 257, 1974.
329. **Haussler, M. R., Bursac, K. M., Bone, H., and Pak, C.Y.C..** Increased circulating 1α,25-dihydroxyvitamin D_3 in patients with primary hyperparathyroidism, *Clin. Res.,* 23, 322a, 1975.
330. **Haussler, M. R. and McCain, T. A.,** Basic and clinical concepts related to vitamin D metabolism and action, *N. Engl. J. Med.,* 297, 974, 1977.
331. **Haussler, M. R., Wasserman, R. H., McCain, T. A., Peterlik, M., Bursac, K. M. and Hughes, M. R.,** 1,25-Dihydroxyvitamin D/glycoside identification of a calcinogenic principle of solanum malacoxylan, *Life Sci.,* 18, 1049, 1976.
332. **Hayashi, T.,** Mechanism of death by massive dose of thiamine, *Vitamins,* 26, 76, 1962.
333. **Hayashi, T.,** Blocking action of thiamine on neuromuscular transmission, *Vitamins,* 26, 76, 1962.
334. **Hayashi, T.,** Comparison of action of thiamine and curare on neuromuscular transmission, *Vitamins,* 27, 227, 1963.
335. **Hayashi, T.,** Action of thiamine and oxythiamine on neuromuscular transmission, *Vitamins,* 27, 476, 1963.
336. **Hayashi, T.,** Effect of pyrithiamine on neuromuscular transmission, *Vitamins,* 28, 119, 1963.
337. **Hayashi, T.,** Effect of thiamine derivatives on neuromuscular transmission, *Vitamins,* 28, 195, 1963.
338. **Hayashi, T.,** Effect of carbalkoxy thiamine administration into cerebrospinal fluid, *Vitamins,* 25, 526, 1962.
339. **Hayashi, T.,** Comparison of convulsive effect of thiamine and curare, *Vitamins,* 27, 228, 1963.
340. **Hayashi, T.,** Convulsion due to administration of oxythiamine and pyrithiamine, *Vitamins,* 32, 183, 1965.
341. **Hayashi, T.,** Convulsive effect of thiamine and its derivatives after administration into cerebrospinal fluid, *Vitamins,* 31, 426, 1965.
342. **Hayashi, T., Kurahashi, Y., and Takeuchi, M.,** Blocking action of thiamine and its derivatives upon neuromuscular transmission of cold blooded animals, *J. Vitaminol.,* 11, 30, 1965.
343. **Hayes, K. C. and Hegsted, D. M.,** Toxicity of the vitamins, in *Toxicants Occurring Naturally in Foods,* National Research Council, National Academy of Sciences, Washington, D.C., 1973, 235.
344. **Hazzard, D. G., Woelfel, C. G., Calhoun, M. C., Rousseau, J. E., Jr., Eaton, H. D., Nielsen, S. W., Grey, R. M., and Lucas, J. J.,** Chronic hypervitaminosis A in Holstein male calves, *J. Dairy Sci.,* 47, 391, 1964.
345. **Hecht, G. and Weese, H.,** Pharmakologisches uber Vitamin B_1, *Klin. Wochenschr.,* 16, 414, 1937.
346. **Hellriegel, K. P. and Reuter, H.,** Side effects of vitamins, in *Meyler's Side Effects of Drugs,* Vol. 8, Dukes, M. N. G., Ed., Excerpta Medica, Amsterdam, 1975.

347. **Hellström, L.**, Lack of toxicity of folic acid given in pharmacological doses to healthy volunteers, *Lancet*, 1, 59, 1971.
348. **Henderson, R. G., Russel, R. G. G., Ledingham, J. G. G., Smith, R., Oliver, D. O., Walton, R. J., Small, D. G., Preston, C., Warner, G. T., and Norman, A. W.**, Effects of 1,25-dihydroxycholecalciferol on calcium absorption, muscle weakness, and bone disease in chronic renal failure, *Lancet*, 1, 379, 1974.
349. **Henninger, G. and Bavers, M.**, Adverse effects of niacin in emergent psychosis, *JAMA*, 204, 1010, 1968.
350. **Henneman, P. H., Dempsey, E. F., Carroll, E. L., and Albright, F.**, The cause of hypercalcemia in sarcoid and its treatment with cortisone and sodium phytate, *J. Clin. Invest.*, 35, 1229, 1956.
351. **Herbert, V.**, The rationale of massive-dose vitamin therapy (megavitamin therapy: hot fictions versus cold facts), Proc. West. Hemi. Nut Cong, 4, 84, 1974.
352. **Herbert, V. and Jacob, E.**, Destruction of vitamin B_{12} by ascorbic acid, *JAMA*, 230, 241, 1974.
353. **Hillman, R. W.**, Hypervitaminosis A: experimental induction in human subjects, *Am. J. Clin. Nutr.*, 4, 603, 1956.
354. **Hillman, R. W.**, Tocopherol excess in man. Creatinuria associated with prolonged ingestion, *Am. J. Clin. Nutr.*, 5, 597, 1957.
355. **Hoffer, A. and Callbeck, M. J.**, The hypocholesterolemic effect of nicotinic acid and its relationship to the autonomic nervous system, *J. Ment. Sci.*, 103, 810, 1957.
356. **Hoffer, A.**, Ascorbic acid and toxicity, *N. Engl. J. Med.*, 285, 635, 1972.
357. **Hoffer, A., Osmond, H., Callbeck, M. J., and Kahar, I.**, Treatment of schizophrenia with nicotinic acid and nicotinamide, *J. Clin. Exp. Psychopathol.*, 18, 131, 1957.
358. **Hoigné, R.**, Araphylactic shock due to drugs, *Int. Arch. Allergy*, 28, 27, 1965.
359. **Holz, P. and Palm, D.**, Pharmacological aspects of vitamin B_6, *Pharmacol. Rev.*, 16, 113, 1964.
360. **Hommes, O. R. and Obbens, E. A. M. T.**, Liver function and folate epilepsy in the rat, *J. Neurol. Sci.*, 20, 269, 1973.
361. **Hommes, O. R., Obbens, E. A. M. T., and Wijffels, C. C. B.**, Epileptogenic activity of sodium folate and the blood-brain barrier in the rat, *J. Neurol. Sci.* 19, 63, 1973.
362. **Hook, E. B., Healy, K. M., Niles, A. M., and Skalko, R. C.**, Vitamin E: teratogen or anti-teratogen?, *Lancet*, 1, 809, 1974.
363. **Høviing, C.**, Anaphylactic reaction after injection of vitamin B_{12}, *Br. Med. J.*, 3, 102, 1968.
364. **Horrobin, D. C.**, D. V. T. after vitamin C, *Lancet*, 2, 317, 1973.
365. **Horwitt, M. K.**, Vitamin E: a reexamination, *Am. J. Clin. Nutr.*, 29, 569, 1976.
366. **Hossain, M.**, Vitamin D treatment in hypothyroidism, *Lancet*, 2, 615, 1970.
367. **Hottinger, A., Berger, H. and Krauthammer, W.**, Klinische beobachtunger zum problem des vitamin B_6 metabolismus, *Schweiz, Med. Wochenschr.*, 94, 221, 1964.
368. **House, H. L.**, Nutrition, in *the Physiology of Insecta*, Vol. 2, 2nd ed., Rockstein, M., Ed., Academic Press, New York, 1974, 1.
369. **Hövels, O. and Stephan, U.**, Das krankheitsbild der "idiopathischen" hypercalcamie, eine chronische vitamin D-intoxikation, *Ergenb. Inn. Med. Kinderheilkd.*, 18, 118, 1962.
370. **Hruban, Z., Russell, R. M., Boyer, J. L., Glagov, S., and Bagheri, S. A.**, Ultrastructural changes in livers of two patients with hypervitaminosis, *Am. J. Pathol.*, 76, 451, 1974.
371. **Hughes, M. R., Baylink, D. J., Jones, P. G., and Haussler, M. R.**, Radioligand receptor assay for 25-hydroxyvitamin D_2/D_3 and 1α25-dihydroxyvitamin D_2/D_3. Application to hypervitaminosis D, *J. Clin. Invest.*, 58, 61, 1976.
372. **Hunt, A. D., Stokes, W. W., McCrory, W. W., and Stroud, H. H.**, Pyridoxine dependency: report of a case of intractable convulsions in an infant controlled by pyridoxine, *Pediatrics*, 13, 140, 1954.
373. **Hunter, R., Barnes, J., and Matthews, D. M.**, Effect of folic acid supplement on serum vitamin B_{12} levels in patients on anticonvulsants, *Lancet*, 2, 666, 1969.
374. **Hunter, R., Barnes, J., Oakley, H. F., and Matthews, D. M.**, Toxicity of folic acid given in pharmacological doses to healthy volunteers, *Lancet*, 1, 61, 1970.
375. **Iliescu, C. C., Iliescu, M., Roman, L., Jacobini, P., Constantinescu, S., and Nutu, S.**, Effect of nicotinic acid on blood lipids in atherosclerosis, *Med. Interna*, 15, 39, 1963.
376. **Inoue, K. and Itokawa, Y.**, Metabolism of O-benzoylthiamine in animal body, *Biochem. Med.*, 8, 450, 1973.
377. **Inoue, K., Katsura, E., and Kariyone, S.**, Secondary riboflavin deficiency, *Vitamins*, 10, 69, 1956.
378. Institute of Food Technologists Expert Panel on Food Safety and Nutrition and the Committee on Public Information, Vitamin E, *Nutr. Rev.*, 35, 57, 1977.
379. **Ireland, A. W., Clubb, J. S., Neale, F. C., Posen, S., and Reeve, T. S.**, Calciferol requirements of patients with surgical hypoparathyroidism, *Ann. Intern. Med.*, 69, 81, 1968.
380. **Irneil, L.**, Metastatic calcification of soft tissue on overdosage of vitamin D, *Acta Med. Scand.*, 185, 147, 1969.

381. **Itokawa, Y.**, Effect of nutrition toxicities in animals and man: thiamine, in *Handbook in Nutrition and Food*, Rechcigl, M., Jr., Ed., CRC Press, Boca Raton, Fla., 1978, 3.

382. **Itokawa, Y., Tanaka, C., and Kimura, M.**, Effect of thiamine on serotonin levels in magnesium-deficient animals, *Metabolism*, 21, 375, 1972.

383. **Itokawa, Y., Schulz, R. A., and Cooper, J. R.**, Thiamine in nerve membranes, *Biochim. Biophys. Acta*, 266, 293, 1972.

384. **Itokawa, Y. and Cooper, J. R.**, Thiamine release from nerve membrane by tetrodotoxin, *Science*, 166, 759, 1969.

385. **Itokawa, Y. and Cooper, J. R.**, On a relationship between ion transport and thiamine in nervous tissue, *Biochem. Pharmacol.*, 18, 545, 1969.

386. **Itokawa, Y. and Cooper, J. R.**, Ion movements and thiamine in nervous tissue. I. Intact nerve preparations, *Biochem. Pharmacol.*, 19, 985, 1970.

387. **Itokawa, Y. and Copper, J. R.**, Ion movement and thiamine. II. The release of the vitamin from membrane fragments, *Biochim. Biophys. Acta*, 196, 274, 1970.

388. **Itokawa, Y.**, Is calcium deficiency related to thiamine-dependent neuropathy in pigeon?, *Brain Res.*, 94, 475, 1975.

389. **Itokawa, Y.**, Role of thiamine in excitable membrane of nerve, *Vitamins*, 49, 415, 1975.

390. **Itokawa, Y.**, Thiamine and nerve membrane, *J. Nutr. Sci. Vitaminol.*, Suppl. 22, 17, 1976.

391. **Iwata, H. and Inoue, A.**, Antagonistic effect of thiamine and its derivatives to nicotinie on atria of guinea pig, *Folia Pharmacol. Jpn.*, 64, 46, 1968.

392. **Jablonska, A. and Cadamska, T.**, Niedokrwistösc hemolityczna u noworodka po przedawkowania witaminy K., *Med. Wiejska*, 1, 79, 1973.

393. **Jaffe, R. M., Kasten, B., Young, D. S., and Maclowry, J. D.**, Fake negative stool occult blood tests caused by ingestion of ascorbic acid (vitamin C), *Ann. Intern. Med.*, 83, 824, 1975.

394. **Jaffee, S. J. and Filer, L. J.**, The use and abuse of vitamin A, *Nutr. Rev.*, 32, (Suppl. 1), 41, 1974.

395. **Jakovlieu, N.**, Scurvy following nutritional stress, *Lanachrunestorschune*, 3, 446, 1958.

396. **James, J. and Warin, R. P.**, Sensitivity to cyancocobalamin and hydroxycobalamin, *Br. Med. J.*, 2, 262, 1971.

397. **Jaros, S. H., Wnuck, A. L., and DeBeer, E. J.**, Thiamine intolerance, *Ann. Allergy*, 10, 291, 1952.

398. **Jauregg-Wagner, J.**, Riboflavin, in *The Vitamins: Chemistry, Physiology, Pathology, Methods*, Vol. 5, 2nd ed., Sebrell, W. H., Jr. and Harris, R. S., Eds., Academic Press, New York, 1972.

399. **Jeghers, H. and Marraro, H.**, Hypervitaminosis A: its broadening spectrum, *Am. J. Clin. Nutr.*, 6, 335, 1958.

400. **Jenkins, D. and Spector, R. G.**, The action of folate and phenytoin on the rat heart in vivo and in vitro, *Biochem. Pharmacol.*, 22, 1813, 1973.

401. **Jenkins, M. Y.**, Effect of nutrient toxicities (excess) in animals and man: vitamin A, in *Handbook in Nutrition and Food*, Rechcigl, M., Jr., Ed., CRC Press, Boca Raton, Fla., 1978, 73.

402. **Jenkins, M. Y. and Mitchell, G. V.**, Influence of excess vitamin E on vitamin A toxicity in rats, *J. Nutr.*, 105, 1600, 1975.

403. **Jirâsek, L. and Schwank, R.**, Berufskontaktekzem durch vitamin K, *Hautarzt*, 16, 351, 1955.

404. **Jones, B.**, Negative effects of vitamin K preparations on glucuronyl transferase activity, *Pediatrics*, 40, 993, 1967.

405. **Jolliffe, N., Colbert, C. N., and Joffe, P. M.**, Observations on the etiologic relationship of vitamin B (B_1) to polyneuritis in the alcohol addict, *Am. J. Med. Sci.*, 191, 515, 1936.

406. **Jolliffe, N.**, Treatment of neuropsychiatric disorders with vitamins, *JAMA*, 117, 1496, 1941.

407. **Josephs, H. W.**, Hypervitaminosis A and carotenemia, *Am. J. Dis. Child.*, 67, 33, 1944.

408. **Jowsey, J. and Riggs, B. L.**, Bone changes in a patient with hypervitaminosis A, *J. Clin. Endocrinol. Metab.*, 28, 1833, 1968.

409. **Jukes, T. M.**, Perosis in turkeys. I. Experiments related to choline, *Poult. Sci.*, 20, 251, 1941.

410. **Jusko, W. L. and Levy, G.**, Absorption, protein binding, and elimination of riboflavin, in *Riboflavin*, Rivlin, R. S., Ed., Plenum Press, New York, 1975, 99.

411. **Kahler, R. L., Braunwald, E., Plauth, W. H., Jr., and Morrow, A. G.**, Familial congenital heart disease, *Am. J. Med.*, 40, 384, 1966.

412. **Kaidbey, K. H., Kligman, A. M., and Yoshida, H.**, Effects of intensive application of retinoic acid on human skin, *Br. J. Dermatol.*, 92, 693, 1975.

413. **Kallner, A.**, Serum bile acids in man during vitamin C supplementation and restriction, *Acta Med. Scand.*, 202, 283, 1977.

414. **Kanis, J. A. and Russell, R. G. G.**, Rate of reversal of hypercalcaemia and hypercalciuria induced by vitamin D and its 1α-hydroxylated derivatives, *Br. Med. J.*, 1, 78, 1977.

415. **Kapp, J., Mahaley, M. S., and Odom, G. L.**, Experimental evaluation of poten tial spasmolytic drugs, *J. Neurosurg.*, 32, 468, 1970.

416. **Kaserer, H. P., Gibitz, H. J., and Witontky, O.,** Uber eine todliche vitamin D-intoxikation beim erwachsenen, *Wien. Klin. Wochenschr.*, 78, 463, 1966.
417. **Katz, C. M. and Tzagournis, M.,** Chronic adult hypervitaminosis A and hypercalcemia, *Metabolism*, 21, 1171, 1972.
418. **Kawasaki, T., Asano, T., and Makita, S.,** A case of anaphylactic shock due to intravenous thiamine hydrochloride, *J. Jpn. Soc. Int. Med.*, 51, 246, 1962.
419. **Khera, K. S.,** Teratogenicity study in rats given high doses of pyridoxine (vitamin B_6) during organogenesis, *Experientia*, 31, 469, 1975.
420. **Klemetti, L.,** Is the vitamin B_{12} treatment of pernicious anemia a predisposing factor for thromboses in aged patients?, *Acta Med. Scand.*, 176, 121, 1964.
421. **Klevay, L.,** Hypercholesterolemia due to ascorbic acid, *Proc. Soc. Exp. Biol. Med.*, 151, 579, 1976.
422. **Klopp, C. T., Abels, J. C., and Rhods, C. P.,** The relationship between riboflavin intake and thiamine excretion in man, *Am. J. Med. S.*, 205, 852, 1943.
423. **Knudson, A. G. and Rothman, P. E.,** Hypervitaminosis A. A review with discussion of vitamin A, *Am. J. Dis. Child.*, 85, 316, 1953.
424. **Kochhar, D. M.,** Limb development in mouse embryos. I. Analysis of teratogenic effects of retinoic acid, *Teratology*, 7, 289, 1973.
425. **Kohn, R. M. and Montes, M.,** Hepatic fibrosis following long acting nicotinic acid therapy: a case report, *Am. J. Med. Sci.*, 250, 94, 1969.
426. **Kooh, S. W., Fraser, D., DeLuca, H. F., Holick, M. F., Belsey, R. E., Clark, M. B., and Murray, T. M.,** Treatment of hypoparathyroidism and pseudohypoparathyroidism with metabolites of vitamin D: evidence for impaired conversion of 25-hydroxyvitamin D to 1α,25-dehydroxyvitamin D, *N. Engl. J. Med.*, 293, 840, 1975.
427. **Kordylas, J. M.,** Vitamin A and fat combination in cholesterol biosynthesis and atherosclerosis, *Lancet*, 2, 606, 1972.
428. **Korner, W. F. and Vollm, J.,** New aspects of the tolerance of retinol in humans, *Int. J. Vitam. Nutr. Res.*, 363, 1975.
429. **Korner, W. F. and Weber, F.,** Zur toleranz hoher ascorbinsautredosen, *Int. Z. Vit. Eru. Forsch.*, 42, 528, 1972.
430. **Kotaki, A., Sakurai, T., Kobayashi, M., and Yagi, K.,** Studies on myo-inositol. IV. Effect of myo-inositol on the cholesterol metabolism of rats suffering from experimental fatty liver, *Vitamins*, 14, 87, 1968.
431. **Kraft, H. G., Fiebig, L., and Hotovy, R.,** Zur Pharmakologie des Vitamin B_6 and seiner Derivate, *Arzneim. Forsch.*, 11, 922, 1961.
432. **Krause, R. F.,** Liver lipids in a case of hypervitaminosis A, *Am. J. Clin. Nutr.*, 16, 455, 1965.
433. **Kritchevsky, D. and Tepper, S. A.,** Influence of nitocinic acid, picolinic and pyridine-3-sulfonic acids on cholesterol metabolism in the rat, *J. Nutr.*, 82, 157, 1964.
434. **Krjukova, L. V., Ulasevia, I. I., and Medvedskaja, V. S.,** Interrelation of vitamins A and E in the live animal, *Byull. Eksp. Biol. Med.*, 67, 59, 1969.
435. **Kuhn, R. and Boulanger, P.,** Uber die Giftigkeit der Flavine, *Hoppe-Seylers Z. Physiol. Chem.*, 241, 233, 1936.
436. **Kusin, J. A., Reddy, V., and Sivakumar, B.,** Vitamin E supplements and the absorption of a massive dose of vitamin A, *Am. J. Clin. Nutr.*, 27, 774, 1974.
437. **Lagerholm, B.,** Hypersensitivity to phenylcarbinol preservative in vitamin B_{12} for injection, *Acta Allergol.*, 12, 295, 1958.
438. **Lamden, M. P. and Chrystowski, G. A.,** Urinary oxalate excretion by man following ascorbic acid ingestion, *Proc. Soc. Exp. Biol. Med.*, 85, 100, 1954.
439. **Lane, B. P.,** Hepatic microanatomy in hypervitaminosis A in man and rat, *Am. J. Pathol.*, 53, 591, 1968.
440. **Langman, J. and Welch, G. W.,** Excess vitamin A and development of the cerebral cortex, *J. Comp. Neurol.*, 131, 15, 1967.
441. **Lascari, A. D. and Bell, W. E.,** Pseudotumour cerebri due to hypervitaminosis A, *Clin. Pediatr.*, 9, 627, 1970.
442. **Laurance, B.,** Danger of vitamin K — analogues to newborn, *Lancet*, 1, 819, 1955.
443. **Laws, C. L.,** Sensitisation to thiamine hydrochloride, *JAMA*, 117, 176, 1941.
444. **Lecombe, J., Deleixhe, A., and Lapiere, C.,** Sur la nature du choc provoque par la thiamine, *Acta Allergol.*, 12, 51, 1958.
445. **Lee, K. W., Abelson, D. M., and Kwon, Y. O.,** Nicotinic acid and ^{14}C metabolism in man, *Am. J. Clin. Nutr.*, 21, 223, 1968.
446. **Leelaprute, V., Boonpucknavig, V., Bhamarapravati, N., and Weerapradist, W.,** Hypervitaminosis A in rats. Varying responses due to different forms, doses, and routes of administration , *Arch. Pathol.*, 96, 5, 1973.

447. **Lesson, P. M. and Fourman, P.**, Increased sensitivity to vitamin D after vitamin D poisoning, *Lancet*, 1, 1182, 1966.
448. **Leeson, P. M. and Fourman, P.**, Acute pancreatitis from vitamin D poisoning in a patient with parathyroid deficiency, *Lancet*, 1, 1185, 1966.
449. **Leicht, E., Strunz, J., von Seebach, H. B., Mäusle, E., and Meiser, R. J.**, Akute vitamin A intoxikation mit hamolytischer anamie, hyperkälzamie und toxischer hepatose, *Med. Klin.*, 68, 54, 1973.
450. **Leitner, Z. A.**, Untoward effects of vitamin B_1, *Lancet*, 2, 474, 1943.
451. **Leitner, Z. A., Moore, T., and Sharman, I. M.**, Fatal self medication with retinol and carrot juice, *Proc. Nutr. Soc.*, 34, 44A, 1975.
452. **Leoni, A.**, Quadri acneiformi da vitamina B_{12}, *G. Ital. Derm.*, 47, 210, 1972.
453. **Levander-Lindgren, M.**, Hypersensitivity to folic acid in a case of erythroblastomatosis, *Acta Med. Scand.*, 157, 233, 1957.
454. **Levander, O. A., Morris, V. C., Higgs, D. J., and Varma, R. N.**, Nutritional interrelationships among vitamin E, selenium, antioxidants and ethyl alcohol in the rat, *J. Nutr.*, 103, 536, 1973.
455. **Levinson, H. Z. and Cohen, E.**, The action of overdose biotin on reproduction of the hide beetle, *Dermestes maculatus, J. Insect Physiol.*, 19, 551, 1973.
456. **Levy, H. and Boas, E. P.**, Vitamin E in heart disease, *Ann Intern. Med.*, 28, 1117, 1948.
457. **Linden, V. and Seeleg, M. S.**, Multiple factors in the hyperlipidaemia of hypervitaminosis D, *Br. Med. J.*, 4, 166, 1975.
458. **Lippi, U., Pulido, E., and Guidi, G.**, Lesioni lisosomiali da dosi elevate di acido ascorbico, *Acta Vitaminol.*, 5, 177, 1966.
459. **Lipton, M. M. and Steigman, A. J.**, Tuberculin type of hypersensitization to crystalline vitamin B_{12} (cyanocobalamin) induced in guinea pigs, *J. Allergy*, 34, 362, 1963.
460. **Lofland, H. B., Jr., Goodman, H. O., Clarkson, T. B., and Prichard, R. W.**, Enzyme studies in thiamine-deficient pigeons, *J. Nutr.*, 79, 188, 1963.
461. **Logan, J.**, Thyroid depression following high doses of vitamin A, *N. Z. Med. J.*, 56, 249, 1957.
462. **Lombaert, A. and Carton, H.**, Benign intracranial hypertension due to A-hypervitaminosis in adults and adolescents, *Eur.Neurol.*, 14, 340, 1976.
463. **Loomis, W. F.**, Rickets, *Sci. Am.*, 223, 76, 1970.
464. **Lucey, J. F.**, Hazards to the newborn infant from drugs administered to the mother, *Pediatr. Clin. North Am.*, 8, 413, 1961.
465. **Lucey, J. F. and Dolan, R. G.**, Hyperbilirubinaemia of newborn infants associated with the parenteral administration of vitamin K analogue to the mothers, *Pediatrics*, 23, 553, 1959.
466. **Nater, J. P. and Doeglas, H. M. G.**, Halibut liver poisoning in 11 fishermen, *Acta Derm. Venereol.*, 50, 109, 1970.
467. **Lumish, S. H., Blyn, C., and Nodine, J. H.**, Inositol nicotinate as a peripheral vasodilator, *Cur. Ther. Res. Clin. Exp.*, 4, 243, 1962.
468. **Lunaas, T., Bladwin, R. L., and Cupps, P. T.**, Ovarian activities of pyridine nucleotide dependent dehydrogenases in the rat during pregnancy and lactation, *Acta Endocrinol. (Copenhagen)*, 58, 521, 1968.
469. **Maddox, G. W., Foltz, F. M., and Nelson, S. R.**, Effect of vitamin A intoxication on intracranial pressure and brain water in rats, *J. Nutr.*, 104, 478, 1974.
470. **Maeda, T.**, The Physiological Functions of Myo-Inositol: The Alteration of Lipid Metabolism Due to Myo-Inositol Deficiency in Rats, Ph. D. thesis, Shizuoka College of Pharmaceutical Sciences, Shizuoka, Japan, 1976.
471. **Maeda, T.**, Clinical and experimental studies on the effect of the administration of large amount of B vitamins on the blood sugar and insulin content of blood, *Vitamins*, 37, 439, 1968.
472. **Mallia, A. K., Smith, J. E., and Goodman, D. S.**, Metabolism of retinol-binding protein and vitamin A during hypervitaminosis A in the rat, *J. Lipid Res.*, 16, 180, 1975.
473. **Mallick, N. P. and Berlyne, G. M.**, Arterial calcification after vitamin D therapy in hyperphosphataemic renal failure, *Lancet*, 2, 1316, 1968.
474. **Mallick, N. and Deb, C.**, Effect of different doses of ascorbic acid on thyroid activity in rats at different levels of protein intake, *Endocrinologie*, 65, 333, 1975.
475. **March, B. E., Wong, E., Seier, L., Sim, J and Biely, J.**, Hypervitaminosis E in the chick, *J. Nutr.*, 103, 371, 1973.
476. **Marie, J. and See, G.**, Acute hypervitaminosis A of the infant, *Am. J. Dis. Child.*, 87, 731, 1954.
477. **Marie, J., Hennequet, A., Marandian, H., and Momenzadeh, A.**, Vitamin D_2 excess due to daily ingestion of an excessive dose of a water-alcohol solution containing 400 IU per drop. Recovery on a diet free of Ca and with corticosteroid, *Sem. Hop. Ann. Pediatr.*, 45, 24, 1969.
478. **Marin-Padilla, M. and Ferm, V. H.** Somite necrosis and developmental malformations induced by vitamin A in the golden hamster, *J. Embryol. Exp. Morphol.*, 13, 1, 1965.

479. **Markiewicz, M. and Uss, B.,** Zapanlenie mózgu w nasterpstwie uczulenia na witamine B_1, *Pol. Tyg. Lek.*, 25, 1661, 1970.
480. **Martin, G. J., Thompson, M. R., and de Carvajal-Forero, J.,** Influence of inositol and other B-complex factors on the motility of the gastrointestinal tract, *Am. J. Dig. Dis.*, 8, 290, 1941.
481. **Wade, A. and Reynolds, J. E. F., Eds.,** *Martindale — The Extra Pharmacopoeia,* The Pharmaceutical Press, London, 1665.
482. **Marxs, W., Marks, L., Meserve, E. R., Shimoda, F., and Deuel, H. J.,** Effects of the administration of a vitamin E concentrate and of cholesterol and bile salts on the aorta of the rat, *Arch. Pathol.*, 47, 440, 1947.
483. **Mathews-Roth, M. M.,** β-Carotene as an oral photoprotective agent in erythropoietic protoporphyria, *JAMA*, 228, 1004, 1974.
484. **Mathur, B. P.,** Sensitivity of folic acid, *Indian J. Med. Sci.*, 20, 133,1966.
485. **Mayfield, H. L. and Roehm, R. R.,** Influence of ascorbic acid and the source of B vitamins on the utilization of carotene, *J. Nutr.*, 58, 203, 1956.
486. **Mayson, J. J.,** False negative tests for urinary glucose in the presence of ascorbic acid, *Am. J. Clin. Pathol.*, 58, 297, 1972.
487. **Melhorn, D. K. and Gross, S.,** Relationships between iron-dextran and vitamin E in iron-deficiency anemia in children, *J. Lab. Clin. Med.*, 74, 789, 1969.
488. **Mellette, S. J. and Leone, L. A.,** Influence of age, sex, strain of rat, and fat soluble vitamins on hemorrhagic syndromes in rats fed irradiated beef, *Fed. Proc. Fed. Am. Soc. Exp. Biol.*, 19, 1045, 1960.
489. **Mengel, C. E. and Green, H. L., Jr.,** Ascorbic acid effects on erythrocytes, *Ann. Intern. Med.*, 84, 490, 1976.
490. **Merritt, A. D., Palmer, C. G., Lurie, P. R., and Petry, E. L.,** Supravalvular aortic stenosi: genetic and clinical studies, *J. Lab. Clin. Med.*, 62, 995, 1963.
491. **Meyer, T. C. and Angus, J.,** The effect of large doses of synkavit in the newborn, *Arch. Dis. Child.*, 31, 212, 1956.
492. **Metildi, P. F.,** The treatment of tabetic lightning pain with thiamine chloride, *Am. J. Syph. Gonorrhea Vener. Dis.*, 23, 1, 1939.
493. **Mickelson, O., Caster, W. O., and Keys, A.,** A statistical evaluation of the thiamine and pyramin excretions of normal young men on controlled intakes of thiamine, *J. Biol. Chem.*, 168, 415, 1947.
494. **Miettinen, T. A.,** Effect of nicotinic acid on catabolism and synthesis of cholesterol in man, *Clin. Chim. Acta*, 20, 43, 1968.
495. **Miettinen, T. A.,** Effect of nicotinic acid on the fecal excretion of neutral sterols and bile acids, in *Metabolic Effects of Nicotinic Acid and its Derivatives,* Gey, K. F. and Carlson, L. A., Eds., Hans Huber, Bern, Switzerland, 1971, 677.
496. **Miettinen, T. A.,** Influence of nicotinic acid on cholesterol synthesis in man. Effect of nicotinic acid on the fecal excretion of neutral sterols and bile acids, in *Metabolic Effects of Nicotinic Acid and its Derivatives,* Gey, K. F. and Carlson, L. A., Eds., Hans Huber, Bern, 1971, 649.
497. **Miettinen, T. A., Taskinen, M. R., Pelkonen, R., and Nikkila, E. A.,** Glucose tolerance and plasma insulin in man during acute and chronic administration of nicotinic acid, *Acta Med. Scand.*, 186, 247, 1969.
498. **Mikkelsen, B., Ehlers, N., and Thomsen, H. G.,** Vitamin A intoxication causing papilloedema and simulating acute encephalitis, *Acta Neurol. Scand.*, 50, 642, 1974.
499. **Miller, O. N., Hamilton, J. G., and Goldsmith, G. A.,** Studies on the mechanism of effects of large doses of nicotinic acid and nicotinamide on serum lipids of hypercholesterolaemic patients, *Circulation*, 18, 489, 1958.
500. **Miller, O. N. and Hamilton, J. G.,** Nicotinic acid and derivatives, in *Lipid Pharmacology,* Paoletti, R., Ed., Academic Press, New York, 1964, 275.
501. **Miller, O. N., Gutierrez, M., Sullivan, A., and Hamilton, J. G.,** The effect of nicotinic acid and ring-substituted analogues on the *in vitro* biosynthesis of cholesterol and fatty acids in rat liver, in *Metabolic Effects of Nicotinic Acid and Its Derivatives,* Gey, K. F. and Carlson, L. A., Eds., Hans Huber, Bern, Switzerland, 1971, 609.
502. **Mills, C. A.,** Discussion on vitamin therapy, *JAMA*, 117, 1500, 1941.
503. **Mills, C. A.,** Thiamine overdosage and toxicity, *JAMA*, 116, 2101, 1941.
504. **Minesita, K. and Ueda, M.,** On the acute toxicity of large doses of thiamine, *Vitamins*, 26, 77, 1962.
505. **Minesita, K., Ueda, M., and Matsumura, S.,** Pharmacological effect of carbalkoxythiamine, *Vitamins*, 25, 522, 1962.
506. **Minkin, W., Cohen, H. J., and Frank, S. B.,** Contact dermatitis from deodorants, *Arch. Dermatol.*, 107, 774, 1973.
507. **Minz, B.,** La role de la vitamin B_1 dans la regulation humorale du systeme nerveux, *Presse Med.*, 76, 1406, 1938.

508. **Minz, B.**, Sur la libération de la vitamine B_1 par le tronc isolé du nerf pneumogastrique soumis l'excitation électrique, *C. R. Soc. Biol.*, 127, 1251, 1938.
509. **Minz, B.**, L'influence de l'aneurine sur l'activite des elements cholinergiques, *C. R., Soc. Biol.*, 131, 1156, 1939.
510. **Misson, R., Guenard, C., Garrel, J., and Millet, P.**, Placards sclerodermiformes des régions ilio-trochanteriénnes paraissant consecutifs a des injections i.m. contenant de la vitamine K_1, *Bull. Soc. Franc. Derm. Syph.*, 79, 581, 1972.
511. **Mistry, S. P. and Dakshinamurti, K.**, Biochemistry of biotin, in *Vitamins and Hormones*, Vol. 22, Harris, R. S., Wool, I. G., and Loraine, J. A., Eds., Academic Press, New York, 1964, 1.
512. **Mitchell, D. C., Vilter, R. W., and Vilter, C. F.**, Hypersensitivity to folic acid, *Ann. Intern. Med.*, 31, 1102, 1949.
513. **Mitrani, M. M.**, Vitamin B_1 hypersensitivity with desensitisation, *J. Allergy*, 15, 150, 1944.
514. **Mittelholzer, E.**, Absence of influence of high dose of biotin on reproductive performance in female rats, *Int. J. Vitamin. Nutr. Res.*, 46, 33, 1976.
515. **Miura, O.**, Allergy by vitamins, *Sogo Rinsho*, 6, 985, 1956.
516. **Miyagawa, K., Ikehata, H., and Murata, K.**, Studies on the administration of large amounts of thiamine propyldisulfide, *Vitamins*, 23, 103, 1961.
517. **Molitor, H.**, Vitamins as pharmacological agents, *Fed. Proc. Fed. Am. Soc. Exp. Biol.*, 1, 309, 1942.
518. **Molitor, H. and Robinson, H. J.**, Oral and parenteral toxicity of vitamin K_1, phyticol and 2-methyl-1,4 naphthoquinone, *Proc. Soc. Exp. Biol. Med.*, 43, 125, 1940.
519. **Molnar, G. D., Berge, K. G., Rosevear, J. W., McGuckin, W. F., and Achor, R. P.**, The effect of nicotinic acid in diabetes mellitus, *Metabolism*, 13, 181, 1964.
520. **Moore, T. and Sharman, I. M.**, Danger of vitamin K analogues to newborn, *Lancet*, 1, 819, 1955.
521. **Moore, T.**, Vitamin A, Elsevier, London, 1957, 340.
522. **Moroz, L. A. and Gilmore, N. J.**, Inhibition of plasmin-mediated fibrinolysis by vitamin E, *Nature*, 259, 235, 1976.
523. **Morrice, G., Havener, W. H., and Kapetansky, F.**, Vitamin A intoxication as a cause of pseudo-tumor cerebri, *JAMA*, 173, 1802, 1960.
524. **Morrison, A. B. and Sarett, H. P.**, Effect of excess thiamine and pyridoxine on growth and reproduction in rats, *J. Nutr.*, 69, 111, 1959.
525. **Morrison, A. B. and Sarett, H. P.**, Studies on B vitamin interrelationships in growing rats, *J. Nutr.*, 68, 473, 1959.
526. **Morrison, R. C. and McNally, H. E.**, The spectrum of abnormalities in supravalvular aortic stenosis, *Am. J. Cardiol.*, 19, 143, 1967.
527. **Morriss, G. M.**, Morphogenesis of the malformations induced in rat embryos by maternal hypervitaminosis A, *J. Anat.*, 113, 241, 1972.
528. **Mosher, L. T.**, Nicotinic acid side effects and toxicity, *Am. J. Psychiatry*, 126, 1290, 1970.
529. **Mouriquand, G. and Edel, V.**, Hypervitaminosis C, *C. R. Soc. Biol (Paris)*, 147, 1432, 1953.
530. **Muenter, M. O.**, Hypervitaminosis A, *Ann. Intern. Med.*, 80, 105, 1974.
531. **Muenter, M. O., Perry, H. O., and Ludwig, J.**, Chronic vitamin A intoxication in adults, *Am. J. Med.*, 50, 129, 1971.
532. **Murakami, M.**, Studies on the effects of excessive administration of thiamine upon the living body. I. Study on habituation of *Uroloncha striata* var. *domestica* in the metabolism of thiamine as a result of successive profuse administration of thiamine, *Vitamins*, 11, 211, 1956.
533. **Murakami, M.**, Studies on the effects of excessive administration of thiamine upon the living body. II. Study on habituation of albino rats in the metabolism of thiamine as a result of successive profuse administration of the vitamin, *Vitamins*, 11, 216, 1956.
534. **Murakami, U. and Kameyama, Y.**, Malformations to the mouse foetus caused by hypervitaminosis A of the mother during pregnancy, *Arch. Environ. Health*, 10, 732, 1965.
535. **Murata, T.**, Influence of thiamine on the amount of lactic acid in urine, *J. Vitaminol.*, 4, 109, 1958.
536. **Murata, K., Suzuki, S., Irimajiri, S., Miyagawa, K., Miyamato, T., and Ikehata, H.**, Studies on the oral administration of large amounts of thiamine to men, *Vitamins*, 27, 282, 1963.
537. **Murata, K., Suzuki, S., Miyatake, K., Miyamoto T., and Ikehata, H.**, Studies on the oral administration of large amounts of thiamine in men, *Vitamins*, 30, 33, 1964.
538. **Murphy, B. F.**, Hypervitaminosis E, *JAMA*, 227, 1381, 1974.
539. **Murphy, J. V., Craig, L. J., and Glew, R. H.**, Leigh disease. Biochemical characteristics of the inhibitor, *Arch. Neurol.*, 31, 220, 1974.
540. **Muting, D.**, Uber die Verhutung von Müchenstichen duren Einnahme von Vitamin B_1, *Med. Klin. (Munich)*, 53, 1023, 1958.
541. **McCuaig, L. W. and Motzok, I.**, Excessive dietary vitamin E: its alleviation of hypervitaminosis A and lack of toxicity, *Poult. Sci.*, 49, 1050, 1970.

542. **McMichael, A. J.,** Kidney stone hospitalisations in relation to changes in vitamin C consumption in Australia 1966—76, *Comm. Health Stud.,* 2, 13, 1978.
543. **McNicholl, B.,** Vitamin D as a public health problem, *Br. Med. J.,* 2, 245, 1964.
544. **Naha, P. N.,** Vitamin A, *J. Assoc. Physicians India,* 22, 860, 1974.
545. **Najjar, S. S. and Yazigi, A.,** Abuse of vitamin D, *J. Med. Liban,* 25, 113, 1972.
546. **Nakao, T., Nitta, T., and Gotooda, T.,** Idiopathic hypercalcemia of infancy, *Tohoku. J. Exp. Med.,* 71, 363, 1960.
547. **Nastuk, W. L. and Kahn, N.,** Blocking action of thiamine on neuromuscular transmission and membrane conduction, *Fed. Proc. Fed. Am. Soc. Exp. Biol.,* 19, 260, 1960.
548. **Nath, N., Harper, A. E., and Elvehjem, C. A.,** Diet and cholesterolemia. Four effects of carbohydrate and nicotinic acid, *Proc. Soc. Exp. Biol. Med.,* 102, 571, 1959.
549. National Academy of Sclences, National Research Council, Hazards of overuse of vitamin D, *Nutr. Rev.,* 33, 61, 1975.
550. National Research Council, National Academy of Sciences, Food and Nutrition Board, Hazards of overuse of vitamin D, *JADA,* 66, 453, 1975.
551. National Research Council, National Academy of Sciences, Recommended Dietary Allowances, 8th ed., (rev.), National Academy of Sciences, Washington, D.C., 1974, 74.
552. **Natume, K.,** Studies on myo-inositol. III. Effect of the excess dosage of myo-inositol on the pattern of lipids in the liver of young rat, *Vitamins,* 32, 363, 1965.
553. **Nayakama, Y.,** Studies on the relationship between thiamine propyl disulfide and riboflavin, I. Effect of oral administration of TPD on riboflavin content in the urine of normal children and in the urine as well as blood of healthy infants under artificial feeding, *Vitamins,* 16, 294, 1959.
554. **Nayakama, Y.,** Studies on the relationship between thiamine propyl disulphide and riboflavin. II. Effect of parenteral administration of TPD and esters of riboflavin on the urinary riboflavin content and its forms in healthy infants under artificial feeding, *Vitamins,* 16, 300, 1959.
555. **Neuweiler, W.,** Hypervitaminosis and its relation to pregnancy, *Int. Z. Vitaminforsch.,* 22, 392, 1951.
556. **Ngai, S. H., Ginsburg, S., and Katz, R. L.,** Action of thiamine and its analog on neuromuscular transmission in the cat, *Biochem. Pharmacol.,* 7, 256, 1961.
557. **Nieman, H. J. and Klein, O.,** The biochemistry and pathology of hypervitaminosis A, *Vitam. Horm.,* 12, 69, 1954.
558. **Nikkila, E. A.,** Effect of nicotinic acid on hepatic lipogenesis and triglyceride synthesis and release, in *Metabolic Effects of Nicotinic Acid and Its Derivatives,* Gey, K. F. and Carlson, L. A., Eds., Hans Huber, Bern, 1971, 471.
559. **Nikkila, E. A. and Mieltinen, T. A.,** NIcotinic acid and gluconeogeneses in man, in *Metabolic Effects of Nicotinic Acid and its Derivatives,* Gey, K. F. and Carlson, L. A., Eds., Hans Huber, Bern, 1971, 753.
560. **Nitzschner, H. and Liebsch, F.,** Vitamin B_{12}-Akne, *Med. Bild,* 13, 100, 1970.
561. **Nolen, G. A.,** Variations in teratogenic response to hypervitaminosis A in three strains of albino rat, *Food Cosmet. Toxicol.,* 7, 209, 1969.
562. **Norcia, L. N., Brown, H. J., and Furman, R. H.,** Non-hypocholesterolemic action of nicotinic acid in dogs, *Lancet,* 1, 1255, 1959.
563. **Norkus, E. P., and Rosso, P.,** Changes in ascorbic acid metabolism of the offspring following high maternal intake of this vitamin in the pregnant guinea pig. Second Conference on Vitamin C, *Ann. N.Y. Acad. Sci.,* 258, 401, 1975.
564. **Nunn, S. L., Tauxe, W. N., and Juergens, J. L.,** Effect of nicotinic acid on human cholesterol biosynthesis, *Circulation,* 24, 1099, 1961.
565. **Nurmio, P., Roine, K., Juokslahti, T., and Loman, A.,** A study of the effects of nicotinic acid in cattle, with special reference to ketosis therapy, *Nord. Veterinaermed.,* 26, 370, 1974.
566. **Nye, E. R. and Buchanan H.,** Short term effect of nicotinic acid on plasma level and turnover of free fatty acids in sheep and man, *J. Lipid Res.,* 10, 193, 1969.
567. **Obbens, E. A. M. T. and Hommes, O. R.,** The epileptogenic effects of folate derivatives in the rat, *J. Neurol. Sci.,* 20, 223, 1973.
568. **Oberndorfer, L.,** Reaccion toxica a la dosis terapeutica unica masiva de vitamina A, *Rev. Colomb. Pediatr.,* 23, 57, 1968.
569. **Odom, G. and McEachern, D.,** Subarachnoid injection of thiamine in cats; unmasking of brain lesions by induced thiamine deficiency, *Proc. Soc. Exp. Biol. Med.,* 50, 28, 1942.
570. **Oguro, Y.,** Mechanism on the effect of intraspinal injection of thiamine in treatment of nervous disease, *Vitamins,* 16, 196, 1959.
571. **Oliver, T. K.,** Chronic vitamin A intoxication: report of a case in an older child and review of the literature, *Am. J. Dis. Child.,* 95, 57, 1958.
572. **Oliver, T. K. and Havener, W. H.,** Eye manifestations of chronic vitamin A intoxication, *Arch. Ophthal.,* 60, 19, 1958.

573. **Olson, J. A.**, The prevention of childhood blindness by the administration of massive doses of vitamin A, *Isr. J. Med. Sci.*, 8, 1199, 1972.

574. **Ong, D. E., Page D. L., and Chytil, F.**, Retinoic acid binding protein: occurrence in human tumors, *Science*, 190, 60, 1975.

575. **O'Reilly, P. O., Demay, M., and Kotlowski, K.**, Cholesteremia and nicotinic acid, *Arch. Int. Med.*, 100, 797, 1957.

576. **Page, H. L., Jr., Vogel, J. H. K., Pyror, R., and Blount, S. G., Jr.**, Unusual observations in supravalvular aortic stenoisis, *Circulation*, 32, 11, 1965.

577. **Pardue, W. O.**, Severe liver dysfunction during nicotinic acid therapy, *JAMA*, 175, 137, 1961.

578. **Parent, G.**, Hypervitaminose A and oedema papillair, *Bull. Soc. Belge Ophtalmol.*, 152, 596, 1969.

579. **Parfitt, A. M.**, Vitamin D treatment in hypoparathyroidism, *Lancet*, 2, 614, 1970.

580. **Parsons, W. B.**, Reduction in serum cholesterol levels and other metabolic effects of large doses of nicotinic acid, *Circulation*, 20, 747, 1959.

581. **Parsons, W. B.**, Activation of peptic ulcer by nicotinic acid, *JAMA*, 173, 1466, 1960.

582. **Parsons, W. B.**, Treatment of hypercholesteremia by nicotinic acid, *Arch. Intern. Med.*, 107, 639, 1961.

583. **Parsons, W. B.**, Studies of nicotinic acid use in hypercholesterolemia, Arch. Intern. Med., 107, 653, 1961.

584. **Parsons, W. B.**, Reduction in hepatic synthesis of cholesterol from C^{14}-acetate in hypercholesterolemia patients by nicotinic acid, *Circulation*, 24, 1099, 1961.

585. **Parsons, W. B.**, Use of nicotinic acid to reduce serum lipid levels, *J. Am. Geriatr. Soc.*, 10, 850, 1962.

586. **Parsons, W. B.**, The effect of nicotinic acid on the liver: evidence favoring functional alteration of enzymatic reactions without hepatocellular damage, in *Niacin in Vascular Disorders and Hyerplipaemia*, Altschul, R., Ed., Charles C Thomas, Springfield, Ill., 1964, 263.

587. **Parsons, W. B., Achor, R. W. P., Berge, K. G., McKenzie, B. F., and Barker, N. W.**, The effect of large doses of nicotinic acid on the plasma and serum lipids of human beings with hypercholesterolemia, *Circulation*, 14, 495, 1956.

588. **Parsons, W. B. and Finn, J. H.**, Reduction in serum cholesterol levels by nicotinic acid (including studies pertaining to the mechanism of action), *Circulation*, 18, 489, 1958.

589. **Parsons, W. B. and Finn, J. H.**, Reduction of serum cholesterol levels by nicotinic acid, *Arch. Intern. Med.*, 103, 783, 1959.

590. **Patlan, B. D., Lebendinsky, R. I., and Petukh, M. I.**, Anaphylactic shock after injection of vitamin B_{12}, *Br. Med. J.*, 3, 102, 1968.

591. **Patterson, J. A.**, Diabetogenic effect of dehydroascorbic and dehydroisoascorbic acid, *J. Biol. Chem.*, 183, 81, 1950.

592. **Patterson, J. A.**, Course of diabetes and development of cataracts after injecting dehydroascorbic acid and related substances, *Am. J. Physiol.*, 165, 61, 1951.

593. **Paul, P. K.**, Effect of nutrient toxicities in animals and man: biotin, in *Handbook in Nutrition and Food*, Rechcigl, M., Jr., Ed., CRC Press, Boca Raton, Fla., 1978, 47.

594. **Paul, P. K., Duttagupta, P. N., and Agarwal, H. C.**, Effect of an acute dose of biotin on reproductive organs of the female rat, *Curr. Sci.*, 42, 206, 1973.

595. **Paul, P. K., Duttagupta, P. N., and Agarwal, H. C.**, Antifertility effect of biotin and its amelioration by estrogen in the female rat, *Curr. Sci.*, 42, 613, 1973.

596. **Paul, P. K. and Kuttagupta, P. N.**, The effect of an acute dose of biotin at the pre-implantation stage and its relation with female sex steroids in the rat, *J. Nutr. Sci. Vitaminol.*, 21, 89, 1975.

597. **Paul, P. K. and Duttagupta, P. N.**, The effect of an acute dose of biotin at the post-implantation stage and its relation with the female sex steroids in the rat, *J. Nutr. Sci. Vitaminol.*, 22, 181, 1976.

598. **Paunier, L., Conen, P. E., Gibson, A. A. M., and Fraser, D.**, Renal function and histology after long term vitamin D therapy of vitamin D refractory rickets, *J. Paediatr.*, 73, 833, 1968.

599. **Pease, C. N.**, Focal retardation and arrestment of growth of bones due to vitamin A intoxication, *JAMA*, 182, 980, 1962.

600. **Pereira, J. N.**, The plasma free fatty acid rebound induced by nicotinic acid, *J. Lipid Res.*, 8, 238, 1967.

601. **Pelner, L.**, Anaphylaxis to the injection of nicotinic acid (niacin); successful treatment with epinephrine, *Ann. Intern. Med.*, 26, 290, 1947.

602. **Pereda, J. M., Arnal, P., Cavaniltes, J. M., Gonzales, F. M., and Facal, J. L.**, Vitamin D poisoning with radiologically visible nephrocalcinosis, *Rev. Clin. Esp.*, 110, 61, 1968.

603. **Perlzweig, W. A., Rosen, F., and Pearson, P. B.**, Comparative studies in niacin metabolism: the fate of niacin in man, rat, dog, pig, rabbit, guinea pig, goat, sheep, and calf., *J. Nutr.*, 40, 453, 1950.

604. **Perov, M. L.**, Congenital supravalvular aortic stenosis, *Arch. Pathol.*, 71, 453, 1961.
605. **Persson, B., Tunell, R., and Ekengren, K.**, Chronic vitamin A intoxication during the first half year of life, *Acta Paediatr. Scand.*, 54, 49, 1965.
606. **Petropulos, S. F.**, The action of an antimetabolite of thiamine on single myelinated nerve fibers, *J. Cell. Comp. Physiol.*, 56, 7, 1960.
607. **Pfaffenschlager, F.**, Late effects of vitamin D_2 poinsoning, *Wien. Klin. Wochenschr.*, 76, 935, 1964.
608. **Pick, E. P. and Unna, K.**, Blockade of the nicotine action on the blood pressure by thiazole compounds sulfathiazole and thiamine, *J. Pharmacol. Exp. Ther.*, 87, 138, 1946.
609. **Pierce, W. F.**, Hypervitaminosis A., *Ill. Med. J.*, 122, 591, 1962.
610. **Pierides, A. M.**, Variable response to long-term 1α-hydroxycholecalciferol in haemodialysis osteodystrophy, *Lancet*, 1, 1092, 1976.
611. **Pillai, M. K. K. and Madhukar, B. V. R.**, Effects of biotin on the fertility of the yellow fever mosquito, *Aedes aegypti, Naturwissenschaften*, 4, 218, 1969.
612. **Pilotti, G. and Scorta, A.**, Hypervitaminosis A during pregnancy and neonatal malformations of the urinary tract, *Minerva Ginecol.*, 17, 1103, 1965.
613. **Pincus, J. H., Cooper, J. R., Itokawa, Y., and Gumbinas, M.**, Subacute necrotizing encephalomyelopathy, *Arch. Neurol.*, 24, 511, 1971.
614. **Pincus, J. H., Cooper, J. R., Murphy, J. V., Rabe, E. F., Lonsdale, D., and Dunn, H. G.**, Thiamine derivatives in subacute necrotizing encephalomyelopathy, *Pediatrics*, 51, 716, 1973.
615. **Pincus, J. H., Itokawa, Y., and Cooper, J. R.**, Enzyme-inhibiting factor in subacute necrotizing encephalomyelopathy, *Neurology*, 19, 841, 1969.
616. **Pinelli, A., Pozza, G., Formento, M. L., Favalli, L., and Coglio, G.**, Effect of vitamin E on urine porphyrin and steroid profiles in porphyria cutanea tarda; report of four cases, *Eur. J. Pharmacol.*, 5, 100, 1972.
617. **Pinto, J., Huang, Y. P., Chaudhuri, R., and Rivlin, R. S.**, Riboflavin excretion and turnover in an unusual case of multiple myeloma, *Clin. Res.*, 23, 426A, 1975.
618. **Pljaskova, L. M. O.**, Damage to the cardiovascular system by excess vitamin D_2 in infants, *Pediatrija*, 7, 39, 1966.
619. **Pljaskova, L. M. O.**, Anaemia of vitamin D excess in young children, *Pediatrija*, 12, 49, 1968.
620. **Plum, P., Dam, H., Dyggve, H., and Hjalmar Larsen, E.**, Administration of vitamin K antepartum. Prophlyaxis against haemorrhagic disease of the newborn, *Dan. Med. Bull.*, 1, 21, 1954.
621. **Pohle, F. J. and Stewart, J. K.**, Observations on the plasma prothrombin and the effects of vitamin K in patients with liver or biliary tract disease, *J. Clin. Invest.*, 19, 365, 1940.
622. **Pollack, H., Ellenberg, M., and Dolger, H.**, Excretion of thiamine and its degradation products in man, *Proc. Soc. Exp. Biol. Med.*, 47, 414, 1941.
623. **Pollitt, N. T.**, Large intravenous dosage of thiamine, *JAMA*, 203, 175, 1968.
624. **Post, M. and Smith, J. A.**, Bronchoconstrictor effect of thiamine, *Am. J. Physiol.*, 163, 742, 1950.
625. **Powers, B. R.**, Circulatory collapse following intravenous administration of nicotinic acid (niacin), *Ann. Intern. Med.*, 29, 558, 1948.
626. **Preuss, H. G.**, Effect of nutrient toxicities — excess in animals and man: folic acid, in *Handbook in Nutrition and Food*, Rechcigl, M., Jr., Ed., CRC Press, Boca Raton, Fla., 1978, 61.
627. **Preuss, H. G., Weiss, F. R., Janicki, R. H., and Goldin, H.**, Studies on the mechanism of folate induced growth in rat kidneys, *J. Pharmacol. Exp. Ther.*, 180, 754, 1972.
628. **Price, J. D. E. and Sookochoff, M. M.**, Arterial calcification after vitamin D, *Lancet*, 1, 416, 1969.
629. **Primbs, E. R., Sinnhuber, R. O., and Warren, C. E.**, Hypervitaminosis A in the rainbow trout: counteraction of vitamin C, *Int. J. Vitam. Nutr. Res.*, 41, 331, 1971.
630. **Harris, R. S. and Wool, I. G., Eds.**, Vitamins and Hormones, *Proc. Int. Symp. Vitamin B_6 in honor of Professor Paul Gyorgy*, Academic Press, New York, 1964.
631. **Puissant, A., Monfort, J., and Vanbremeersch, F.**, Les acnés provoquées par la vitamine B_{12}, *Gaz. Med. Fr.*, 76, 4535, 1969.
632. **Raaflaub, J.**, Zur pharmakokinetic von Rocinal retard, *Med. Pharmacol. Exp.*, 16, 393, 1967.
633. **Rahm. U.**, Besitzt Vitamin B_1 insektenabhaltende Eigenschaften?, *Schweiz. Med. Wochenschr.*, 26, 634, 1958.
634. **Ralston, A. J., Snaith, R. P., and Hinley, J. B.**, Effects of folic acid on fit frequency and behaviour in epileptics on anticonvulsants, *Lancet*, 1, 867, 1970.
635. **Reingold, I. M. and Webb, F. R.**, Sudden death following intravenous injection of thiamine hydrochloride, *JAMA*, 130, 491, 1946.
636. Report of Suspected Adverse Reactions No. 4, 1964—1976, Australian Drug Evaluation Committee, Australian Government Public Service, Canberra 1978.
637. **Restak, R. M.**, Pseudotumor cerebri psychosis, and hypervitaminosis A, *J. Nerv. Ment. Dis.*, 155, 7205, 1972.

638. **Rhead, W. J. and Schrauzer, G. N.**, Risks of long term ascorbic acid overdosage, *Nutr. Rev.*, 11, 262, 1971.
639. **Ricci, C. and Pallini, V.**, Occurrence of free nicotinic acid in the liver of nicotinamide-injected rats, *Biochem. Biophys. Res. Commun.*, 17, 34, 1964.
640. **Richards, M. B.**, Imbalance of vitamin B factors. Pyridoxine deficiency caused by additions of aneurin and chalk, *Br. Med. J.*, 1, 4395, 1945.
641. **Richards, R. K. and Shapiro, S.**, Experimental and clinical studies on the action of high doses of hykinone and other menadione derivatives, *J. Pharmacol. Exp. Ther.*, 84, 93, 1945.
642. **Rivin, A. V.**, Jaundice occurring during nicotinic acid therapy for hypercholesterolemia, *JAMA*, 170, 2088, 1959.
643. **Rivin, A. V.**, Hypercholesterolemia — use of niacin and niacin combinations in therapy, *Calif. Med.*, 96, 267, 1962.
644. **Rivlin, R. S.**, Effect of nutrient toxicities (excess) in animals and man: riboflavin, in *Handbook in Nutrition and Food*, Rechcigl, M., Jr., Ed., CRC Press, Boca Raton, Fla., 1978, 25.
645. **Rivlin, R. S., Huang, Y. P., Pinto, J., and McConnell, R. J.**, Increased excretion of riboflavin due to boric acid intoxication, *Clin. Res.*, 24, 424A, 1976.
646. **Rivlin, R. S., Menedez, C. E., and Langdon, R. G.**, Biochemical similarities between hypothyroidism and riboflavin deficiency, *Endocrinology*, 83, 461, 1968.
647. **Robens, J. F.**, Teratogenic effects of hypervitaminosis A in the hamster and the guinea pig, *Toxicol. Appl. Pharmacol.*, 16, 88, 1976.
648. **Robie, T. R.**, The administration of vitamin B_1 by intraspinal injection, *Am. J. Surg.*, 48, 398, 1940.
649. **Rodahl, K.**, Hypervitaminosis A in the rat, *J.Nutr.*, 41, 399, 1950.
650. **Rodahl, K. and Moore, T.**, The vitamin A content and toxicity of bear and seal liver, *Biochem. J.*, 37, 166, 1943.
651. **Rodriguez, J. A., Robinson, C. A., Smith, M. S., and Frye, J. H.**, Evaluation of an automated glucose-oxidase procedure, *Clin. Chem.*, 21, 1513, 1975.
652. **Rodriguez, M. S. and Irwin, M. I.**, Hypervitaminosis A, *J. Nutr.*, 102, 919, 1972.
653. **Roe, D. A.**, *Drug Induced Nutritional Deficiencies*, AVI Press, Westport, Conn., 1976.
654. **Roe, D. A., McCormick, D. B., and Lin, R. T.**, Effects of riboflavin on boric acid toxicity, *J. Pharm. Sci.*, 61, 1081, 1972.
655. **Rosenberg, D., Roux, J. A., Frederich, A., Philippe, N., Peytel, J., and Monnet, P.**, Facial paralysis as a result of vitamin D poisoning, *Pediatr.*, 23, 565, 1968.
656. **Rosenstreich, S. J., Rich, C., and Volwider, W.** Deposition and release of vitamin D_3 from body fat: evidence for a storage site in the rat, *J. Clin. Invest.*, 50, 679, 1971.
657. **Rosenthal, G.**, Interaction of ascorbic acid and warfarin, *JAMA*, 215, 1671, 1971.
658. **Rubin, E., Florman, A. L., Degnan, T., and Diaz, J.**, Hepatic injury in chronic hypervitaminosis A, *Am. J. Dis. Child.*, 119, 132, 1970.
659. **Ruby, L. K. and Mital, M. A.**, Skeletal deformities following chronic hypervitaminosis A, *J. Bone Jt. Surg. Am.*, 56A, 1283, 1974.
660. **Ruiter, M. and Meyler, L.**, Skin changes after therapeutic administration of nicotinic acid in large doses, *Dermatologica*, 120, 139, 1960.
661. **Russell, R. G. G., Smith, R., Walton, R. J., Preston, C., Basson, R., Henderson, R. G., and Norman, A. W.**, 1,25-Dihydroxycholecalciferol and 1α-hydroxycholecalciferol in hypoparathyroidism, *Lancet*, 2, 14, 1974.
662. **Russell, R. G. G., Smith, R., Preston, C., Walton, R. J., Woods, C. G., Henderson, R. G., and Norman, A. W.**, The effect of 1,25-dihydroxycholecalciferol on renal tubular reabsorption of phosphate, intestinal absorption of calcium and bone histology in hypophosphataemic renal tubular rickets, *Clin. Sci. Mol. Med.*, 48, 177, 1975.
663. **Russell, R. M., Boyer, J. L., Bagheri, S. A., and Hruban, Z.**, Hepatic injury from chronic hypervitaminosis A resulting in portal hypertension and ascites, *N. Engl. J. Med.*, 435, 1974.
664. **Russell, R. M., Smith, V. C., Multack, R., Krill, A. E., and Rosenberg, I. H.**, Dark adaptation testing for diagnosis of subclinical vitamin A deficiency and evaluations of therapy, *Lancet*, 2, 1161, 1973.
665. **Saller, S. and Bolzano, K.**, The action of nicotinic acid on the esterification rate of plasma free fatty acids to plasma triglycerides, in *Metabolic Effects of Nicotinic Acid and Its Derivatives*, Gey, K. F. and Carlson, L. A., Eds., Hans Huber, Bern, 1971, 479.
666. **Samborskaia, E. P.**, Effect of large doses of ascorbic acid on course of pregnancy in the guinea pig, *Bull. Exp. Biol. Med. (Moskva)*, 57, 105, 1966.
667. **Sasa, M., Takemoto, I., Nishino, K., and Itokawa, Y.**, The role of thiamine on excitable membrane of crayfish giant axon, *J. Nur. Sci. Vitaminol.*, Suppl. 22, 21, 1977.
668. **Schacter, D., Kowarski, S., and Finkelstein, J. D.**, Vitamin D_3: direct action on the small intestine of the rat, *Science*, 143, 1964.

669. **Schiff, L.**, Collapse following parenteral administration of solution of thiamine hydrochloride, *JAMA*, 117, 609, 1941.

670. **Schiff, L.**, Discussion on vitamin therapy, *JAMA*, 117, 1501, 1941.

671. **Schmidt, U.and Dubach, U. C.**, Acute renal failure in the folate-treated rat: early metabolic changes in various structures of the nephron, *Kidney Int.*, 10, S39, 1976.

672. **Schrauzer, G. N., Ishmael, D., and Kiefer, G. W.**, Some aspects of current vitamin C usage: diminished high altitude resistance following overdosage, *Ann. N.Y. Acad. Sci.*, 258, 377, 1975.

673. **Schrauzer, G. N. and Rhead, W. J.**, Ascorbic acid abuse: effects of long-term ingestion of excessive amounts on blood levels and urinary excretion, *Int. J. Vit. Nutr. Res.*, 43, 201, 1973.

674. **Schroll, M., Petersen, B., and Christiansen, C.**, Is hypocalcaemia protective against hyperlipidaemia?, *Br. Med. J.*, 3, 226, 1975.

675. **Schubert, G. E.**, Folic acid-induced acute renal failure in the rat: morphological studies, *Kidney Int.*, 10, S46, 1976.

676. **Schumacher, M. F., Williams, M. A., and Lyman, R. L.**, Effect of high intakes of thiamine, riboflavin and pyridoxine on reproduction in rats and vitamin requirements of offspring, *J. Nutr.*, 86, 343, 1965.

677. **Schwieter, U., Tamm. R., Weiser, H., and Wiss, O.**, Synthesis and vitamin E activity of tocopheramines and their N-alkyl derivatives, *Helv. Chim. Acta*, 49, 2297, 1966.

678. **Scipicyna, L. P.**, Effect of toxic doses of vitamin D_2 on the dissociation of oxidative phosphorylation, *Bjull. Eksp. Biol. Med.*, 9, 60, 1966.

679. **Seawright, A. A., English, P. B., and Gartner, R. J. W.**, Hypervitaminosis A and deforming cervical spondylosis of the cat, *J. Comp. Pathol.*, 77, 29, 1967.

680. **Seawright, A. A., English, P. B., and Gartner, R. J. W.**, Hypervitaminosis A of the cat, *Adv. Vet. Sci.*, 14, 1, 1970.

681. **Sebrell, W. H. and Harris, R. S.**, Eds., Tocopherols, in *The Vitamins: Chemistry, Physiology, Pathology*, Vol. 5, Academic Press, New York, 1972, chap. 16.

682. **Seelig, M. S.**, Vitamin D and cardiovascular, renal and brain damage in infancy and childhood, *Ann. N.Y. Acad. Sci.*, 147, 537, 1969.

683. **Seelig, M. S. and Heggtveit, H. A.**, Magnesium interrelationships in ischemic heart disease: a review, *Am. J. Clin. Nutr.*, 27, 59, 1974.

684. **Sehgal, S. S., Agarwal, H. C., and Pillai, M. K. K.**, Sterilizing effect of a dietary surplus of biotin in *Trogoderma granarium everts, Curr. Sci.*, 39, 551, 1970.

685. **Serge, A.**, Cocarboxylase. I., Mitteilung, *Arzneim. Forsch.*, 9, 1, 1959.

686. **Serge, A.**, Cocarboxylase. II. Mitteilung, *Arzneim. Forsch.*, 9, 102, 1959.

687. **Seusing, J.**, Allergisches verhalten gegen Vitamin B_1 (aneurin, thiamin), *Klin. Wochenschr.*, 29, 394, 1951.

688. **Shapero, W. and Gwinner, M. W.**, Sensitivity to thiamine hydrochloride, *Ann. Allergy*, 5, 349, 1947.

689. **Shaw, E. W. and Niccoli, J. Z.**, Hypervitaminosis A: report of a case in an adult male, *Ann. Intern. Med.*, 39, 131, 1953.

690. **Sheehy, T.**, How much folic acid is safe in pernicious anaemia?, *Am. J. Clin. Nutr.*, 9, 708, 1961.

691. **Shen, F., Baylink, D., Nielsen, R., Sherrard, D., and Haussler, M.**, A study of the pathogenesis of idiopathic hypercalciuria, *Clin. Res.*, 24, 157a, 1976.

692. **Shenefelt, R. E.**, Animal model: treatment of various species with a large dose of vitamin A at known stages in pregnancy, *Am. J. Pathol.*, 66, 589, 1972.

693. **Shetty, K. R., Ajlouni, K., Rosenfeld, P. S., and Hagen, T. C.**, Protracted vitamin D intoxication, *Arch. Int. Med.*, 135, 986, 1975.

694. **Shimkin, M. B.**, Toxicity of naphthoquinones with vitamin K activity in mice, *J. Pharmacol. Exp. Ther.*, 71, 210, 1941.

695. **Shinagawa, T.**, Influence of the over-dose of thiamine and thiamine allyldisulfide on the urinary excretion of riboflavin and niacin, *Vitamins*, 11, 175, 1956.

696. **Shinagawa, T.**, Effect of the over-dose of diphosphothiamine on the urinary excretion of B-vitamins, and the effect of over-dose of thiamine on the urinary excretion of niacin and methylnicotinamine, *Vitamins*, 12, 243, 1957.

697. **Shintani, S.**, On the relationship between hydroxymethylpyrimidine and pyridoxine. II. On the agumentation of convulsant effect of hydroxymethylpyrimidine by pyridoxine-deficient synthetic diet in mice, *J. Pharm. Soc. Jpn.*, 76, 15, 1956.

698. **Shintani, S.**, On the relationship between hydroxymethylpyrimidine and pyridoxine. III. The antipyridoxine activity of hydroxymethylpyrimidine in rats, *J. Pharm. Soc. Jpn.*, 76, 18, 1956.

699. **Shintani, S.**, Studies on the change of hydroxymethylpyrimidine in vivo. I. Determination of hydroxymethylpyrimidine in urine and its excretion in rabbit, *J. Pharm. Soc. Jpn.*, 77, 781, 1957.

700. **Shintani, S.**, On the relationship between hydroxymethylpyrimidine and pyridoxine. I. Inhibitory effect of pyridoxine on the convulsion by hydroxymethylpyrimidine in mice, *J. Pharm. Soc. Jpn.*, 76, 13, 1956.
701. **Shintani, S.**, On the relationships between the analogs of hydroxymethylpyrimidine and vitamin B_6 II. Accelerating effect of the analogs of hydroxymethylpyrimidine upon the development of vitamin B_6 deficiency syndrome in rat, *J. Pharm. Soc. Jpn.*, 77, 993, 1957.
702. **Shioda, K., Yamada, K., Ohgai, Y., Matsumoto, Y., Yamamoto, Y., and Maeda, Y.**, Anaphylactic shock following injection of thiamine, *Nippon Rinsho*, 18, 543, 1960.
703. **Shrifter, H. and Steigmann, F.**, The effect of large doses of synthetic vitamin K and K_1 on the prothrombin time of patients with liver disease, *J. Lab. Clin. Med.*, 44, 930, 1954.
704. **Shrifter, H. and Steigmann, F.**, Use of large doses of vitamin K in liver disease, *J. Lab. Clin. Med.*, 46, 951, 1955.
705. **Shute, E.**, Wheat-germ oil therapy. I. Dosage idiosyncrasy, *Am. J. Obstet. Gynecol.*, 35, 249, 1938.
706. **Siegel, N. J. and Spackman, T. J.**, Chronic hypervitaminosis A with intracranial hypertension and low cerebrospiral fluid concentration of protein, *Clin. Pediatr.*, 2, 580, 1972.
707. **Silver, J., Shvil, Y., and Fainaru, M.**, Vitamin D transport in an infant with vitamin D toxicity, *Br. Med. J.*, 93, 1978.
708. **Singh, H. P., Herbert, M. A., and Gault, M. H.**, Effect of some drugs on clinical laboratory values as determined by the Technicon SMA-12/60, *Clin. Chem.*, 18, 137, 1972.
709. **Singh, V. N., Singh, M., and Venkitasubramanian, T. A.**, Early effects of feeding excess vitamin A. Mechanism of fatty liver production in rats, *J. Lipid Res.*, 10, 395, 1969.
710. **Sise, H. S.**, Warfarin: a new coumarin anticoagulant, *Practitioner*, 181, 98, 1958.
711. **Smith, A. M. and Custer, P. P.**, Toxicity of vitamin K, *JAMA*, 173, 502, 1960.
712. **Smith, D. W., Blizzard, R. M., and Harrison, H. E.**, Idiopathic hypercalcemia, *Pediatr.*, 24, 258, 1959.
713. **Smith, E. C.**, Interaction of ascorbic acid and warfarin, *JAMA*, 221, 1166, 1972.
714. **Smith, F. R. and Goodman, D. S.**, Vitamin A transport in human vitamin A toxicity, *N. Engl. J. Med.*, 294, 805, 1976.
715. **Smith, J. H.**, Hypervitaminosis A: report of a case, *Oral Surg.*, 17, 305, 1964.
716. **Smith, J. J., Ivy, A. C., and Foster, R. H. K.**, The pharmacology of two water soluble vitamin K-like substances, *J. Lab. Clin. Med.*, 28, 1667, 1943.
717. **Smith, J. A., Foa, P. P., and Weinstein, H. R.**, Some toxic effects of thiamine, *Fed. Proc. Am. Soc. Exp. Biol.*, 6, 204, 1947.
718. **Smith, J. A., Foa, P. P., and Weinstein, H. R.**, The curare-like action of thiamine, *Science*, 108, 412, 1948.
719. **Smith, J. A., Foa, P. P., Weinstein, H. R., Ludwig, A. S., and Wertheim, J. M.**, Some aspects of thiamine toxicity, *J. Pharmacol. Exp. Ther.*, 93, 294, 1948.
720. **Smith, J. A., Foa, P. P., and Weinstein, H. R.**, Observations on the curare-like action of thiamine, *Am. J. Physiol.*, 155, 469, 1948.
721. **Smith, J. A. and Sohn, H.**, Dual role of thiamine in producing bronchoconstriction, *Fed. Proc. Fed. Am. Soc. Exp. Biol.*, 10, 128, 1951.
722. **Snyderman, S. E. and Holt, L. E.**, Nutrition in infancy, in *Modern Nutrition in Health and Disease*, Goodhart, R. S. and Shils, M. E., Eds., Lea & Febiger, Philadelphia, 1973, 650.
723. **Solder-Bechara, J. and Soscia, J. L.**, Chronic hypervitaminosis A: report of a case in an adult, *Arch. Intern. Med.*, 112, 462, 1963.
724. **Soliman, M. K**, Vitamin A-uberdosie-rung: Mogliche teratogene wirkungen, *Int. J. Vitam. Nutr. Res.*, 42, 389, 1977.
725. **Soliman, M. K.**, Vitamin A overdosing. II. Cytological and biochemical changes in blood of rats treated with high doses of vitamin A and alpha-tocopherol, *Int. J. Vitam. Nutr. Res.*, 42, 576, 1972.
726. **Soloshenko, E. N. and Brailorski, A.**, Allergic skin reactions caused by group B vitamins, *Sov. Med.*, 10, 141, 1975.
727. **Sommer, H.**, Nicotinic acid levels in the blood and fibrinolysis under the influence of the hexa nicotinic acid ester of mesoinositol., *Arzneim. Forsch.*, 15, 1337, 1964.
728. **Spaulding, W. B. and Yendt, E. R.**, Prolonged vitamin D intoxication in a patient with hypoparathyroidism, *Can. Med. Assoc. J.*, 90, 1049, 1964.
729. **Spiegel, H. E. and Pinili, E.**, Effects of vitamin C on SGOT, SOPT, LDH and bilirubin, *Med. J. Aust.*, 2, 117, 1974.
730. **Spirichev, W. B. and Blazheievich, N. V.**, Mechanism of toxicity of vitamin D, *Int. Z. Vitaminforsch.*, 39, 30, 1969.
731. **Srikantia, S. G. and Reddy, V.**, Effect of a single massive dose of vitamin A on serum and liver levels of the vitamin, *Am. J. Clin. Nutr.*, 23, 114, 1970.

732. **Stamp, T. C. B., Haddad, J. G., and Twigg, C. A.,** Comparision of oral 25-hydroxycholecalciferol, vitamin D and ultraviolet light as determinants of circulating 25-hydroxyvitamin D, *Lancet,* 1, 1341, 1977.
733. **Stanbury, S. W.,** Bone disease in uraemia, *Am. J. Med.,* 44, 712, 1968.
734. **Stanbury, S. W.,** In round table discussion on renal osteodystrophy, *Arch. Int. Med.,* 124, 674, 1969.
735. **Steigmann, F., Firestein, R., and De La Huerga, J.,** Intravenous choline therapy, *Pharmacol. Exp. Ther.,* 11, 393, 1952.
736. **Stein, H. B., Hassan, A., and Fox, I. H.,** Ascorbic acid-induced uricosuria: a consequence of megavitamin therapy, *Ann. Intern. Med.,* 84, 385, 1976.
737. **Stein, W. and Morgenstern, M.,** Sensitisation to thiamine hydrochloride: report of another case, *Ann. Intern. Med.,* 20, 826, 1944.
738. **Stell, P. M. and McLoughlin, M. P.,** Vitamin A and chronic laryngitis, *Lancet,* 1, 147, 1972.
739. **Stenflo, J.,** Vitamin K, prothrombin and γ-carboxyglutamic acid, *N. Engl. J. Med.,* 296, 624, 1977.
740. **Stern, E. L.,** The intraspinal (subarachnoid) injection of vitamin B_1 for the relief of intractable pain, and for inflammatory and degenerative diseases of the central nervous system, *Am. J. Surg.,* 45, 495, 1938.
741. **Stich, H. F., Karim, J., Koropatnick, J., and Lo, I.,** Mutagenic action of ascorbic acid, *Nature (London),* 260, 722, 1976.
742. **Stickler, G. B., Beabout, J. W., and Riggs, B. L.,** Vitamin D resistant rickes: clinical experience with 41 typical familial hypophosphatemic patients and two atypical nonfamilial cases, *Mayo Clin. Proc.,* 45, 197, 1970.
743. **Stiles, M. H.,** Hypersensitivity to thiamine chloride, with a note on sensitivity to pyridoxine hydrochloride, *J. Allergy,* 12, 507, 1941.
744. **Stimson, W. H.,** Vitamin A intoxication in adults, *N. Engl. J. Med.,* 265, 369, 1961.
745. **Sulzberger, M. B. and Lazar, M. P.,** Hypervitaminosis A: report of a case in an adult, *JAMA,* 146, 788, 1951.
746. **Sunaga, I., Tadokoro, S., and Takeuchi, S.,** Studies on prolonged administration of vitamin K_3 (menadione), *Gunma J. Med. Sci.,* 8, 357, 1959.
747. **Svedmyr, N., Harthon, L., and Lundholm, L.,** Dose response relationship between concentration of free nicotinic acid concentration of plasma and some metabolic and circulatory effects after administration of nicotinic acid and pentoery thritol tetranicotinate in man, in *Metabolic Effects of Nicotinic Acid and Its Derivatives,* Gey, K. F. and Carlson, L. A., Eds., Hans Huber, Bern, 1971, 1085.
748. **Swoboda, W. and König, I.,** Gefahren der vitamin-D-prophylaxe (intoxikation), *Pediatr. Prax.,* 13, 351, 1973/1974.
749. **Takada, R. and Katsura, E.,** *Japanese Literature on Vitamins,* The Vitamin Society of Japan, Kyoto, 1962, 1385.
750. **Tamori, Y.,** Studies on the anti-nicotine effect of thiamine, *Folia Pharmacol. Jpn.,* 54, 571, 1958.
751. **Taussig, H. B.,** Possible injury to the cardiovascular system from vitamin D, *Ann. Int. Med.,* 65, 1195, 1966.
752. **Taylor, T. G., Morris, K. M. L., and Kirkley, J.,** Effects of dietary excesses of vitamins A and D on some constituents of the blood of chicks, *Br. J. Nutr.,* 22, 713, 1968.
753. **Taylor, W. E. and McKibbin, J. M.,** Effect of choline on yeast bioassay of inositol., *Proc. Soc. Exp. Biol. Med.,* 79, 95, 1952.
754. **Telford, I. R.,** The effects of hypo- and hyper-vitaminosis E on lung tumor growth in mice, *Ann. N.Y. Acad. Sci.,* 52, 132, 1949.
755. **Teo, S. T., Newth, J., and Pascoe, B. J.,** Chronic vitamin A intoxication, *Med. J. Aust.,* 2, 324, 1973.
756. **Terroine, T.,** Physiology and biochemistry of biotin, in *Vitamins and Hormones,* Vol. 18, Harris, R. S. and Ingle, D., J., Eds., Academic Press, New York, 1960, 1.
757. **Tetrault, A. F. and Beck, I. A.,** Anaphylactic shock following intramuscular thiamine chloride, *Ann. Intern. Med.,* 45, 134, 1956.
758. **Texier, L., Gauthier, Y., Gauthier, O., Surleve Bazeille, J. E., and Boineau, D.,** Sclerodermies lombo-fessières consécutives à des injections intramusculaires de vitamine K_1, *Bordeaux Med.,* 7, 1571, 1974.
759. **Texier, L., Gendre, P., and Gauthier, O.,** Localised dermatosclerosis of the buttock due to intramuscular injection of vitamin K, *Ann. Dermatol. Syph.,* 99, 363, 1972.
760. **Thompson, J. N.,** Vitamin A in development of the embryo, *Am. J. Clin. Nutr.,* 22, 1063, 1969.
761. **Thompson, J. N. and Pitt, G. A. J.,** Vitamin A acid and hypervitaminosis A, *Nature, (London),* 19, 672, 1960.
762. **Thornton, P. A. and Omdahl, J. L.,** Further evidence of skeletal response to exogenous ascorbic acid, *Proc. Soc. Exp. Biol. Med.,* 132, 618, 1969.

763. **Threlfall, G., Taylor, D., and Buck, A. T.,** The effect of folic acid on growth and deoxyribonucleic acid synthesis in the rat kidney, *Lab. Invest.*, 15, 1477, 1966.
764. **Tilson, M. O.,** A dissimilar effect of folic acid upon growth of the rat kidney and small bowel, *Proc. Soc. Exp. Biol. Med.*, 134, 95, 1970.
765. **Toone, W. M.,** Effects of vitamin E: good and bad, *N. Engl. J. Med.*, 289, 979, 1973.
766. **Torda, C. and Wolff, H. G.,** Effect of vitamin B_1 and cocarboxylase on synthesis of acetylcholine, *Proc. Soc. Exp. Biol. Med.*, 56, 88, 1944.
767. **Traina, V.,** Toxicity studies on vitamin B_{12} in albino mice, *Arch. Pathol.*, 49, 278, 1950.
768. **Tromovitch, T. A., Jacobs, P. H., and Kern, S.,** Acarthosis nigricans-like lesions from nicotinic acid, *Arch. Dermatol.*, 89, 222, 1964.
769. **Tuchweber, B., Garg, B. D., and Salas, M.,** Microsomal enzyme inducers and hypervitaminosis A in rats, *Arch. Pathol. Lab. Med.*, 100, 100, 1976.
770. **Turner, W. M. L.,** Acute pancreatitis after vitamin D, *Lancet*, 1, 1423, 1966.
771. **Turtz, C. A. and Turtz, A. I.,** Vitamin A intoxication, *Am. J. Ophthal.*, 50, 165, 1960.
772. **Unger, P. N. and Shapiro, S.,** The prothrombin response to parenteral administration of large doses of vitamin K in subjects with normal liver function and in cases of liver disease: a standardized test for the estimation of hepatic function, *J. Clin. Invest.*, 27, 39, 1948.
773. **Unger, P. N., Weiner, M., and Shapiro, S.,** the vitamin K tolerance test, *Am. J. Clin. Pathol.*, 18, 835, 1948.
774. **Ungley, C. C.,** The chemotherapeutic action of vitamin B_{12}, *Vitam. Horm. N.Y.*, 13, 137, 1955.
775. **Unna, K.,** Studies on the toxicity and pharmacology of nicotinic acid, *J. Pharmacol., Exp. Ther.*, 65, 95, 1939.
776. **Unna, K. and Antopol, W.,** Toxicity of vitamin B_6, *Proc. Soc. Exp. Biol. Med.*, 43, 116, 1940.
777. **Unna, K.,** Studies on the toxicity and pharmacology of vitamin B_6 (2-methyl-3-hydroxy-4, 5-bis (hydroxymethyl)-pyridine), *J. Pharmacol. Exp. Ther.*, 70, 400, 1940.
778. **Unna, K. and Clark, J. D.,** Effect of large amounts of single vitamins of the B group upon rats deficient in other vitamins, *Am. J. Med., Sci.*, 204, 364, 1942.
779. **Unna, K. and Greslin, J. G.,** Studies on the toxicity and pharmacology of riboflavin, *J. Pharmacol. Exp. Ther.*, 75, 80, 1942.
780. **Vedrova, I. N.,** Vitamin A in the treatment of Palmar and Plantar hyperkeratosis, *Abstr. Wildl. Med.*, 41, 388, 1967.
781. **Van der Kolk, W. F. J.,** Hypervitaminose A: iatrogene oorzaak van pseudotumor-cerebri-syndroom met passagere nierfunctiestoornis bij een zesjarig kind, *Maandschr. Kindergeneeskd.*, 41, 81, 1973.
782. **Ventura, U., Ceriani, I., Zelaschi, F., and Rindi, G.,** Action potential modifications in rat myocardial cells induced by hypervitaminosis A, *Q. J. Exp. Physiol.*, 56, 147, 1971.
783. **Vest, M.,** Vitamin K in medical practice: paediatrics, *Vitam. Horm. N.Y.*, 24, 649, 1966.
784. **Vickery, R. E.,** Unusual complication of excessive ingestion of vitamin C tablets, *Int. Surg.*, 58, 422, 1973.
785. **Vietti, T. J., Murphy, T. P., James, J. A., and Pritchard, J. A.,** Observations on the prophylactic use of vitamin K in the newborn infant, *J. Pediatr.*, 56, 343, 1960.
786. **Vitter, C. F., Vitter, R. W., and Spres, T. D.,** Treatment of pernicious anaemia and related anaemias with synthetic folic acid, *J. Lab. Clin. Med.*, 32, 262, 1947.
787. **Vogelsang, A. B., Shute, E. V., and Shute, W. E.,** Vitamin E in heart disease, *Med. Rec.*, 160, 279, 1947.
788. **von Frohberg, H., Gleich, J., and Kieser, H.,** Reproduktionstoxikologische studien mit ascorbinsaure, au mausen und ratten, *Arzneim. Forsch.*, 23, 1081, 1973.
789. **von Kunz, H. A.,** Uber die Wirkung von antimetaboliten des aneurins auf die einzelne markhaltige nervenfaser, *Helv. Physiol. Pharmacol. Acta*, 14, 411, 1956.
790. **von Muralt, A.,** The role of thiamine (vitamin B_1) in nerve excitation, *Exp. Cell Res. Suppl.*, 5, 72, 1958.
791. **Vorhees, C. V.,** Some behavioral effects of maternal hypervitaminosis A in rats, *Teratology*, 10, 269, 1974.
792. **Wagner, A. F. and Folkers, K.,** *Vitamins and Coenzymes*, Interscience, New York, 1964, 138.
793. **Warkany, J., Guest, G. M., and Grahill, F. J.,** Vitamin D in human serum during and after periods of ingestion of large amounts of vitamin D, *J. Lab. Clin. Med.*, 27, 557, 1941.
794. **Warkany, J. and Mahon, H. E.,** Estimation of vitamin D in blood serum, *Am. J. Dis. Child.*, 60, 606, 1940.
795. **Warkany, J. and Schraffenberger, E.,** Congenital malformations induced in rats by maternal nutrition deficiency, *J. Nutr.*, 27, 477, 1944.
796. **Warren, A. G.,** Letter to editor, *Lepr. Rev.*, 44, 220, 1973.
797. **Waterman, R. A.,** Nutrient toxicities in animals and man: niacin, in *Handbook in Nutrition and Food*, Rechcigl, M., Jr., Ed., CRC Press, Boca Raton, Fla., 1978, 29.

798. **Waterman, R. and Schultz, L. H.,** Carbon^{14}-labeled palmitic acid metabolism in fasted, lactating goats following nicotinic acid administration, *J. Dairy Sci.*, 56, 1569, 1973.
799. **Waterman, R. and Schultz, L. H.,** Nicotinic acid loading of normal cows: effects of blood metabolites and excretory forms, *J. Dairy Sci.*, 55, 1511, 1972.
800. **Waterman, R., Schwalm, J. W., and Schultz, L. H.,** Nicotinic acid treatment of bovine ketosis. I. Effects on circulatory metabolites and interrelationships, *J. Dairy Sci.*, 55, 1447, 1972.
801. **Waters, W. J., Dunham, R., and Bowen, W. R.,** Inhibition of bilirubin conjugation in vitro, *Proc. Soc. Exp. Biol. Med.*, 99, 175, 1958.
802. **Wehinger, H.,** Spatrachitis nach vitamin-D-uberempfindlichkeit bei adiponekrosis subcutanea in der neugeborenenperiode, *Kinderheilk*, 107, 42, 1969.
803. **Weigand, C. G.,** Reactions attributed to administration of thiamin chloride, *Geriatrics*, 5, 574, 1950.
804. **Weigand, C., Eckler, C., and Chen, K. K.,** Action and toxicity of vitamin B_6 hydrochloride, *Proc. Soc. Exp. Biol. Med.*, 44, 147, 1940.
805. **Weiland, R. G., Hendricks, F. H., Leon, F. A., Gutierrez, L., and Jones, J. C.,** Hypervitaminosis with hypercalcaemia, *Lancet*, 1, 698, 1971.
806. **Wendel, O. W.,** A study of urinary lactic acid levels in humans. I. Influence of thiamine and pyrithiamine, *J. Vitaminol.*, 6, 16, 1960.
807. **Weissman, G., Uhr, J. W., and Thomas, L.,** Acute hypervitaminosis A in guinea pigs. I. Effects on acid hydrolases, *Proc. Soc. Exp. Biol. Med.*, 112, 284, 1963.
808. **Wild, J., Schorah, C. J., and Smithells, R. W.,** Vitamin A, pregnancy and oral contraceptives, *Br. Med. J.*, 1, 57, 1974.
809. **Williams, J. C., Barratt-Boyes, B. G., and Lowe, J. B.,** Supravalvular aortic stenosis, *Circulation*, 1311, 1961.
810. **Williamson, D. H., Mayor, F., Veloso, D., and Page, M. A.,** Effects of nicotinic acid and related compounds on ketone body metabolism, in *Metabolic Effects of Nicotinic Acid and Its Derivatives*, Gey, K. F., and Carlson, L. A., Eds., Hans Huber, Bern, 1971, 227.
811. **Winberg, J. and Zetterström, R.,** Cortisone treatment of vitamin D intoxication, *Acta Paediatr.*, 45, 96, 1956.
812. **Winter, S. L. and Boyer, J. L.,** Hepatic toxicity from large doses of vitamin B_3 (nicotinamide), *N. Engl. J. Med.*, 289, 1180, 1973.
813. **Winterborn, M. H., Mace, P. J., Heath, D. A. and White, R. H. R.,** Impairment of renal function in patients on 1α-hydroxycholecalciferol, *Lancet*, 2, 150, 1978.
814. **Wisse Smit, J.,** Vitamine-A-intoxicatie bij volwassenen, *Ned. Tijdschr. Geneeskd.*, 110, 10, 1966.
815. **Witte, F. C.,** Antihyperlipidaemic agents, in *Progress in Medicinal Chemistry*, Vol. 11, Ellis, G. P. and West, G. B., Eds., North-Holland, Amsterdam, 1975, 119.
816. **Wittenborne, J. R., Weber, E. S. P., and Braun, M.,** Niacin in the long term treatment of schizophrenia, *Arch. Gen. Psychiatry*, 28, 308—15, 1973.
817. **Wolbach, S. B. and Hegsted, D. M.,** Hypervitaminosis A in young ducks, *Arch. Pathol.*, 55, 47, 1953.
818. **Wolfson, S. K., Jr. and Ellis, S.,** Thiamine toxicity and ganglionic blockage, *Fed. Proc. Fed. Am. Soc. Exp. Biol.*, 13, 418, 1954.
819. **Wolke, R. E., Nielsen, S. W., and Rousseau, J. E., Jr.,** Bone lesions of hypervitaminosis A in the pig, *Am. J. Vet. Res.*, 29, 1009, 1968.
820. **Wollenweber, J., Kottke, B. A., and Owen, C. A.,** The effect of nicotinic acid on pool size and turnover of taurocholic acid in normal and hypothyroid dogs, *Proc. Soc. Exp. Biol. Med.*, 122, 1070, 1966.
821. **Woodard, W. K., Miller, L. J., and Legart, O.,** Acute and chronic hypervitaminosis A in a 4-month old infant, *J. Pediatr.*, 59, 260, 1961.
822. **Woodliff, H. J. and Davis, R. E.,** Allergy to folic acid, *Med. J. Aust.*, 351, 1966.
823. Vitamin A Deficiency and Xerophthalmia, Report of Joint W.H.O./U.S.A.I.D. Meeting, Tech. Rep. Ser. No. 590, World Health Organization, Geneva, 1976.
824. **Wynn, R.,** Relationship of menadiol tetrasodium diphosphate (Synkavite) to bilirubinaemia and haemolysis in the adult and newborn rat, *Am. J. Obstet. Gynecol.*, 86, 495, 1963.
825. **Yaffe, S. J. and Filer, L. J.,** The use and abuse of vitamin A, *Paediatrics*, 48, 655, 1971.
826. **Yamamoto, I.,** Thiamine as a nicotine antagonist, *Jpn. J. Pharmacol.*, 13, 240, 1963.
827. **Yamamoto, I., Iwata, H., Tamori, Y., and Hirayama, M.,** On the nicotine antagonistic effect of thiamine. I, *Folia Pharmacol. Jpn.*, 52, 429, 1956; II, Folia Pharmacol. Jpn., 53, 307, 1957.
828. **Yarington, C. T., Jr. and Stivers, F. E.,** Lathyrogenic effects of vitamin A in the rat embryo, *Laryngoscope*, 84, 1310, 1974.
829. **Yatzidis, H., Digenis, P., and Fountas, P.,** Hypervitaminosis A accompanying advanced chronic renal failure, *Br. Med. J.*, 2, 352, 1975.

830. **Young, C. M.**, Overnutrition, *World Rev. Nutr. Diet.*, 16, 187, 1973.
831. **Young, W. C., Ulrich, C. W., and Fouts, P. J.**, Sensitivity to vitamin B_{12} concentrate, *JAMA*, 143, 893, 1950.
832. **Zbinden, G.**, Therapeutic use of vitamin B_1 in diseases other than beriberi, *Ann. N.Y. Acad. Sci.*, 98, 550, 1962.
833. **Zeller, W.**, A long term study of the effect of nicotinic acid and Ronical retard on serum lipids and fatty liver, in *Progress in Biochemical Pharmacology*, Vol. 2, Kritchevsky, D., Paoletti, R., and Steinberg, D., Eds., S. Karger, Basel, 1967, 401.
834. **Zetterström, R.**, Discussion following paper by Hsia, D. Y., Dowben, R. M., and Riabov, S., "Inhibitors of glucuronyl transferase in the newborn", *Ann. N.Y. Acad. Sci.*, 111, 333, 1963.
835. **Zinkham, W. H.**, An in vitro abnormality of glutathione metabolism in erythrocytes from normal newborns: mechanisms and clinical significance, *Pediatrics*, 23, 18, 1959.
836. *Martindales Extra Pharmacopoedia*, 27th ed., Pharmaceutical Press, London, 1977.
837. *Meyhler's Side Effects of Drugs*, Vol. 1, Elsevier, Amsterdam, 1952.
838. **Dukes, M. N., Ed.**, *Side effects of Drugs Annual*, Vol. 1, Elsevier, Amsterdam, 1977.

INDEX

A

Abdominal symptoms
from nicotinic acid, 87
from vitamin C overdose, 196
Abnormal calcification, 201
Abortions, 198, 205
AC, see Activation coefficient
Acanthosis nigricans, 87
Acetylated p-aminobenzoic acid (PABA), 20
Acetylation of sulfanilamide, 20
Acetylcholine, 120—122
Acetylkynurenine, 41
Acid mucopolysaccharides, 200
Acid phosphatases, 201
Acne, 164, 207
Acne vulgaris, 165—169
Actinic keratoses, 170
Activation coefficient (AC), 4
Acute hypotension, 210
Acute mental alertness, 210
Acute pancreatitis, 202
Adipose tissue, 201
Alanine aminotransferase, 4, 15, 38
Alcoholism
in elderly, 74
nicotinic acid in, 86
pantothenic acid and, 20
thiamin in, 74
Wernicke-Korsakoff-like syndromes in, 124
Wernicke's encephalopathy in, 8
Alkaline phosphatase, 49
serum, 21
Allergic contact dermatitis, 206
Allergic reactions
to vitamin B_{12}, 212
to vitamin C, 196
to vitamin E, 206
Alloxan, 195, 196
Altered state of consciousness, 109
Aluminum nicotinate, 86, 88
Amino acid metabolism, 119—120
p-Aminobenzoic acid (PABA), 20
Aminotransferase, 41
Amnesia, 110
Anaphylactic shock, 210, 212, 213
Anemia, 13, 69, 190, 205, 208, 209
dyshaemopoietic, 70
hemolytic, 208
iron deficiency, 13, 207
macrocytic, 44
megaloblastic, 17, 19, 43, 45, 46, 67, 70, 71, 73
pernicious, 10, 19, 72, 213
sideroblastic, 198
symptoms of, 73
Angina pectoris, 83, 86
Angular stomatitis, 2, 67
Anorexia, 19, 75, 202, 210
Antibiotics, 3, 11, 22, 67, 213
Anticonvulsants, 67, 71
Antilipolysis, 82
Antimony trichloride, 7
Antinauseant, 212
Antithiamin substances, 8
Antivitamin K compounds, 22
Aortic abnormalities, 200, 204
Apoenzyme levels, 4
Apoproteins, 83
Appetite loss, 210
Arrhythmia, 88
Arterial cholesterol, 83
Arterial disease, 200
Arthritis, 87, 202
Ascorbate, 49
hypolipemic action of, 99—102
oxidase activity, 48
Ascorbic acid, see Vitamin C
L-Ascorbic acid, 195
Aspartate aminotransferase (EASPAT), 4, 15, 38—40
Ataxia, 109, 210
Atheroma, 200
Atheromata, 200
Atherosclerosis, 101
Atherosclerotic vascular disease, 82, 83
Atopic eczema, 181
Atrial fibrillation, 88
Atromid-S, 83

B

Backache, 68
Back stiffness, 212
Basal cell layers, 199
BCDHC, see Branched chain dehydrogenases
Behavioral changes, 13
Behavioral disorders, 153
Benzoyl peroxide, 166
Beriberi, 8, 74
Bilateral nephrectomy, 202
Bile acids, 102
Bilirubin, 208
Biotin, 211
toxicity of, 214
Birth weight, 203
Blood-brain barrier, 139—141, 213
Blood cells, 4
Blood-cerebrospinal fluid barriers, 139—141
Blood clotting, 51
disorders of, 22
intravascular, 198
time of, 205, 209
Blood flow, 86
coronary, 86
obstruction of, 194
Blood levels of pyridoxine, 14
Blood levels of vitamins, 4
Blood pressure, 209, 213

Blood thiamin, 9
Blood transaminase activities, 15—16
Blood vessel walls, 199
Bones
  density of, 69
  growth of, 193
  malformation of, 191
  matrix of, 193, 194
  metabolism of, 195
  mineral mobilization in, 201
  physiology of, 194
  tenderness of, 68
  vitamin D and, 69
Boric acid, 214
Brain barrier systems, 139—142
  transport through 142—143
Brain disorders, 139
  megavitamin therapy for, 152—153
Brain functions, 143, 147, 154
Branched chain dehydrogenases (BCDHC), 115—116, 127
"Burning feet syndrome", 20

C

Calcareous deposition, 200
Calcification of soft tissue, 204
Calcitonin, 49
Calcium, 49, 50, 68, 195, 198, 200, 203
  absorption of, 201
  balance of, 194
  concentration of, 202
  deposits of, 194, 200
  metabolism of, 194
  mobilization of, 201
  transport of, 200
Calcium-dependent binding of prothrombin, 22
Calcium phosphate, 201
Cancer, 10, 15
  of the lung, 204
  of the urinary tract, 41, 42
Carbohydrates, 42
  metabolism of, 42, 117—119
  oxidation of, 117—118
Carboxylation, 22
γ-Carboxylation, 22
Cardiac conditions, see also Heart; Heart disease, 206, 213
Cardiovascular abnormalities, 52, 109, 203, 205, 209
β-Carotene, 188
Carotenes, 32
Carotenoids, 188
Catecholamines, 121
Cateracts, 196
Cell membranes, 200
Cellular damage, 195, 208
Central nervous system manifestations of thiamin deficiency, 108
Cerebrospinal fluid (CSF), 139—141, 144—152, 191
Ceruloplasmin, 48, 49
Cervical malignancy, 46
Cheilosis, 67
Chenodeoxycholate concentration, 197
Chest pains, 209
Children, see also Infants
  anemia treatment in, 22
  hypercalcemia in, 203
  hyperkinetic, 153
  hypervitaminosis A in, 191, 193
  mentally retarded, 15
  nicotinic acid in, 86
  pyridoxine and, 13, 14
Cholecalciferol, see Vitamin D
Cholecystectomy, 88
Cholelithiasis, 88
Cholesterol, 22, 50, 83, 99, 200, 202, 205, 207, 215
  niacin and, 82
  oxidation of, 82
  vitamin C and, 96, 102
Cholesterolemia, 100
Cholestyramine, 83, 86, 89
Choline, 216
  deficiency of, 217
Choroid plexus, 146
Chromium, 87
Chronic constipation, 67
Chronic hypervitaminosis A, 191
Chronic renal disease, 68
Cimetidine, 87
Circulatory disorders, 215
Cirrhosis, 10, 13
Clinical lesions, 12
Clinical scurvy, 69
Clofibrate, 83, 86, 206
Clotting, see Blood clotting
Coagulation, see Blood clotting
Coagulopathy, 205
Cocarboxylase, 209
Coenzyme adenosyl cobalamine, 22
Coenzyme A mutase, 22
Coenzymes, 4
Colestipol, 89
Colitis, 22
Collagenous tissue, 204
Collagen turnover 195
Colloid emulsion, 209
Colorimetric procedures, 7, 16
Confabulation, 110
Confabulatory psychosis, 110
Congenital malformations, 190
Conjugase, 45—47, 71
Conjunctivitis, 210
Constipation, 67, 202
Contaminant allergins in vitamin B_{12}, 213
Convulsive seizures, 13
Cooking techniques, 70, 71
Copper, 48, 198
Coronary blood flow, 86
Coronary Drug Project, 86—88
Corticosteroids, 202, 203

Cosmetic sprays and creams, 206
Coumarin, 22
Cranio-facial complex, 203
Creatine, 206
Creatinine
  clearance of, 197
  excretion of, 3
CSF, see Cerebrospinal fluid
Cutaneous physiology, 159
Cutaneous vasodilation, 86
Cyanide, 212
Cyanocobalamin, see Vitamin B_{12}
Cyanosis, 208
Cystathionine, 40
L-Cysteine sulphinic acid, 40
Cystic fibrosis, 22
Cystoid macular edema, 88
Cytochrome P-450, 96
Cytoplasmic inclusion, 208

D

Daily quantities
  of niacin (nicotinic acid), 138
  of thiamin, 138
  of vitamin B_6, 138
  of vitamin C, 195
Darier's disease, 158, 176
DCP, see Dichlorophenol
DEAE Sephadex, 89
Deep vein thombosis, 197
Deficiencies of vitamins, see also under specific vitamins 5—6
  causes of, 66—67
Definitions of vitamin, 138
Degenerative atrophy, 190
Degenerative neurological disease, 66
Dehydro-ascorbic acid, 143, 195, 196
7-Dehydro-cholesterol, 198
Dehydrogenases, 114—116
Dementia, 71, 75
Deoxyuridine suppression test, 18—19
Depressed growth, see Growth inhibition
Depressive illness, 42, 46
Dermatitis, 13, 75, 206
Dermatoses, 41, 158, 180
Desquamation of skin, 191
Diabetes, 10, 42, 87, 196
Diarrhea, 75, 87, 196, 202, 216
2,6-Dichlorophenol (DCP), 21
Dietary protein, 14
Dietary vitamin A, 7
Diet surveys, 3
Dihydro-tachysterol, 199
1α-Dihydroxycholecalciferol, 49
1,25-Dihydroxy-cholecalciferol- glycoside, 200
1,25-Dihydroxy D_2 or D_3, 21
Di-hydroxylated derivatives, 199
Dinitrophenylhydrazine reagents, 21
Diphosphate, 147
Disturbed sleep, 213
Disulfide bonding, 198
Diverticulosis, 67
DNA synthesis, 213
Dopamine, 143
Dopamine-β-hydroxylase, 143
Double vision, 74
Dreams, 213
Drug-metabolizing enzymes, 214
Drugs, 6, 67, 71, 89, 198
  clearance rates for, 214
Dyshaemopoietic anemia, 70
Dyspnoea, 209

E

EASPAT, see Aspartate aminotransferase
Ecchymoses, 206
Eczema, 181
EGOT, see Erythrocyte glutamate oxaloacetate transaminase
EGR, see Erythrocyte glutathione reductase
Elderly
  folacin and, 70—71
  niacin and, 75
  nutritional status of, 66
  thiamin and, 8, 74—75
  vitamin B_{12} and, 72—74
  vitamin C and, 69—70, 197
  vitamin D and, 67—69
Encephalopathy, 74, 148—152, 201
Endogenous calcium, 194
Endogenous synthesis rate, 195
Endoplasmic reticulum, 88
Enlarged liver, 191
Enzymatic action of TPP, 113—116
Enzymatic tests, 4, 15—16
Epidermal differentiation, 162
Epilepsy, 202, 213
Ergocalciferol, 199, 201, 204
Erythema, 162, 164
Erythrocyte enzyme function test, 75
Erythrocyte glutamate oxaloacetate transaminase (EGOT), 15, 16
9-Erythrocyte glutathione reductase (EGR), 12—13, 35
Erythrocyte hemolysis, 51
Erythrocytes, 4, 34, 38, 43, 46
  folate in, 17—18, 44, 47
  riboflavin in, 11—12, 35, 37
  vitamin C in, 197, 198
Erythrocyte transketolase test (ETK), 10—11
Estradiol, 30
Estrogens, 15, 30, 32, 33, 37, 38, 41, 46, 48—51, 68
Ethinyl, 30
ETK, see Erythrocyte transketolase test
Euglena gracilis, 19
Exencephaly, 204
Exogenous ascorbic acid, 195
Exophthalmos, 208
Extraskeletal tissues, 199, 201

F

FABP, see Folic acid-binding protein
FAD, see Flavin adenine dinucleotide
Familial hypercholesterolemia, 86
Fatigue, 205
Fatty liver, 216, 217
Fermented tea leaves, 8
Fertility, 198
Fever, 202, 210
FFA, see Free fatty acids
Fibrinolysis, 207
FIGLU, see Formiminoglutamic acid
Fish, 199
Flavin adenine dinucleotide (FAD), 12, 35, 37
Flavoproteins, 214
Flexion contractures, 191
Fluorometric methods, 7, 11
5-Fluoronicotinate, 86
Flushing, 86—87, 209, 215
Folacin, 70—71
  deficiency of, 70—71
  vitamin B_{12} and, 72
Folate, see Folic acid
Folic acid (folate), 16—19, 35, 43, 52, 67, 148—149, 154
  daily quantities of, 138
  deficiency of 10, 12, 16
  erythrocyte, 17—18, 44
  oral contraceptives and, 44—47
  serum, 17—18
  toxicology of, 213
  vs. vitamin B_{12}, 19
Folic acid-binding protein (FABP), 47
Folic acid (folate) deficiency encephalopathy, 149
Food preparation, 70, 71
Foot drop, 74
Formiminoglutamic acid (FIGLU), 17, 18, 44
Fractures of bone, 68
Free cortisol, 42
Free fatty acids (FFA), 82, 88
Fruits, 70
Functional tests, 4—5
  for folate, 18—19
  for pantothenic acid, 20
  for pyridoxine, 14—16
  for thiamin, 9—11
  for vitamin C, 21

G

Gallbladder bile, 102
Gallstones, 102
Ganglionic blocking, 210
Gastrectomy, 67
Gastric atrophy, 66, 67
Gastric mucosa, 201
Gastrointestinal disturbances, 10, 72, 87, 213, 215—217
Genetic aspects of Wernicke-Korsakoff syndrome, 125—128
Geriatric patients, see Elderly
Giddiness, 206
Glomeralur filtration rate, 197
Glossitis, 2
Gluconeogenesis, 42
Glucose, synthesis of vitamin C from, 195
Glucose-fatty acid cycle, 87
Glucose load test, 9
Glucose-6-phosphate dehydrogenase, 36, 198
  deficiency of, 13
Glucose tolerance, 41, 42, 87, 215
Glucose tolerance factor (GTF), 87
Glucuronide conjugation, 208
Glutamic acid residues, 22
Glutathione, 49, 209
Glutathione reductase test for riboflavin, 4
Glycoprotein fibrils, 198
Gout, 213, 216
Gouty arthritis, 87
Growth inhibition, 190, 191, 193, 214, 217
GTF, see Glucose tolerance factor
Guidelines for laboratory tests, 5

H

Hair follicle hyperkeratosis, 69
Hair loss, 190
HDL, see High-density lipoproteins
Headaches, 202, 206, 210
Heart
  calcification influence on, 201
  glucose tolerance and, 87
Heart disease, see also Cardiac conditions; Cardiovascular abnormalities, 202, 205
Heinz body, 208
Hematological abnormalities, 16, 197, 207
Hematoma, 207
Hematopoiesis, 44, 45
Hemochromotosis, 198
Hemoglobin levels, 73
Hemoglobinuria, 207—209
Hemolysins, 208
Hemolytic anemia, 208
Hemorrhages, 69, 190
Hemosiderosis, 208
Hemostasis, 207
Hepatectomy, 213
Hepatic fibrosis, 194
Hepatic portal vein ligature, see also Liver, 213
High altitude resistance, 196
High-density lipoproteins (HDL), 82, 101
Histidine load, 17—19, 44
Homeostatis of vitamins, 143
Homocysteine, 40, 46
Hormone-sensitive neoplasms, 207
Hormone therapies, 6
H-thymidine incorporation, 18
5-HT, see 5-Hydroxytryptamine
Hydrogen peroxide, 198
3-Hydroxyanthranilic acid, 37
1α-Hydroxycholecalciferol, 202

25-Hydroxycholecalciferol, 49, 50, 68, 69, 199
Hydroxycobalamin, 212
3-Hydroxykynurenine, 37
7α-Hydroxylation, 102
Hydroxylation reactions, 20
6-Hydroxy-niacin, 215
6-Hydroxy-niacinamide, 215
Hydroxyproline, 195
5-Hydroxytryptamine (5-HT), 42
5-Hydroxytryptophan decarboxylase, 42
25-Hydroxy-vitamin D, 200, 201
Hyperactivity, 13
Hypercalcemia, 194, 199, 201—203
Hypercholesterolemia, 86, 99, 102, 195
Hyperchylomicronemia, 89
Hyperexcitability, 213
Hyperkeratosis, 69, 158
Hyperkinetic children, 153
Hyperlipidemia, 86, 202, 215
Hyperlipoproteinemia, 84—85
Hyperphosphatemia, 200
Hyperplasia, 203
Hypertension, 194, 201, 216
Hyperthyroid-like symptoms, 210
Hypertriglyceridemia, 89, 102
Hyperuricemia, 86, 87, 213
Hypervitaminosis A, 188, 190, 191, 193, 194, 200
Hypervitaminosis D, 194, 199—202
Hypocholesterolemic effects
  of niacin, 215
  of vitamin C, 101
Hypolipemic action of ascorbate, 99—102
Hypolipidemia, 202
Hypoparathyroidism, 203
Hypophosphatemic vitamin D resistant rickets, 203
Hypoprothrombinemia, 52, 209
Hypotension, 209, 210
Hypovitaminosis A, 7, 159
Hypovitaminosis C, 102

I

Iatrogenic conditions, 19
Ichthyosiform dermatoses, 158
Ichthyosis, 171—176
Ichthyosis vulgaris, 158
ICNND guidelines, 12
Idiopathic hemochromotosis, 198
Idiopathic hypercalcemia, 203
Idiosyncratic reactions to vitamins, 188, 202
Ileocecal valve, 196
Immunological disturbances, 73, 213
Indanedione, 22
Indomethacin, 87
Infants, see also Children, 208
  hypercalcemia in, 203
  premature, 22, 208
Infections, 3
Inositol, 216—217
Institutional catering, 67
Institutionalized elderly, 67, 69—71
Insulin, 42, 87, 216
Intercranial pressure, 190
Intestines, 200
  absorption of riboflavin in, 214
  pain in, 206
  stones in, 196
Intimal sclerosis, 204
Intracellular osmotic pressure, 209
Intrauterine development of idiopathic hypercalcemia, 203
Intravascular coagulation, 198
Intrinsic factor, 19, 74
Iron, 44, 52, 198
  absorption of, 198
Iron deficiency anemia, 13, 207
Iron dextran, 207
Irritability, 213
Ischemic heart disease, 202, 205
Isoascorbic acid, 145
Isoenzyme GdA, 198
Itching, 215

J

Jaundice, 88, 209

K

Keratinization, 159
Keratosis follicularis, 176
Keratosis pilaris, 176
Kernicterus, 208
Ketoglutarate dehydrogenases, 114, 127
Ketone production, 87
Kidneys, 146, 199, 201
  damage to, 208
  stones in, 197
  tubule damage in, 193
  weight of, 213
Kinetic methods, 16
Korsakoff's psychosis, see also Wernicke-Korsakoff syndrome, 74, 110—111
Kynurenine, 41
Kynurenine transaminase, 38
Kynurinase, 6, 14

L

Laboratory methods, 2—6
Lacrimation, 208
Lactate, 87
Lactobacillus casei, 18
Lactobacillus leichmanni, 19
Lamellar ichthyosis, 158
LDL, see Low density lipoprotein
Learning disabilities, 110, 153
Leg pains, 212

Leigh's disease, 124
Lesions, 12, 123, 190, 194, 204
 mammillary bodies, 110
 neurological, 19
Lethargy, 210
Leukocytes, 4, 34, 48, 49
 pyruvate decarboxylase in, 11
 transketolase in, 35
 vitamin C in, 20—21, 70
Lichen planus, 180
Lipids, 22, 200, 202, 207
 deposition of, 194
 lowering of, 82
 metabolism of, 119
Lipoproteinemia, 22
Lipoprotein lipase, 83, 97
Lipoproteins, 22, 50, 51, 82, 200
β-Lipoproteins, 22
Liver, 201, 208
 ascorbate in, 99
 cholesterol in, 205
 cirrhosis of, 13
 diseases of, 10, 22, 68, 86, 203, 209
 dysfunction of, 87—88, 194, 209, 213, 215
 enlarged, 191
 failure of, 194
 fatty infiltration of, 216, 217
 function of, 201
 insufficiency of, 201, 202
 problems in, 203
 size of, 194
 tubular damage in, 208
 vitamin A in, 6
 vitamin C in, 138
 vitamin D in, 69, 199
Load retention tests, 3
Load return tests, 8
Low birth weight, 203
Low density lipoprotein (LDL), 82, 101
 calcification influence on, 201
 cancer of, 204
Lysine, 20
Lysis, 198
Lysosomes, 193
Lytic sensitivity, 198

M

Macrocytic anemia, 44
Maculopaphy, 88
Magnesium deficiency, 202
Maize, 16
Malabsorption, 22, 44, 67, 69, 71, 74, 205
 in elderly, 66
 of vitamin E, 22
Malaise, 213
Malignant conditions, 170—171
 cervical, 46
Malnutrition, 66, 69
Mandibular hyperplasia, 203
Manganese, 198
Maple-Syrup-Urine Disease (MSUD), 127
Matrix dissolution, 201
Medial calcific degeneration, 200
Medroxyprogesterone acetate, 32, 48, 49
Megadosage
 of niacin, 82
 of vitamin A, 190
 of vitamin C, 196
Megaloblastic anemia, 17, 19, 43, 45—67, 70, 71, 73
Megaloblastic changes, 17
Megavitamin therapy, 138—139, 207
 for brain disorders, 152
Melanin, 199
Membrane function, 121—122
Menadione, 208
Menaphthone, 207
Meningeal irritation, 210, 212
Meningitis, 154
Menstrual cycle, 44
Mental alertness, 210
Mental changes, 71, 73, 216
Mental defects, 203
Mental frailty, 66
Mental retardation, 13, 15
Mesenchymal reactions, 200
Mestranol, 30, 32, 35
Metabolism
 disorders of, 197
 errors in, 6
 inhibitors of, 128
 tests of, 4
 thiamin and, 116—117
Methemoglobinuria, 208
Methionine, 37
Methionine load, 16, 40
N′-Methoylnicotinamide, 16
Methylamine, 216
Methylmalonic acid (MMA), 20, 22
Methylmalonyl coenzyme A mutase, 22
N-Methyl-niacinamide, 215
N′-Methyl-2-pyridone-5-carboxylamide, 16
Microbiological assays, 11, 14, 18, 19
Micrognathia, 204
Microsomal enzymes, 210
Milk production, 212
Mineralization, 201
Minimal brain dysfunction, 153
Mitochondria, 88
MMA, see Methylmalonic acid
Mobilization of bone mineral, 201
Mobilization of calcium, 201
Molluscum contagiosum, 180
Monoclonal immunoglobulin, 214
Monohydroxylated metabolite, 199
Monophosphate, 147
MSUD, see Maple-Syrup-Urine Disease
Mucus membranes, 191
Multiple myeloma, 214
Multivitamin preparations, 194, 214
Muscles
 damage to, 206

stiffness in, 210
weakness of, 206
Muscular dystrophy, 205
Mutagenicity of ascorbic acid, 198
Myeloma, 214
Myocardial infarction, 86
Myocarditis, 200
Myoinositol, 217

N

NAD, see Nicotinamide-adenine dinucleotide
NADP, see Nicotinamide-adenine dinucleotide phosphate
Naturally occurring vitamins, 189
Nausea, 87, 202, 210
Negative calcium balance, 194
Negative feedback control, 200
Negative nitrogen balance, 3, 11
Neonatal period, 203
Neoplasia, 190
Nephrectomy, 202
Nerve conduction, 147
Neurological damage, 74
Neurological disorders, 13, 66, 108, 153
Neurological lesions, 19
Neuromuscular blocking, 210
Neuromuscular paralysis, 210
Neuropathy, 2, 10, 19, 73, 74, 109
Neuropsychiatric disorders, 153
Neurotransmitters, 120, 121
Neutrophilic polymorphonuclear leucocytes (PMN), 16—17
Newborn infants, 208
Niacin (nicotinic acid), 16, 37, 41, 151—152, 154
cholesterol lowering properties of, 82
daily quantities of, 138
deficiency of, 16, 75
derivatives and analogues of, 88—89
effectiveness of, 83—86
elderly and, 75
mechanism of, 82—83
side effects from, 86—88
toxicology of, 214—216
Niacin (nicotinic acid) deficiency encephalopathy, 152
Niacin (nicotinic acid) pathway, 6, 15, 16
Niacianamide, 214
Nicalex, 86, 88
Niceritrol, 83, 87, 88
Nicotinamide-adenine dinucleotide (NAD), 215
Nicotinamide-adenine dinucleotide phosphate (NADP), 215
Nicotine, 212
Nicotinic acid, see Niacin
Nicotinyl alcohol, 88
Nitrogen balance, 3, 11
N′-methoylnicotinamide, 16
N-methyl-niacinamide, 215
N′-methyl-2-pyridone-5-carboxylamide, 16
Noradrenaline, 143
Normocalcemia, 202
Nucleic acids, 120
degradation of, 213
Nutritional deficiencies, 5—6
"Nutritional melalgia", 20
Nutritional status of elderly, 66
Nystagmus, 74

O

Ochromonas danica, 9
25-(OH) D_3, 21
Open eye, 204
Opthalmoplegia, 109
Oral contraceptives, 4, 6, 11, 15, 30—31
and folic acid, 44—47
and riboflavin, 35—37
and thiamin, 33
and vitamin A, 32—33
and vitamin B_1, 33—35
and vitamin B_6, 37—43
and vitamin B_{12}, 43—44
and vitamin C, 47—49
and vitamin D, 49—50
and vitamin E, 50—51
and vitamin K, 51—52
Oral vitamin A, 165
Organic matrix of bone, 194
Orthomolecular psychiatry, 139
Osteomalacia, 67, 68
Osteoporosis, 68
Overdose
of vitamin C, 196
of vitamin D, 202
Over pigmented skin, 75
Oxalate formations, 197
Oxalic acid, 196
Oxidation of carbohydrates, 117—118
Oxidative phosphorylation, 199

P

PABA, see p-Aminobenzoic acid
Packed cell volume, 201
Pain with fractures, 68
Palmar plantar hyperkeratosis, 158
Palmoplantar keratodermas, 176
Pancreas, 201
damage to, 196
Pancreatitis, 202
Pantothenic acid, 20, 214
Paralysis, 208, 210, 216
Parathyroid deficiency, 202
Parathyroid hormone, 49
Parietal cells, 66
Partial hepatectomy, 213
Pathogenesis of fracture, 68
Pathologic lesions, 123
Pathophysiology, 128
Pediococcus cerevisiae, 18

Pellagra, 16, 75, 152, 214
Pentaerythritol tetranicotinate, 87, 88
Pentose phosphate pathway, 118—119, 214
Peptic ulcer, 67, 87
Percutaneous absorption of vitamin A, 159
Pericyt, 87, 88
Periferal vascular collapse, 209
Perioral dermatosis, 41
Peripheral nerve degeneration, 74
Peripheral neuropathy, 2, 73, 74, 109
Peripheral vasodilatory effects, 215
Pernicious anemia, 10, 19, 72, 213
Perosis, 217
Peroxides, 22
Pharmacokinetics, 83
Pharmacology of vitamin A acid, 160—165
Phenylcarbinol, 213
Phosphate, 201
Phosphate ions, 200
Phosphoenolpyruvate carboxykinase, 42
Phosphorous, 49, 195
Phosphorylated derivatives, 13
Physiatric emergencies, 210
Phytate content of the diet, 21
Phytomenadione, 207
Phytonadione, 208, 209
Placental transfer of vitamins, 203
Plant tissues, 199
Plasma, 201
Plasma binding globulin, 200
Plasma calcium, 201
Plasma carotenoid concentration, 207
Plasma cholesterol, 82
Plasma half-life, 202
Plasma lipids, 83
  and vitamin C, 97—102
Plasma lysosomal enzymes, 201
Plasma prothrombin, 205
Plasma tocopherol, 50
Plasma triglyceride, 82
Platelet, 32, 33, 48
  disturbances of, 49
  structure of, 198
PLP, see Pyridoxal phosphate
PMN, see Polymorphonuclear leukocytes
Polished rice, 74
Polyglutamate, 46
Polyglutamyl folate, 45—47
Polymorphonuclear leukocytes (PMN), 16—17
Polyneuritis, 74
Polysaccharides, 200
Polyuria, 202
Porphyria cutanea tarda, 207
Portal blood flow obstruction, 194
Portal hypertension, 194
Postoperative patients, 75
Potassium, 88
Poverty, 75
Pregnandiol excretion, 207
Pregnant women
  megaloblastic anemia in, 17
  thiamin deficiency in, 8
  vitamin C and, 198
  vitamin D and, 203
  xanthurenic acid excretion in, 15
Premalignant conditions, 170—171
Premature epiphyseal closing, 191
Premature infants, 22, 208
Primary lesions, 194
Procarcinogenicity, 204
Progesterone, 49
Progestogens, 30—32, 35, 37, 38, 43, 48, 50, 51
Prolactin secretion, 212
Proline, 20
Prophylactic vitamin D, 69
Prophylactic vitamin E, 205
Prostaglandin, 82, 87
Protein, 14, 32, 190, 195, 216
  copper transport, 48
  deficiencies of, 10
  synthesis of, 120
Proteolytic enzymes, 193
Prothrombin, 22, 51, 205, 206, 208
  synthesis of, 209
Provitamins, 188
Pruritus, 87, 215
Pseudo fractures, 68
Pseudohypoparathyroidism, 203
Psoriasis, 158, 169—170
Psychiatric conditions, 41, 214
Psychosis, 74
Pterolpolyglutamate hydrolase, 71
Purine-free diet, 197
Pyridine compounds, 82
2-Pyridones, 215
4-Pyridones, 215
Pyridoxal, see also Vitamin B_6, 13
Pyridoxal hydrochloride, 212
Pyridoxal phosphate (PLP), 13, 35, 37, 38, 41
Pyridoxal phosphate dependent enzymes, 37
Pyridoxamine, see also Vitamin B_6, 13
4-Pyridoxic acid, 13, 37, 38
Pyridoxine, see also Vitamin B_6, 4, 13—16, 42, 43
  blood levels of, 14
  deficiency of, 10, 12, 15, 42, 46
  urinary, 13—14
Pyridoxine deficiency encephalopathy, 150—151
Pyridoxine hydrochloride, 37
β-Pyridyl carbinol, 83
Pyrithiamine, 123
Pyrophosphokinase, 116
Pyruvate, 87
  oxidation of, 117
Pyruvate dehydrogenase, 114, 126—127

Q

Quality control, 5
Quaternary ammonium compounds, 210, 216
Questran, 83
Quinolinic acid, 42

R

Radio-iron absorption, 198

Radio ligand receptor assay, 201
Rami of the pelvis, 68
Raw fish, see Thiaminase
RBP, see Retinol-binding protein
Red blood cells, see Erythrocytes
Redox systems, 209
Renal tubule epithelia cells, 146
Renal tubule lumen, see also Liver, 201
Resin, 89
Resorcinol, 166
Respiratory paralysis, 208, 216
Retardation of growth, see Growth inhibition
Retinal, 158
Retinoic acid, 158
  clinical studies with, 165—169
  pharmacology of, 160—165
Retinoic acid-binding protein, 190
Retinol, 6, 32, 33, 158
Retinol-binding protein (RBP), 6, 32, 33, 195
Retinol transport system, 195
Retinyl esters, 32, 33
Retinyl palmitate, 188
Rheumatoid arthritis, 202
Riboflavin (vitamin B_2), 11—13, 35, 52, 152, 212
  glutathione reductase test, 4
  oral contraceptives and, 35—37
  toxicology of, 214
  urinary, 11, 12
Rice, 74
Rickets, 21, 201, 203
Ronicol, 88

S

Saccharomyces carlsbergensis, 14
Saccharamyces uvarum, 14
Salts of ascorbic acid, 195
Sarcoidosis, 204
Scaling, 162, 164
Schilling test, 19, 44
Schizophrenia, 153, 215
Sclerosis, 194, 204
Scorbutic diet, 197
Scurvy, 69, 144, 197
Seasonal variation in vitamin D status, 68
Secholex, 89
Sedatives, 67
Sensitization, 210
Seromucoid levels, 200
Serotonin, 121
Serum alkaline phosphatase, 21
Serum ascorbic acid, 20—21
Serum bile acids, 197
Serum calcium, 194, 200, 203
Serum ceruloplasmin, 49
Serum cholesterol, 202, 207
Serum creatine kinase activity, 206
Serum folate, 17—18
Serum lipids, 202, 207
Serum magnesium, 202
Serum tocopherol, 22
Serum triglyceride, 202
Serum uric acid, 197, 213, 216
Serum vitamin A, 6, 7
Serum vitamin B_{12}, 19
Serum vitamin C, 70
Sideroblastic anemia, 198
Skeletal disorders, 191
Skeletal pains, 194
Skin disorders, 158, 190, 191, 196, 206, 209, 213
Sleep disturbance, 213
Small intestine, 200
Smoking, 70
Soft tissue calcification, 194, 204
Solanum malocoxylon, 200
Solemness, 210
Soreness of tongue, 19
Sorghum, 16
Sorghum vulgare, 16
Spectrophotometric method of measurement of serum vitamin A, 7
Sperm penetration, 198
Spleen size, 194
Sprue, 214
Standardization of laboratory methods, 5
Steroid balance, 82
Steroid B ring, 198
Stillbirths, 195
Stomatitis, 2
  angular, 67
Streptococcus faecalis, 18
Streptomyces griseus, 213
Stressful states, 70
Subacute combined degeneration of the cord, 73
Subcorneal pustular dermatoses, 180
Sulfanilamide, 20
Sulfur, 166
Sunlight, 203
  and folacin, 71
  and over pigmented skin, 75
  and vitamin D, 67
Supravalvular aortic stenosis, 200, 203
Synthesis, 189
  of vitamin C, 195
  of vitamin D, 199
Systolic murmers, 203

T

Tachyphylaxis, 164
Taurine, 40
TDP, see Thiamin pyrophosphate (TPP)
Tea leaves, 8
Teeth, 203
Tetany, 202
Tetrahymena pyriformis, 14
Thalassaemia major, 198
Thiamin, 8—11, 147—148, 154
  blood, 9
  daily quantities of, 138
  deficiency of, 8, 12, 74—75, 108, 111—112, 122—125

elderly and, 74—75
functions of, 112—123
metabolic roles of, 116—117
oral contraceptives and, 33
role of in brain, 147
therapy with, 111—112
toxicology of, 209—212
transketolase test for, 4
urinary, 8—9
Wernicke-Korsakoff syndrome, 111—112
Thiaminase, 8
Thiamin deficiency encephalopathy, 148
Thiamin-deficient ganglia, 122
Thiamin monophosphate (TMP), 113
Thiamin propyl disulfide, 212
Thiamin pyrophosphate (TPP), 9, 10, 35, 112, 119
enzymatic actions of, 113—116
Thiamin pyrophosphokinase, 116
Thiamin triphosphate (TTP), 113, 122
Thiochrome method of estimating urinary thiamin, 9
Thiozolium ring, 210
Thromboembolic disorders, 33, 46, 49, 51
Thrombosis, 52, 197, 213
Thyroid
function of, 205
suppression of, 191
TMP, see Thiamin monophosphate
α-Tocopherol, 204, 205
Tocopherol amines, 204
Tocopherols, see also Vitamin E, 204
levels of, 22
Tongue soreness, 19
Toxicology
of biotin, 214
of choline, 216
of folic acid, 213
of inositol, 216—217
of niacin, 214—216
of riboflavin, 214
of thiamin, 209—212
of vitamin A, 188—195
of vitamin B_6, 212
of vitamin B_{12}, 211—213
of vitamin C, 195—198
of vitamin D, 198—204
of vitamin E, 204—207
of vitamin K, 207—210
TPP, see Thiamin pyrophosphate
Transaminases, 35, 37, 38
Transcobalamins, 43, 44
Transketolase, 4, 10, 33, 35, 114—115, 118, 125—127
Transport through brain barrier systems, 142—143
Tricarboxylic acid cycle, 119
Trichostatis spinulosa, 177—180
Trifluoroacetic acid, 7
Triglyceride, 51, 82, 202, 205
metabolism of, 96—97
vitamin C and, 96—97
Tri-hydroxylated derivatives, 199
Triphosphate, 147
Tryptophan, 37, 38
metabolism of, 42
Tryptophan load, 14—15
Tryptophan metabolites, 6
Tryptophan-niacin pathway, 6, 14, 16
Tryptophan-nicotinic acid, 41
Tryptophan oxygenase, 6, 15
Tryptophan pyrrolase, 38, 41
TTP, see Thiamin triphosphate
Tubular ascorbic acid, 153
Tubular function, 197
Tyrosine, 21

U

UBBC, see Unsaturated vitamin B_{12} binding capacity
Ulceration, 215
Ulcerative colitis, 22
Ulcers, 67
peptic, 87
varicose, 69
Ultraviolet light, 68, 69, 198
Unsaturated vitamin B_{12} binding capacity (UBBC), 43, 44
Urate-binding, 197
Urate stones, 197
Uremia, 13
Uremic neuropathy, 10
Uric acid clearance, 197
Uric acid stones, 197
Uricosuria, 197
Urinary excretion
of calcium, 194
of copper, 198
of formiminoglutamate, 18
of metabolite of tyrosine, 21
of riboflavin, 12
of urocanate, 18
of vitamin C, 197
of vitamins, 3
of xanthurenic acid, 15
of zinc, 198
Urinary oxalate, 196
Urinary pyridoxin 13—14
Urinary riboflavin, 11
Urinary stone formation, 196
Urinary thiamin, 8—9
Urinary tract cancer, 41, 42
Urobilinuria, 207
Urocanate, 18
Urogilinoguria, 207
Urticaria, 213
Uterine cervic mucus, 198

V

Vagans, 19
Valine load test, 20

Varicose ulcers, 69
Vasodilation, 86
Vasopressin, 202
Vegans, 74
Vegetables, 70
Vegetarians, 19
Verrucous nevus, 177
Vertebral irregularities, 191
Very low density lipoprotein (VLDL) triglyceride, 82
Vitamin A, 6—7, 158, 201, 205
  deficiency of, 6
  dietary or administered vs. serum, 7
  esters of, 188
  megadoses of, 190
  oral, 165
  oral contraceptives and, 32—33
  percutaneous absorption of, 159
  synthetic derivatives of, 189
  toxicology of, 188—195
Vitamin A acid, see Retinoic acid
Vitamin A alcohol, see Retinol
Vitamin A aldehyde, see Retinal
Vitamin B, synthetic derivatives of, 189
Vitamin B_1, 52
  and oral contraceptives, 33—35
Vitamin B_2, see Riboflavin
Vitamin B_6, see also Pyridoxal; Pyridoxamine; Pyridoxine, 52, 149—151, 154, 216
  daily quantities of, 138
  deficiency of, 13, 42, 216
  oral contraceptives and, 37—43
  toxicology of, 212
Vitamin B_{12} (cyanocobalamin), 19—20, 22, 52, 152
  absorption of, 67
  binding capacity of, 43
  deficiency of, 19, 67, 73—74
  elderly and, 72—74
  oral contraceptives and, 43—44
  serum, 19
  toxicology of, 211—213
  vs. folic acid, 19
Vitamin B_{12} binders, 19
Vitamin C (ascorbic acid), 20—21, 52, 143—147
  cholesterol and, 96, 102
  daily quantities of, 138
  deficiency of, 10, 12, 21, 69, 70
  dosage problems, 97—99
  elderly and, 69—70
  erythrocyte, 198
  homeostatis of, 146—147, 153—154
  hypocholesterolemic effect of, 101
  in tissues, 21
  leukocyte, 20—21, 70
  optimum dosage of, 97—99
  oral contraceptives and, 47—49
  plasma lipids and, 97—102
  serum, 20—21, 70
  synthetic derivatives of, 189
  toxicology of, 195—198
  triglyceride metabolism and, 96—97
Vitamin D (cholecalciferol), 21—22, 68, 194, 199, 201, 204, 205
  deficiency of, 67—69
  elderly and, 67—69
  oral contraceptives and, 49—50
  prophylactic, 69
  synthetic derivatives of, 189
  toxicology of, 198—204
Vitamin D-binding protein, 50
Vitamin E, see also Tocopherols, 22, 209
  deficiency of, 20, 22, 205
  fatigue associated with, 205
  oral contraceptives and, 50—51
  synthetic derivatives of, 189
  toxicology of, 204—207
Vitamin H, see Biotin
Vitamin K, 22, 205, 209
  deficiency of, 22, 190
  oral contraceptives and, 51—52
  synthetic derivatives of, 189
  toxicology of, 207—210
Vitamin-responsive inborn metabolism errors, 6
VLDL, see Very low density lipoprotein
Vomiting, 75, 87, 202, 210

W

Warfarin, 206
Warts, 181
Weakness, 202, 210
Weight, 88
Wernicke-Korsakoff-like syndromes, 123—124
Wernicke-Korsakoff syndrome, 108—112, 119, 123
  genetic aspects of, 125—128
Wernicke's encephalopathy, 8, 74
Wernicke's syndrome, 111
Wet beriberi, 74
Wheat-germ oil, 206
White cell count, 88

X

Xanthomata, 83
Xanthurenic acid, 15, 37, 42

Y

Yeast apo pyruvate decarboxylase, 9

Z

Zinc, 47, 52, 198